MERCER UNIVERSITY
MAIN LIBRARY

# Current Problems in
Condensed Matter

# Current Problems in Condensed Matter

Edited by

## J. L. Morán-López
*Universidad Autónoma de San Luis Potosí*
*San Luis Potosí, Mexico*

Plenum Press • New York and London

Library of Congress Cataloging-in-Publication Data

On file

Proceedings of an International Workshop on Current Problems in Condensed Matter: Theory and
Experiment, held January 5–9, 1997, in Cocoyoc, Morelos, Mexico

ISBN 0-306-45915-9

© 1998 Plenum Press, New York
A Division of Plenum Publishing Corporation
233 Spring Street, New York, N.Y. 10013

http://www.plenum.com

10 9 8 7 6 5 4 3 2 1

All rights reserved

No part of this book may be reproduced, stored in a retrieval system, or transmitted in any form or by any
means, electronic, mechanical, photocopying, microfilming, recording, or otherwise, without written
permission from the Publisher

Printed in the United States of America

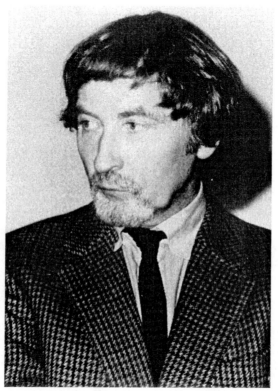

Professor Karl Heinz Bennemann

# Preface

This volume contains the papers presented at the International Workshop on the Current Problems in Condensed Matter: Theory and Experiment, held at Cocoyoc, Morelos, Mexico, during January 5–9, 1997. The participants had come from Argentina, Austria, Chile, England, France, Germany, Italy, Japan, Mexico, Switzerland, and the USA. The presentations at the Workshop provided state-of-art reviews of many of the most important problems, currently under study, in condensed matter. Equally important to all the participants in the workshop was the fact that we had come to honor a friend, Karl Heinz Bennemann, on his sixty-fifth birthday.

This *Festschrift* is just a small measure of recognition of the intellectual leadership of Professor Bennemann in the field and equally important, as a sincere tribute to his qualities as an exceptional friend, college and mentor. Those who have had the privilege to work closely with Karl have been deeply touched by **Karl's** inquisitive scientific mind as well as by his kindness and generosity.

Professor Bennemann has contributed to the understanding of many important issues in condensed matter. Most of the themes included in this volume have been of his interest. The volume begins with a set of papers on the physics of highly correlated systems. A paper on high-$T_c$ superconductors, one of his favorite problems, opens the book. This paper is followed by others on magnetism and superconductivity, intermediate valence systems, spin density waves, and Kondo insulators. Magnetism is other of the issues at Karl's heart. Various aspects are presented, and in particular various papers are devoted to the solution of the Hubbard and related models.

One of the most active fields of research in condensed matter is the study of the physical properties of low-dimensional systems; surfaces, thin films, sandwiches, heterostructures, two- and one-dimensional systems, and small aggregates. It has been observed that the properties of these particular systems differ considerably from those in the the two limits, atomic or molecular and the solid state. Karl has also contributed to the understanding of a large number of properties of those systems. We include various papers that address these important problems: molecular dynamics studies of metallic and semiconductor clusters, electronic properties of fullerenes, and magnetism of low dimensional-systems, within traditional and novel methods.

The surface properties of alloys and their capability to chemisorb atoms, as well as vibrational modes in one-dimensional chains are also addressed in some chapters. The calculation of the bulk properties of alloys and nonperiodic systems, in general, are presented in the last chapters.

This volume is just a small sample of the large variety of problems that have been of Karl's interest and that we have had the opportunity and privilege to share with him.

I wish to thank the participants for their enthusiasm and support in the realization of the Workshop. Special thanks go to the other members of the organizing committee: Professors Michel Avignon, Vijay Kummar, and Akio Okiji. I am particulary grateful to Dr. Alejandro Díaz-Ortiz, Dr. Juan Martín Montejano-Carrizales, and Mr. Teodoro Córdova Fraga, for their invaluable assistance in the preparation of this volume. Finally we acknowledge the financial support of Consejo Nacional de Ciencia y Tecnología (Mexico), Secretaría de Educación Pública (Mexico), Sociedad Mexicana de Física, Centro Latinoamericano de Física (Mexico), Universidad Autónoma de San Luis Potosí, Universidad Nacional Autónoma de México, and United Nations Educational, Scientific, and Cultural Organization (Montevideo).

J. L. Morán-López
San Luis Potosí, S.L.P., Mexico

# Contents

Spin Fluctuation Effects in High-$T_c$ Superconductors ............................. 1
    Sören Grabowski, Jörg Schmalian, and K.H. Bennemann

Superconductivity and Magnetism in $f$ Electronic Systems ...................... 11
    R. Escudero, F. Morales, A. Brigg, and P. Monceau

Intermediate Valence Model for $Tl_2Mn_2O_7$ ....................................... 27
    C. I. Ventura and B. Alascio

Spin Density Waves in Dimerized Systems ........................................ 35
    S. Caprara, M. Avignon, and O. Navarro

Bond-Order-Wave *versus* Spin-Density-Wave Dimerization in Polyacetylene .... 45
    G. M. Pastor and M. B. Lepetit

Impurity States in Kondo Insulators ............................................. 53
    P. Schlottmann

Real-Space Ground-State of a Generalized Hubbard Model ...................... 73
    O. Navarro, E. Flores, and M. Avignon

Coherent and Ultracoherent States in Hubbard and Related Models ........... 79
    K. A. Penson and A. I. Salomon

What is Noncollinear Magnetism? ................................................ 87
    P. Weinberger

Spectral Properties of Transition Metal Compounds and Metal-Insulator
    Transition: A Systematic Approach within the Dynamical
    Mean Field Theory ......................................................... 95
    P. Lombardo, M. Avignon, J. Schmalian, and K.H. Bennemann

Theoretical Study of the Metal-Nonmetal Transition in
    Transition Metal Clusters ............................................... 109
    F. Aguilera-Granja, J. A. Alonso, and J. M. Montejano-Carrizales

First-Principles Langevin Molecular Dynamics Studies of Metallic and
    Semiconductor Clusters .................................................. 119
    Luis C. Balbás

Theoretical Study of the Collective Electronic Excitations of the
    Endohedral Clusters $Na_N@C_{780}$ .................................... 133
  J.M. Cabrera-Trujillo, R. Pis-Diez, J. A. Alonso, M.J. López,
  M.P. Iñiguez, and A. Rubio

Molecular Precursor of Soot and Quantification of the
    Associated Health Risk ............................................... 143
  K. Siegmann and H.C. Siegmann

Magnetism of Transition Metal Clusters: Overview and Perspectives ......... 161
  G. M. Pastor

Magnetic Moments of Iron Clusters: a Simple Theoretical Model ............. 177
  F. Aguilera-Granja, J. M. Montejano-Carrizales, and J.L. Morán-López

Magnetic Anisotropy of $3d$-Transition Metal Clusters, Chains, and
    Thin Films ......................................................... 185
  J. Dorantes-Dávila and G. M. Pastor

Noncollinear Magnetic Structures in Small Compact Clusters ................ 195
  M.A. Ojeda-López, J. Dorantes-Dávila, and G. M. Pastor

Phase Transitions in Ising Square Antiferromagnets:
    A Controversial System ............................................. 203
  E. López-Sandoval, F. Aguilera Granja, and J. L. Morán-López

Interfacial Interdiffusion and Magnetic Properties of Transition Metal
    Based Materials .................................................... 209
  M. Freyss, D. Stoeffler, S. Miethaner, G. Bayreuther, and H. Dreysse

Magnetic Surface Enhancement and the Curie Temperature in Ising
    Thin Films ......................................................... 219
  S. Meza-Aguilar, F. Aguilera-Granja, and J. L. Morán-López

Conductivity Oscillations in Magnetic Multilayers ........................... 227
  Miguel Kiwi, Ricardo Ramírez, Ana María Llois, and Mariana Weissmann

Slave-Boson Approach to Electron Correlations and Magnetism in
    Low-Dimensional Systems .......................................... 239
  E. Muñoz-Sandoval, J. Dorantes-Dávila, and G. M. Pastor

Segregation and Ordering at the Ni-10 at.% Al Surface from
    First Principles .................................................... 247
  T. C. Schulthess and R. Monnier

Study of Deep Level Defects in Polycrystalline Cadmium Sulfide Films ....... 255
  U. Pal, R. Silva González, F. Donado, M. L. Hernández, and
  J. M. Gracia-Jiménez

Intrinsic Localized Modes in the Bulk and at the Surface of
    Anharmonic Chains ................................................. 263
  V. Bortolani, A. Franchini, and R. F. Wallis

Molecular Orientation Dependence of Dynamical Processes on
    Metal Surfaces: Dissociative Adsorption and Scattering,
    and Associative Desorption of Hydrogen ............................. 275
  Ayao Okiji, Hideaki Kasai, and Wilson Agerico Diño

Gaps in the Spectra of Nonperiodic Systems ............................. 283
  R. A. Barrio, Gerardo G. Naumis, and Chumin Wang

Electronic Theory of Colossal Magnetoresistance Materials .............. 291
  R. Allub and B. Alascio

First-Principles Study of Phase Equilibria in the Ni-Cr System ......... 301
  J.M. Sanchez and P. J. Craievich

Theoretical Aspects of Porous Silicon .................................. 315
  M. R. Beltrán, C. Wang, M. Cruz, and J. Tagüeña-Martínez

The Principle of Self-Similarity and Its Applications to the Description
    of Noncrystalline Matter ........................................... 323
  Richard Kerner

Ring Statistics of Glass Networks ...................................... 339
  Matthieu Micoulaut

GaAs Layers Grown by the Closed-Space Vapor Transport Technique
    Using Two Transport Agents ......................................... 347
  E. Gomez, R. Valencia, R. Silva, and F. Silva-Andrade

Index .................................................................. 353

# Spin Fluctuation Effects in High-$T_c$ Superconductors

Sören Grabowski, Jörg Schmalian, and K. H. Bennemann

*Institut für Theoretische Physik*
*Freie Universität Berlin*
*Arnimallee 14, D-14195 Berlin*
*GERMANY*

## Abstract

By using a theory that includes antiferromagnectic short-ranged correlations, recent experiments on the high-transition temperature (high-$T_c$) cuprate superconductors like photoemission, tunneling measurements, and the doping dependence of $T_c$ can be understood. In particular for underdoped compounds, we find the formation of shadows of the Fermi surface, **k**-dependent pseudogap structures in the excitation spectrum and by considering interlayer effects a blocking of the $c$-axis charge transport as precursors of the antiferromagnetic phase transition.

Although there is still no consensus concerning the pairing interaction in the high-$T_c$ superconductors, there is growing evidence that short-ranged antiferromagnetic spin fluctuations are important for their unusual normal state properties[1] and for the Cooper pair formation.[2] However, the quasi-particle excitations, the doping dependence of the superconducting state, the importance and the interplay of the multiple $CuO_2$ layers within materials like $Bi_2Sr_2CaCu_2O_{8+\delta}$ (BSCCO) or $YBa_2Cu_3O_{6+\delta}$ (YBCO) and finally the charge transport properties perpendicular to the $CuO_2$-planes ($c$-axis) are still not well understood.

In this article we show that the dynamical short-range antiferromagnetic correlations in the high-$T_c$ superconductors explain a variety of unconventional properties. In particular they yield a $d$-wave superconducting pairing mechanism and lead to the occurrence of an *optimal doping* where $T_c$ becomes maximal. We demonstrate that significant fingerprints of this dynamical antiferromagnetism were observed especially in angular resolved photoemission experiments (ARPES), where characteristic fine structures like shadows of the Fermi surface (FS),[4] flat single particle bands,[5] excitation gaps in the normal state of *underdoped* cuprates[6] and additional dip features in the superconducting phase[7] were found. Furthermore, by investigating interlayer interactions in the same theoretical framework we argue that the antiferromagnetically correlated $CuO_2$-planes in YBCO[8] or in $La_{2-x}Sr_xCuO_4$ (LSCO)[10] lead to an effective blocking of the single-particle charge transport across the layers for small doping concentractions.

To investigate the short-range antiferromagnetic correlations in the cuprates, the competition between delocalization and correlations effects and the interlayer interactions we use the two dimensional Hubbard Hamiltonian for a mono- and a bilayer $CuO_2$-system with local Coulomb repulsion $U$. As input we use the interaction-free dispersion given by $\varepsilon^0_\pm(\mathbf{k}) \varepsilon^0(\mathbf{k}) \pm t_\perp$ with two-dimensional wave vector $\mathbf{k}$, where $\varepsilon^0(\mathbf{k})$ describes the in-plane contribution. The bare, coherent interlayer coupling $t_\perp$ yields an antibonding $(-)$ and a bonding band $(+)$ with a corresponding bilayer splitting of $2t_\perp$. In the following we will consider $\varepsilon^0(\mathbf{k}) = -2t[\cos(k_x) + \cos(k_y)]$ (Ref. 11) with nearest-neighbor hopping integral $t = 250$ meV and Coulomb repulsion $U = 4t$. For the interplane hopping $t_\perp = 0$ refers to the monolayer with decoupled planes and $t_\perp = 0.4t = 100$ meV to the bilayer.[12,13]

Since we expect spin-fluctuactions to be the dominating low energy excitations in the doping between simple metallic and antiferromagnetic behavior, where superconductivity occurs, we use for the numerical treatment of the Hubbard model the fluctuation exchange approximation (FLEX).[14] This self-consistent, perturbative approach, which sums up all orders of the so-called ladder and bubble diagrams, and the simultaneous application of the strong coupling Eliashberg theory[15,16] allows the study of the superconducting phase transition by including the dynamical short-range order. These correlation processes generate a purely electronic pairing interaction that has a significant energy and momentum dependence for the intra plane $[V_{\parallel}(\mathbf{k},\omega)]$ and interplane contributions $[V_\perp(\mathbf{k},\omega)]$ in the bilayer or in the monolayer $[V_\perp(\mathbf{k},\omega) = 0]$. Here, in constrast to the usual BCS-theory of superconductivity, where two electrons feel an net attraction due to polarizing the medium of phonons, the paired electrons from the pairing medium themself. Furthermore, our method to solve the Eliashberg equations yields directly interesting temperature-, frequency- and momentum-dependent properties of the model that can be compared to experimental measurements. In particular, we are interested in the electronic self-energy $\Sigma_\lambda(\mathbf{k},\omega)$, which includes all correlation effects, the corresponding single particle scattering rate $\tau_\lambda^{-1}(\mathbf{k},\omega) = -\text{Im}\{\Sigma_\lambda(\mathbf{k},\omega)\}$ and the spectral density $\varrho_\lambda(\mathbf{k},\omega)$ with $\lambda = \pm$. Moreover, the superconducting gap function $\Delta_{\parallel}(\mathbf{k},\omega)$ $[\Delta_\perp(\mathbf{k},\omega)]$ describes a Cooper pair formation, where both partners come from one (different) layers and includes the dynamical formation and destruction of Cooper pairs.[17]

In figure 1(a) we present our results for $T_c$ of the mono- and the bilayer Hubbard model within FLEX approximation as schematic phase since its dependence on the doping $x$ is one of the most important characteristic properties of cuprates superconductors. $T_c$ was obtained similar to the method described in Ref. 16 and the order parameter has a $d_{x^2-y^2}$ symmetry for all values of $x$ and $t_\perp$. Interestingly, there are three doping regimes with qualitatively different behavior: the *overdoped*, the *optimally doped* and *underdoped* systems.

In the following we will argue, for simplicity we start with the monolayer, that strong antiferromagnetic correlation and remarkable lifetime effects dominate the quasi-particle properties and lead to an *optimal doping* where $T_c$ becomes largest. In the *overdoped* compounds ($x > 0.13$), we find that the lifetime $\tau(\mathbf{k},\omega)$ of the single

# Spin Fluctuation Effects in High-$T_c$ Superconductors

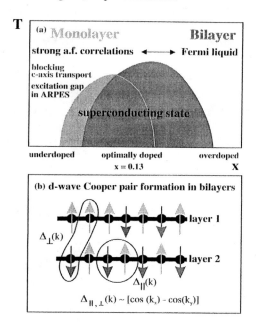

**Figure 1.** (a) Phase diagram of mono- and bilayer high-$T_c$ superconductors. The *underdoped* compounds are characterized by the formation of shadow states and by the blocking of the c-axis hopping in bilayers systems. (b) The in-plane Cooper pair formation $\Delta_\parallel$ in bilayer compounds fits perfectly with the spin-fluctuation induced interaction while interplane pairing $\Delta_\perp$, that can only be present when two in-plane spins are parallel, competes with the short-range antiferromagnetic order.

particles are relatively large with respect to the typical lifetime of the Cooper pairs ($\sim \Delta_o^{-1}$ with $\Delta_o = \mathrm{Re}\{\Delta(\mathbf{k}, \Delta_o)\}$) and $T_c$ increases with decreasing doping, because the pairing interaction $V_\parallel(\mathbf{k}, \omega)$ increases when the antiferromagnetic instability near half-filling is approached. Now, in the *optimally doped* system ($x \approx 0.13$), there is a constructive interplay between $d$-wave superconductivity and dynamical antiferromagnetism due to the comparable time scales of both types of excitation. Furthermore the real space behavior of $V_\parallel(\mathbf{k}, \omega)$ yields in leading order a strong on-site repulsion and a nearest-neighbor attraction. This perfectly fits with the real space Cooper pair formation since the dominate functional contribution to the $d_{x^2-y^2}$ order parameter down to the *optimal doping* is $\cos(k_x) - \cos(k_y)$ that corresponds to a nearest-neighbor pairing. Here, the first electron polarizes the spin system and the resulting spin excitation propagates to a neighboring site. Then, the second electron feels due to the retarded nature of this interaction an attraction leading to the formation of a Cooper pair. Moreover, this interference phenomena leads not only to an optimized superconducting state, but also to an enhancement of the antiferromagnetic correlations characterized by an increased $V_\parallel(\mathbf{k}, \omega)$ with respect to the normal state.[16] By further decreasing the doping concentraction, we observe a dramatic reduction of $\tau(\mathbf{k}, \omega)$ (Ref. 18) compared to $\Delta_o^{-1}$ which change the dynamical character of the spin fluctuations. Thus, in the *underdoped* region $T_c$ is not governed by the still increasing effective pairing interaction, but by the fact that the single-particle lifetime becomes too small to guarantee well defined single-particles during the pairing process. Phys-

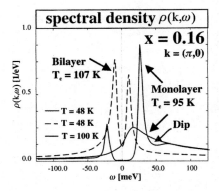

**Figure 2.** Spectral density for the antibonding band of the bilayer with $T_c = 107$ K and for the monolayer with $T_c = 95$ K. Note the suppression of spectral weight with respect to the normal state causing the dip structure at $\omega = 50$ meV below $T_c$.

ically, this signals the tight coupling of neighboring spins, which forces the system to pair single-particles that are further apart in correspondence with our observation that higher harmonics occur in the k-dependence of $\Delta(\mathbf{k},\omega)$ below the *optimal doping*.[19]

So far we only considered the in-plane correlations in an insulated $CuO_2$-plane. However, when various layers are coupled like in the bilayer systems YBCO and BSCCO the influence of the interplane interaction $V_\perp(\mathbf{k},\omega)$ on the superconducting state and $T_c$ is a significant and important open question. By investigating $V_\perp(\mathbf{k},\omega)$ we find in agreement with neutron scattering experiments[8,10] that it has the same k-dependence as $V_\parallel(\mathbf{k},\omega)$, but with an opposite sign yielding antiferromagnetic correlated bilayers. This coupling is strongest for low doping concentractions, where inter- and intralayer interactions become comparable in magnitude although we always find $V_\perp(\mathbf{k},\omega) < V_\parallel(\mathbf{k},\omega)$. By considering the doping dependence of $T_c$, we find that the *optimal doping* shifts to larger $x$, because $T_c$ is suppressed for *underdoped* systems ($x = 0.09$: $T_c^{mono} = 85$ K to $T_c^{bi} = 77$ K; $x = 0.12$: $T_c^{mono} = T_c^{bi} = 92$ K) with respect to the monolayer whereas it is enhanced for *overdoped* compounds ($x = 0.16$: $T_c^{mono} = 95$ K to $T_c^{bi} = 107$ K). This remarkable behavior is directly related to the spin fluctuaction mediated pairing process and in particular to the $d_{x^2-y^2}$ symmetry of superconductivity. By analyzing the different contributions to the gap function we find that interlayer pairing becomes significant in bilayer cuprates and in particular for small doping the d-wave $\Delta_\perp(\mathbf{k},\omega)$ almost equals. $\Delta_\parallel(\mathbf{k},\omega)$. However, the additional energy gain compared to the monolayer has to complete with the corresponding energy loss that is caused by magnetic frustration. In distinction to the in-plane pairing case the interference between the bilayer correlations and the interplane d-wave $[\Delta_\perp(\mathbf{k}) \sim \cos(k_x) - \cos(k_y)]$ is destructive for perfect antiferromagnetic order. Nevertheless as shown in Fig. 1(b), local magnetic fluctuations allow parallel oriented neighboring spins which permits a significant $\Delta_\perp(\mathbf{k},\omega)$. Therefore, the behavior of $T_c$ with increasing $t_\perp$ for a given concentraction depends sensitively on which contribution is dominant. For larger doping, the magnetic correlation length $\xi$ is relatively small ($\sim 2$–$3$ lattice spacings)[24] and thus $T_c$ increases due to the interlayer Cooper pairing, whereas for small doping, $\xi$ and the region of tightly arranged antiferromagnetically correlated spins is much larger leading to a reduced $T_c$.

In figure 2 we plot the spectral density $\varrho(\mathbf{k},\omega)$ of the monolayer and compare them with $\varrho_-(\mathbf{k},\omega)$ of the bilayer for **k** near the FS as typical example of the quasi-particle excitation spectrum in the superconducting state. Here a significant

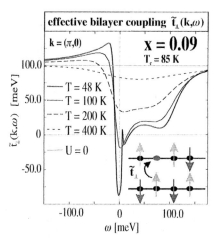

**Figure 3.** Effective interlayer hopping $\tilde{t}_\perp(\mathbf{k},\omega)$ compared with interaction-free case ($U = 0$). Inset: $\tilde{t}_\perp(\mathbf{k},\omega)$ is reduced since a quasi-particle with an up (down) spin that hops to an empty site also hops in the opposite antiferromagnetic environment due to the interlayer coupling. This causes the unfavorable parallel alignment of neighborint spins.

consequence of the magnetic frustration effects in bilayer cuprates can be observed in the single-particle excitation spectrum. It is interesting that although the bilayer system has a higher transition temperature, the physical gap in the monolayer ($\Delta^{mono}(\pi,0) = 24$ meV) is larger than in the bilayer ($\Delta^{bi}(\pi,0) = 10$ meV) resulting from the enhanced scattering processes. Furthermore, the pronounced dip structure above the Fermi energy with respect to the normal state in Fig. 2 is another intrinsic feature of the electronic pairing mechanism that we found also for YBCO or BSCCO-like dispersions, where it appears below the Fermi level. The reason for the dip-structure in $\varrho(\mathbf{k},\omega)$ at $(\pi,0)$ is the interplay between Cooper pairing and dynamical short-range order and is caused by the opening of the superconducting gap that was already discussed in Ref. 16. Most interestingly, the temperature dependence and energetic position of these fine structures agrees well with similar dips found by ARPES[7,20] and by SIS-tunneling measurements,[21] which further supports the importance of spin fluctuactions for high-$T_c$ superconductivity.

Presently, the unusual c-axis transport properties of the layered cuprates are intensively discussed.[9] Here, in particular two key issues are not well understood: (a) the mechanism that reduces the interlayer hopping in comparison with band structure calculations and (b) the experimental observation that *underdoped* LSCO and YBCO systems are characterized by a semiconductor-like temperature-dependence of the c-axis resistivity while *overdoped* cuprates show a metallic-like behavior. In the following we present results indicating that the antiferromagnetic bilayer coupling in the cuprates has important consequences for *underdoped* compounds and cause a blocking of the c-axis hopping.

In figure 3 we plot the renormalized frequency and momentum dependent interplane hopping amplitude $\tilde{t}_\perp(\mathbf{k},\omega)$ given by

$$\tilde{t}_\perp(\mathbf{k},\omega) = t_\perp - \mathrm{Re}\{\Sigma_\perp(\mathbf{k},\omega)\} = t_\perp - \tfrac{1}{2}\mathrm{Re}\{\Sigma_+(\mathbf{k},\omega) - \Sigma_-(\mathbf{k},\omega)\}.$$

This quantity follows directly from the matrix Dyson equation in the layer representation and from the self-energies of the bonding and antibonding band. We find that the sign of Re $\{\Sigma_\perp(\mathbf{k},\omega)\}$ is determined for small $\omega$ by the sign of $V_\perp(\mathbf{k},\omega)$, such that $\tilde{t}_\perp(\mathbf{k},\omega) \ll t_\perp$ for $\omega < t_\perp$. This interesting phenomenon can be found along the entire FS although it is most pronounced at $(\pi,0)$. Physical, this behavior is related to the fact that each interlayer hopping process is accompanied by spin-flips due to the opposite antiferromagnetic environment in the other plane as can be seen in the inset of Fig. 3. Thus, by considering the time scales of this phenomenon, the typical interlayer hopping time $\tau_\perp$ is in a correlated bilayer not determined by $t_\perp^{-1}$ alone but in addition by local redistributions of the parallel oriented spins which has to be compared with the lifetime $\tau_\parallel(\mathbf{k},\omega) = -\text{Im}[\Sigma_\parallel(\mathbf{k},\omega)]$ of the quasi-particles in the plane. Consequently, since $\tau_\parallel < \tau_\perp$ the effective interplane hopping at the Fermi energy is blocked for this *underdoped* compound, which leads to the disappearance of a coherent quasi-particle tunneling process although due to the frequency dependence of Re $\{\Sigma_\perp(\mathbf{k},\omega)\}$ there are still splitted bonding and antibonding bands for high energies. Furthermore, notice also the remarkable temperature dependence of $\tilde{t}_\perp(\mathbf{k},\omega)$. While the blocking of the interlayer hopping disappears for larger temperatures ($T \approx 500$ K) due to decreasing magnetic correlations, we observe that $\tilde{t}_\perp(\mathbf{k},\omega)$ becomes larger and negative below $T_c$ for very small energies. This is due to the fact that the scattering rates are reduced in the region of the superconducting gap since the quasi-particle states are pushed to larger binding energies.

Comparing this incoherent c-axis hopping with the interesting confinement idea of Ref. 3, we find in our calculation that the strong renormalization of $t_\perp$ occurs only in *underdoped* systems whereas for large doping the suppression of $t_\perp$ is rather small and a coherent transport ($\tau_\parallel > \tau_\perp$) is restored. Furthermore, as discussed above for small doping we still observe a coherent interplane Cooper pair formation and one superconducting system with only one global phase of the order parameter. Finally, the remarkable doping and temperature dependence of our results should be of particular importance for c-axis transport properties and the observed crossover from a three-dimensional to a two-dimensional behavior of the resistivity with doping. Our results indicate these interesting properties might be related to the antiferromagnetic bilayer correlations and to a crossover from a coherent, hopping-like dynamics to a phonon or impurity dominated c-axis transport.[22]

In figure 4 we demonstrated another important result of our theory, the occurrence of shadows of the FS[23] as fingerprint of the dynamical antiferromagnetic correlation that were observed by Aebi *et al.*[4] As shown schematically in Fig. 4(a), antiferromagnetic long-range order yields in distinction to a paramagnetic compound two FS crossings on the path from $(0,0)$ to $(\pi,\pi)$ because the long-range order imposes a new, reduced antiferromagnetic Brillouin zone (AF-BZ) and a magnetic gap occurs between the solid and dashed branch of the dispersion. However, since the cuprates are not long-range ordered in the doping region where superconductivity appears, the observation of shadows of the FS are qualitative new features of the excitation spectrum[24] that can be explained in terms of strong antiferromagnetic spin fluctuations.

In figure 4(b) we plot the quasi-particle dispersion $\varepsilon(\mathbf{k})$ for an *underdoped* system and compare it with the interaction-free case ($U = 0$). Comparing these results with Fig. 4(a) we find that a rather short-ranged correlated systems can generate a quasi-particle spectrum which resembles that of a long-range ordered compound. In detail, the first interesting feature of our data is that the quasi-particle band is pushed to the Fermi energy and becomes very flat with respect to the uncorrelation $\varepsilon_0(\mathbf{k})$ which leads to the observed extended van-Hove singularity in various cuprates.[5] More

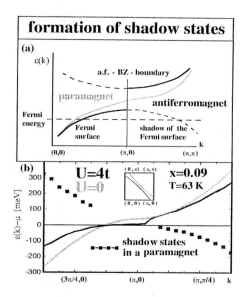

**Figure 4.** Formation of the shadow states. (a) Schematic quasi-particle dispersion of a paramagnetic and an antiferromagnetic systems, where due to the long-range order a magnetic gap and a shadow of the Fermi surface appears. (b) Our calculation show (i) that the dispersion becomes very flat with respect to the interaction-free case near $(\pi, 0)$ and (ii) that shadow states occur in *underdoped* systems although there is no long-range order.

interestingly, the appearance of additional bands that are shifted by $\mathbf{Q} = (\pi, \pi)$ from the main band are the shadow states as precursor of the antiferromagnetic instability in the paramagnetic state. They are caused by a transfer of spectral weight from the FS at $\mathbf{k}$ to its shadow at $\mathbf{k} + \mathbf{Q}$ and becomes observable below $x = 0.13$, *e.g.* the *optimal doping*. One important consequence of the shadow state formation is the occurrence of the $\mathbf{k}$-dependent pseudogap in the spectral density that is very similar to the $d$-wave gap of the superconducting state.

In figure 5 we present results for the spectral density $\varrho(\mathbf{k}, \omega)$ of the monolayer model dispersion for a $\mathbf{k}$-point near $(\pi, 0)$. For temperatures well below $T_c$, there is a large superconducting gap to the $d$-wave symmetry of the pairing state that closes when $T$ is approaching $T_c$. Nevertheless, in the normal state a large pseudogap remains that is due to the shadow band formation. Consequently, it has due to its magnetic origin a different temperature scale and vanishes not at $T_c$ but at $T^* \approx 400$ K. These effects should be observable by ARPES experiments and give a natural explanation of the important observations by Loeser *et al.*[6] Here a pseudogap in the excitation spectrum of the normal state in *underdoped* BSCCO systems was found, that closes at the *optimal doping* and at $T^*_{\exp} \gg T_c$. Moreover it was found that it has an anomalous momentum dependence, *i.e.*, that similar to the magnitude of the $d$-wave superconducting gap it is maximal at the $(\pi, 0)$-point (region A) and minimal at $(\pi/2, \pi/2)$ (region B)[6] as shown in the upper inset of Fig. 5. Furthermore, our theory also explains the anomalous $\mathbf{k}$-dependence of the ARPES measurements since we find evidently no superconducting gap at $(\pi/2, \pi/2)$ and in addition a much weaker pseudogap, respectively shadow state formation. This is due to the fact that the shadow band intensity is not only dependent on the magnetic correlation length $\xi$,

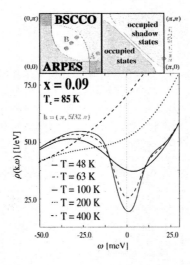

**Figure 5.** Spectral density $\varrho(\mathbf{k},\omega)$ for an occupied shadow state (right, upper inset) in an *underdoped* compound. Region in k-space, where pseudogaps were found in BSCCO by ARPES measurements (left, upper, inset). Below $T_c$: Excitation gap is due to $d$-wave order parameter. Above $T_c$: Pseudogap is related to formation of shadow states. It vanishes at $T^* = 400$ K $\gg T_c$, disappears at the *optimal doping* and can only be found near $(\pi,0)$.

but moreover on the flatness of the quasi-particle dispersion which is most pronounced in region A as experimentally observed[6] and shown in Fig. 4(b). Finally, our results clarify that these new and important experimental observations are unambiguously due to short-range spin fluctuations and are probably not related to the existence of preformed Cooper pair without long-range coherence.[6,25,26]

In this report we have presented properties of the Hubbard Hamiltonian as model for the high-$T_c$ superconductors. It has been demostrated that this approach and its numerical treatment within the FLEX approximation describing the low energy spin dynamics explains a variety of experimental observations. Here we obtained a $d$-wave symmetry of the order parameter and an *optimal doping* for the superconducting state. In this context we discussed important fine structures in the quasi-particle excitation spectrum of mono- and bilayer systems that compare well with recent experiments. Thus our results yield further evidence for magnetically driven high-$T_c$ superconductivity.[27]

# References

1. E. Dagotto, *Rev. Mod. Phys.* **66**, 763 (1994).
2. D. J. Scalapino, *Phys. Rep.* **250**, 331 (1995).
3. S. Chakravarty *et al.*, *Science* **261**, 337 (1993); P. W. Anderson, *Science* **268**, 1154 (1995); P. W. Anderson, *Science* **256**, 1526 (1992).
4. P. Aebi *et al.*, *Phys. Rev. Lett.* **72**, 2757 (1994).
5. Z. X. Shen and D. S. Dessau, *Phys. Rep.* **253**, 1 (1995).
6. D. S. Marshall *et al.*, *Phys. Rev. Lett.* **76**, 4841 (1996); A. G. Loeser *et al.*, to be published in *Science*.
7. D. S. Desau *et al.*, *Phys. Rev. Lett.* **66**, 2160 (1991).

8. J. M. Tranquada *et al.*, *Phys. Rev. B* **46**, 5561 (1992).
9. S. L. Cooper and K. E. Gray, in *Physical Properties of High-$T_c$ Superconductors IV*, edited by D. M. Ginsberg (World Scientific, Singapore, 1994).
10. Although the LSCO system has only one layer within a unit cell, the nearest-neighbor planes from different cells are antiferromagnetically correlated. However, this intercell coupling is much smaller than the bilayer effects in YBCO.
11. The FS of this model dispersion is similar to the LSCO system. Nevertheless we can draw similar physical conclusion by using bare dispersion that are more appropriate for YBCO or BSCCO or by taking a **k**-dependent $t_\perp(\mathbf{k}) = \frac{1}{4} t_\perp [\cos(k_x) - \cos(k_y)]^2$.
12. O. K. Andersen *et al.*, *J. Phys. Chem. Solids* **56**, 1573 (1995).
13. We carefully checked the dependence of our results on $U$ and $t_\perp$, but found no physical significant changes in our data up to values of $U = 6t$ and for $t_\perp = 0.1 - 0.8t$. However, $t_\perp = 0.4t$ was suggested by LDA calculations for YBCO.
14. N. E. Bickers *et al.*, *Phys. Rev. Lett.* **62**, 961 (1989).
15. P. Monthoux *et al.*, *Phys. Rev. Lett.* **72**, 1874 (1994); C. H. Bickers *et al.*, *Phys. Rev. Lett.* **72**, 1870 (1994); T. Dahm *et al.*, *Phys. Rev. Lett.* **74**, 793 (1995).
16. S. Grabowski, M. Langer, J. Schmalian, and K. H. Bennemann, *Europhys. Lett.* **34**, 219 (1996).
17. In bilayer systems we only considered intraband Cooper formation. This refers to even parity pairing with respect to the bilayer inversion symmetry yielding $\Delta_\perp(\mathbf{k}, \omega) \neq 0$ and a larger $T_c$ than the odd (interband) pairing state with $\Delta_\perp(\mathbf{k}, \omega) = 0$. See also J. Maly *et al.*, *Phys. Rev. B* **53**, 6786 (1996). The actual calculations were performed on a $(64 \times 64)$ square lattice with an energy resolution of $0.014t \approx 4$ meV. The numerical procedure is described in J. Schmalian, M. langer, S. Grabowski, and K. H. Bennemann, *Comp. Phys. Comm.* **93**, 141 (1996).
18. Note that $\tau^{-1}(\mathbf{k}, \omega)$ generates naturally an energy and consequently a temperature scale for the magnetic excitations. By comparing the doping dependence and the absolute magnitude of $\tau^{-1}(\mathbf{k}, \omega)$ at the FS and the Fermi energy with the characteristic temperature $T^{AF}$ observed in transport measurements for LSCO by H. Y. Hwang *et al.*, *Phys. Rev. Lett.* **72**, 2636 (1994), we find an excellent agreement with the experimental data.
19. The reason for the *optimal doping* in our results is physically different from the antiferromagnetic van Hove scenario (AFVH) by Dagotto *et al.*, *Phys. Rev. Lett.* **74**, 310 (1995). In the AFVH approach $T_c$ becomes maximal when the peak in the momentum averaged density of states $\varrho(\omega)$ crosses the Fermi level, but is not dependent on lifetime effects and the variation of the pairing interaction on doping.
20. H. Ding *et al.*, *Phys. Rev. Lett.* **76**, 1533 (1996).
21. D. Mandrus *et al.*, *Nature* **351**, 460 (1991).
22. R. J. Radtke *et al.*, *Phys. Rev. B* **53**, R552 (1996).
23. A. P. Kampf *et al.*, *Phys. Rev. B* **42**, 7967 (1990).
24. M. Langer, J. Schmalian, S. Grabowski, and K. H. Bennemann, *Phys. Rev. Lett.* **75**, 4508 (1995).
25. V. J. Emery and S. A. Kivelson, *Nature* **374**, 434 (1995).
26. S. Doniach and M. Inui, *Phys. Rev. B* **41**, 6668 (1990).
27. We acknowledge the financial support of the DFG, thank Z. X. Shen for sending us his papers prior to publication and E. Dagotto for useful discussions.

# Superconductivity and Magnetism in $f$ Electronic Systems

R. Escudero,[1] F. Morales,[1] A. Briggs,[2] and P. Monceau[2]

[1] *Instituto de Investigaciones en Materiales*
*Universidad Nacional Autónoma de México*
*Apartado postal 70-360, México, D. F.*
*MEXICO*

[2] *Centre de Recherches sur les Tres Basses Temperature*
*Centre National de la Recherche*
*Boite Postal 166X, 38042 Grenoble CEDEX*
*FRANCE*

## Abstract

We present two examples of very correlated electronic systems. One related with the behavior of a heavy electron compound, and other related with reentrant superconductivity. The first system; the heavy fermion compound is an alloy formed with transition metals and rare-earth and/or actinides elements whose $f$-like atomic electrons behave as noninteracting spins near room temperature but hybridize with the conduction electrons to form a coherent, strongly correlated electronic system at low temperatures. We analyze the behavior described above that occurs in general in heavy fermion systems, particularly a description of the compound $URu_2Si_2$ is given. This compound presents two types of very correlated phenomena: Spin density waves formation at about 17 K, and the superconducting behavior at about 1.5 K. Secondly, we describe the behavior of a reentrant system produced by the competition of superconductivity and magnetism. This compound $HoMo_6S_8$ is known as a Chevrel phase. One finds a beautiful competition between this two phenomena giving as a result a reentrant superconducting behavior at low temperatures. In this system both ground states compete for the same portions of the Fermi surface. One of the fascinating aspects of heavy fermions and reentrant materials is their unusual superconducting behavior.

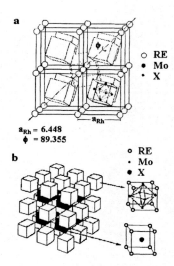

**Figure 1.** Two different aspects of the Chevrel phase structure.

## I. Introduction

Superconductivity and ferromagnetism are two different cooperative phenomena with antagonistic character that compete for parts of the Fermi surface. In some materials as heavy fermions, Chevrel phases, borocarbides, and high-temperature superconductors, magnetism (particularly antiferromagnetism) becomes the precursor of the superconducting behavior.

Superconductivity and magnetic ordering are two electronic behaviors that results from very correlated electronic processes. Both can be observed in the same specimen under different circumstances. The interest of the interplay between superconductivity and magnetism started in 1957 when Ginzburg[1] proposed the existence of a ferromagnetic superconductor theoretically. The first experimental results in this area of research were obtained by Mathias and coworkers in 1958. They studied lanthanum with magnetic impurities of gadolinium and observed that 1% or 2% of gadolinium destroy superconductivity.[2] Abrikosov and Gor'kov studied theoretically the problem of a superconductor with paramagnetic impurities, motivated by Mathias experiments.[3] Their predictions were experimentally confirmed with great success.

The possibility to study a specimen in which superconductivity and long-range magnetic order are present was provided by Chevrel. He and his coworkers synthesized a series of compounds with general formula $RE_xMo_6 X_8$, where RE is a metal or a rare earth metal, X can be one of the three chalcogenides; S, Se, or Te.[4] Figure 1(a) shows the rhombohedric crystalline structure of the Chevrel phases, and Fig. 1(b) shows the basic blocks needed to form the Chevrel structure.

Some of these compounds, now called Chevrel phases, present superconductivity and antiferromagnetism and the singular case, $HoMo_6S_8$, presents ferromagnetism and superconductivity. This latter compound is known as a reentrant superconductor named by its extraordinary temperature behavior that occurs when the two processes are settled, and compete one each other. In figure 2(a) it is shown a schematic plot of the resistance *versus* temperature of $HoMo_6S_8$. This is a drawn to observe with better clarity the different regions of interest of the compound. In figure 2(b) it is shown the

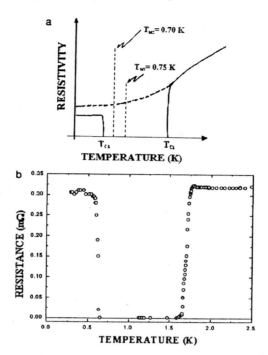

**Figure 2.** (a) Qualitative behavior of the electrical resistence as function of temperature for $HoMo_6S_8$. (b) Experimental measurements. Note the transition temperatures.

experimental part of the same phase. The plots are shown in a region slightly above the superconducting transition temperature. In this material the superconducting transition temperature is $T_{c1} = 1.6$ K. On lowering the temperature the superconducting phase is preserved down to the reentrance phase transition $T_{c2} = 0.65$ K to the normal state. Below of this temperature the compound behaves as a *normal* ferromagnetic system. The ferromagnetic state settles at two temperatures, the first one at $T_{M1} = 0.75$ K and the second at $T_{M2} = 0.70$ K, then due to these transitions a sinusoidal magnetic order appears. These sinusoidal magnetic phases form magnetic domains in which are trapped superconducting regions. These magnetic domains will be more dominant as temperature goes down, squeezing therefore the superconducting phase trapped inside the regions between domains. Nonetheless, the superconducting phase is also increasing its strength as the temperature goes down, and the competition is settled between the two processes. Which will become dominant? It is difficult to predict; but apparently, without further analysis, one can say that magnetism will be the dominant phase. In the schematic illustration of Fig. 2(a), clearly we can see the anomalous behavior that occurs at lower temperatures when the compound is completely in the normal state. At that lower temperature the value of the normal resistance is smaller than the value obtained if it is extrapolated from the initial part of the resistance *vs.* temperature before the onset of $T_{c1}$ (Ref. 5).

In general it is belived that the magnetic order, particularly ferromagnetism, will be the dominant phenomena. However, experimental results indicate that still at very low temperatures, in the normal region of a reentrant superconductor, some small superconducting regions still remain. These remanent superconducting parts are the

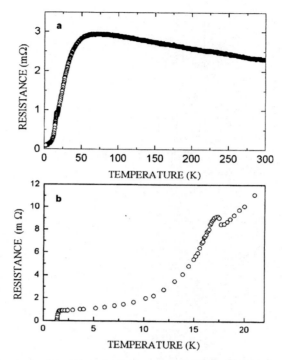

**Figure 3.** Resistance as function of temperature in $URu_2Si_2$ single crystal. It is worth noting the antiferromagnetic and superconducting transitions. The inset is an amplification of the superconducting transition.

reason of the lower value of the electrical resistance at lower temperatures. It is an experimental fact that the superconducting phase persist at temperatures about 100 mK (Ref. 5).

The conclusion about of the interaction between two antagonistic phenomena as ferromagnetism and superconductivity, is that more experimentation and analysis are required in order to understand in more detail the microscopic phenomena that govern the electronic processes occurring in the Fermi surface and the interactions that form the two ground states.

It is interesting to mention that until today not many reentrant superconductors have been discovered. The first and better known compound is $ErRh_4B_4$ (Ref. 6). More recently the Ho-Chevrel phase mentioned here, some Sn-based compounds, and some of the new borocarbides compounds also based in Ho ($HoNi_2B_2C$) have been sintetized. However the phenomena of reentrance to the normal state may be induced in some superconductors when a magnetic field is externally applied.[7]

Other two families of compounds in which superconductivity and magnetic order occur, are the heavy fermion superconductors and the high $T_c$ cuprates. The former compounds are characterized by its high effective electronic masses. The phenomenon in general settles at lower temperatures, where it can be observed that the electronic masses are increasing to values as high as many orders of magnitude of the free electronic mass (it was observed in some heavy fermion compounds enhancement as big as $10^2$ to $10^3$ times the electronic free mass $m_e$). This characteristic confers the name to the heavy fermions systems.

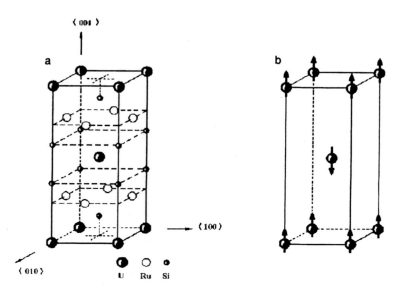

**Figure 4.** (a) Body-centered tetragonal structure of the heavy fermion $URu_2Si_2$. (b) Magnetic structure formed by the magnetic moments of $URu_2Si_2$ below 17 K.

As a typical illustration of the behavior of a heavy fermion alloy we are plotting in Fig. 3 the resistance *vs.* temperature characteristic of a single crystal of $URu_2Si_2$. In figure 3(a) we show the high-$T$ behavior with the anomalous increasing resistance characteristic of the heavy fermions when the temperature decreases. In figure 3(b) we show the behavior at low temperatures where the antiferromagnetic and superconducting transition can be observed.

Figure 4(a) illustrates the simple crystal structure of this compound; simple as compared to the Chevrel phases. The magnetic behavior arises from the uranium atoms, the arrows represent the magnetic ordering that forms the antiferromagnetic transition at a temperature about 17 K [Fig. 4(b)].

In the case of the high temperature superconductors ($Cu-O_2$ based), it is well known that a precursor of the superconducting phenomenon is the antiferromagnetic behavior occurring at higher temperatures than the superconducting transition.

The common aspect of these superconducting families (reentrants, heavy fermions, and high $T_c$ compounds) is the presence of atoms with $4f$ or $5f$ electronic orbitals. These electrons are the responsibles of the magnetic behavior. In the case of the high $T_c$ compounds Cu is mainly the responsible of the magnetic behavior involved with the antiferromagnetic transition at high temperatures, and also is involved with the superconducting transition.

In this paper we present a experimental study performed on $URu_2Si_2$ single crystals, a heavy fermion system, and the $HoMo_6S_8$ a Chevrel phase. To study these systems we mainly used tunneling and point contact spectroscopies as techniques of analysis. Before presenting the experimental part we describe in a very concise manner the differences between BCS and non-BCS superconductors.

## II. Coexistence of Superconductivity and Magnetism

### II.1 BCS Superconductors

Before to the discovery in the last two decades of the heavy fermion superconductors few doubts existed to believe that the basic mechanism of electron-phonon to produce superconductivity was the only existing or the most viable in nature. This physical mechanism was totally explained by the BCS theory in their basic model.

All other proposed mechanisms were thought to be not very probable. If additional improvement was required it was questioned to go further and to use Eliashberg theory. In addition also was believed that the basic symmetry of the order parameter was of the $s$-type, perhaps only with small asymmetries due to reasons not very important but not well understood.

In the BCS theory the Cooper pairs are formed by electrons with opposite momenta and spin. If a magnetic field is applied or a paramagnetic impurity is introduced into the superconducting material, then the interaction with the superconducting electrons will cause breaking of Cooper pairs and therefore depressing the critical temperature. However, Nature provide other systems in which the behavior was not so simple like BCS model; i.e., non-electron-phonon coupling. These other systems are the compounds known as heavy fermions. The evidence of further mechanisms was also reinforced with the discovery of the high-$T_c$ compounds. Here again it is believed that magnetism and pairing of electrons are enough compatible to produce at lower temperature the superconducting transition, and the antiferromagnetism phase at high temperature. It is interesting to mention that the long process to understand superconductivity only as an electron-phonon process took too many theorist to *explain* why the maximum of the transition temperature must be only around 30 K.

### II.2 Non-BCS Superconductors

According to the previous Section now we know that many types of pairing processes can occur in Nature. At least we are confident that two are different to BCS pairing, and that of course can be catalogued as non-BCS processes. This type of pairing with non-phonon mechanism exists in heavy fermions and in high-$T_c$ compounds, and of course in superfluid Helium. Other type of mechanism, called excitonic mechanism, was predicted by Little[8] and could be possible to exist in semiconducting systems or low-dimensional organic compounds. In Table I we show a catalogue of conventional and unconventional superconducting materials.

### II.3 Heavy Fermions

The heavy fermion systems can be mainly distinguished from the ordinary normal metallic state, by the magnitude of the electronic specific heat and the Pauli paramagnetic susceptibility. The behavior of the $T$-dependence of the resistivity is also a different feature to normal metals. For a heavy fermion compound the Sommerfeld constant $\gamma$, may be as large as 8000 mJ/mole K$^2$, as is the case of YbBiPt. If the high electronic specific heat is attributed to the enhanced mass $m^*$, then it may be of the order of $\sim 1000 m_e$.

The heavy fermions are intermetallic compounds or alloys, and are exclusively found in systems with lanthanides and actinides elements (principally Ce and U). They are characterized by a strong hybridization of the $4f/5f$ electrons with the conduction electrons.

## Table I

The superconducting materials shown in this table are divided into two groups named conventional and unconventional superconductors. By conventional superconductors we mean those materials whose behavior is described by BCS theory. The other catalogued as non-conventional, are those in which the pairing mechanism is due to interactions other than electron-phonon coupling. X might be $PF_6$, $ClO_4$, $AsF_6$, etc. A is an alkaline atom or a combination of two.

| Superconductors | | | | |
|---|---|---|---|---|
| Conventional | | Unconventional | | |
| Elements | Alloys | Heavy fermions | Organic | Perovskites |
| Hg (4.5) | $V_3Si$ (15) | $UPt_3$ | $(SN)_x$ | $Na_xWO_3$ |
| Pb (7.2) | $Nb_3Sn$ (18) | $UBe_{13}$ | $(TMTSF)_2X$ | $SrTiO_{3-\delta}$ |
| Nb (9) | $Nb_3Ge$ (23) | $URu_2Si_2$ | $(ET)_2X$ | $Ba(Bi_{1-x}Pb_x)O_3$ |
| Sn (3.7) | $PbMo_6S_8$ | $UNi_2Al_3$ | ET-CuSn | $La_{2-x}Ba_xCuO_{4-\delta}$ |
| Al (1.2) | $HoMo_6S_8$ | $UPd_2Al_3$ | $A_3C_{60}$ | $YBa_2Cu_3O_{7-\delta}$ |
| In (3.4) | $LaMo_6S_8$ | $CeCu_2Si_2$ | TTF-TCNQ | $Bi_2SrCa_2Cu_3O_{10}$ |
| V (5.4) | $ErRh_4B_4$ | | | $Tl_2Ba_2Ca_2Cu_3O_{10}$ |
| | | | | $Ba_{1-x}K_xBiO_3$ |
| | | | | $HgBa_2Ca_2Cu_3O_{8+\delta}$ |
| | | | | $Hg_{0.8}Pb_{0.2}Ba_2Ca_2Cu_3O_x$ |
| | | | | $Sr_2RuO_4$ |

The heavy fermion behavior is delimited by a characteristic temperature $T^*$. This is typically between 10 K and 100 K. At temperatures above $T^*$ the behavior of the electron gas is similar to a ordinary metal, and the electronic density of states $N(E)$ is a smooth function of the energy. The magnetic behavior is Curie-Weiss type with similar magnetic moments to the free ion and the $f$ electrons are localized and weakly coupled with the conduction electrons. Below $T^*$ the hybridization of the $f$-bands takes place and the effective mass and the $N(E)$ are enhanced around the Fermi energy. The magnetic behavior becomes independent of temperature indicating the delocalization of some $f$-electrons.

For heavy fermion systems there is not a unique ground state; it may be a superconducting ground state, a magnetic ground state or both. However, it depends on the competition between the indirect RKKY exchange interaction and the Kondo interaction. The energy associated to these types of interactions have the form $k_B T_{RKKY} \sim (N_F|J|)^2$ and $k_B T^* \sim \exp[(-1)/(N_F|J|)]$, where $N_F$ is the density of states at the Fermi level and $J$ is the exchange integral.

In the superconducting state the heavy fermions show a potential temperature behavior in some of their physical properties. This behavior is a signature of the existence of points or lines in the Fermi surface in which the superconducting energy gap vanishes. The superconducting energy gap in heavy fermions has been studied using point contact spectroscopy giving information about the anisotropic behavior and the interaction strength between quasiparticles.[9-13]

In the particular compound involved with this study, $URu_2Si_2$, it is interesting to mention that this compound does not present a unique ground state; it exhibits two ground states, the antiferromagnetic state below 17.5 K, and the superconducting state below at about 1.2 K. It is important to remark that there are strong evidences that the superconducting behavior in this system is of an unusual character. It might be of anisotropic BCS type, or due to a triplet pairing process. This exotic pairing to form the superconducting state requires more careful experimental studies to unambiguously determinate exactly the type of paring and processes associated.

## III. Experimental

In the next paragraphs and in the rest of the article we describe our experimental procedures performed in the $URu_2Si_2$ and $HoMo_6S_8$.

The single crystals of $URu_2Si_2$ growth by Czochralsky method[14] were used in this study. These have typical dimensions of $2 \times 3 \times 0.5$ mm$^3$. The single crystals of $HoMo_6S_8$ have dimensions of $1 \times 1 \times 2$ mm$^3$ with irregular forms, and were grown as described by Peña et al.[15]

The critical temperature was determined by performing measurements of the electrical resistance $R$ *versus* temperature characteristic at four wires configuration using an AC bridge and a small AC current of the order of 10–100 $\mu$A.

The point contacts junctions (PC) were fabricated using very fine gold wires of high purity (99.999%) with diameters of $\phi = 5$ $\mu$m and Al wires (99.999%) $\phi = 10$ $\mu$m as one of the electrodes of the junctions. The other electrode is the specimen under study. The junction configuration is the standard one, consisting of the single crystal and Au or Al. To form this metallic contact we have to clean the native oxide that grows on the surface of the alloy. It is cleaned with an etching solution of diluted acid, or simply by scrapping material from the surface. The small diameter of the gold wire permits to form junctions of very small area. The junctions are made by crossing the gold wire (Al wire) over the surface of the sample, taking care that the contact area has to be as small as possible. To obtain this small area of contact, we have to cut a piece of the specimen leaving a sharp surface in such a form that when the wire crosses this part of the compound only touches it one edge of the sample. The junctions fabricated in this form have dimensions which in general are smaller than $1 \times 1$ $\mu$m, as observed with a microscope. The junctions were measured at temperatures from 0.3 K to about 30 K depending on the sample. We measured 30 junctions of each specimen and we checked that the results were reproducible. One of the important characteristics of these junctions is the thermal stability, that means stability respect to variations of temperature. This is different to what in general occurs in point contact junctions or scanning tunneling microscopes, where due to thermal contractions when temperature changes, the differential resistance of the junction changes and the part of the area of the crystal that is being sensed.

The tunnel junctions with gold electrodes were fabricated in the same manner. The only difference is that the native oxide was not removed from the surface, in one case. In other case it was removed from the surface with an etching solution and the cleaned piece of the specimen left in the air of the laboratory for different periods of time. For the case where the Al wire was the electrode the natural oxide that grows in the aluminum was used to make the electronic barrier needed in the tunnel junction. In order to improve the thermal contact of the specimen to the thermal bath this is glued on a glass substrate with GE barniz or silver paint.

The tunneling and PC differential resistance $dV/dI$ as function of the bias voltage $V$ was the characteristic that we measured using a standard lock-in and modulation technique. The applied bias voltage in the PCs was limited to prevent heating effects. Once we have collected all the information of $dV/dI$ versus bias voltage at different temperatures, we can obtain the differential conductance by taking the inverse of the differential resistance.

In the case of the URu$_2$Si$_2$ heavy fermion, in order to prevent heating effects when the zero bias resistance of the point contacts were between 0.5 $\Omega$ and 6 $\Omega$ we sweep the voltage between $\pm 2.5$ mV only. Higher values result in heating and distortion of the real characteristic. To estimate the radii of the contacts we used the electronic mean free path $l \simeq 100$ Å (Ref. 11), the resistivity $\rho \sim 40$ $\mu\Omega$ cm measured at 2 K (Ref. 16), and the interpolation Wexler's formula.[17] We obtain values between 320 Å to 3700 Å. The point contacts studied were therefore in the diffusive regime.

In HoMo$_6$S$_8$ the zero bias resistance of the point contacts were between 5 $\Omega$ and 50 $\Omega$, these values give radii values between 43 Å and 254 Å. We obtained these values using the extrapolation Wexler's formula also with $l \simeq 50$ Å (Ref. 18) and the resistivity measured at 2 K of 25 $\mu\Omega$ cm (Ref. 15).

The measurements at lower temperatures were done in a $^3$He refrigerator which permits operation between 0.3 K and helium temperature, and the measurements between 2 and 30 K in a conventional dewar with very precise control of temperature.

## IV. Results and Discussions

### IV.1 URu$_2$Si$_2$

The URu$_2$Si$_2$ single crystal was characterized by measurements of resistance as function of temperature. Figure 3 shows the typical resistance behavior, in which is observed the antiferromagnetic transition $T_N$ at about 17 K and the superconducting transition at 1.37 K. The antiferromagnetic transition is associated with the formation of a spin density wave (SDW). The SDW opens a gap on parts of the Fermi surface. The $R(T)$ around 17 K is similar to the behavior observed in Cr (Ref. 19), typical SDW antiferromagnet, and the exponential decreases below $T_N$ are the initial suggestion of the existence of a gap.

Previous experiments on URu$_2$Si$_2$-Al$_2$O$_3$-Al tunnel junctions[19,20] have shown that an energy gap opens when the heavy fermion transits from the high temperature to the antiferromagnetic state. Using a temperature smearing BCS electronic density of states we can extract the value of the energy gap $2\Delta = (11.7 \pm 0.2)$ meV and $\varepsilon_0 = 12$ meV. The parameter $\varepsilon_0$ is related to the number of electronics states that remain unnested after the Fermi surface has opened an energy gap. It is clear according to this tunnel experiments, that the energy gap is not completely open, result which is in complete agreement with measurements of specific heat and resistivity. They show that only small portions of the Fermi surface open a gap and that only about 40% of the fermi surface is used to form the SDW.[22]

In figure 5 we show a $dV/dI$ vs. $V$ curve measured at 342 mK. In this curve two features can be observed clearly. The two asymmetric minimums at around $\pm 7$ mV are associated with the presence of the energy gap related with the SDW. The second feature, very dim, and situated around zero bias voltage, is related with the superconducting energy gap. It is worth pointing out about the importance of this observation; because we are observing two processes arising around the Fermi energy. Both processes are competing for the maximum number of electrons to strength one of the respective phenomenon (SDW or superconductivity). The presence of the two energy

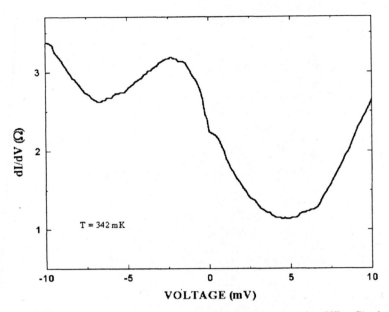

**Figure 5.** Differential resistance as function of bias voltage of a $URu_2Si_2$-Au(W) point contact. The deep at zero bias is related to the superconducting energy gap. The two relative minima are a signature of the spin density wave energy gap.

gaps is one of the stronger evidences that in this case, both, antiferromagnetic order and superconductivity can coexists in the same specimen. Figure 6 shows a scheme of the expected competition between magnetism and superconductivity.

Together to our experiment reported here, Naidyuk et al.[13] also found using point contact spectroscopy that the superconducting energy gap is anisotropic in the $URu_2Si_2$. Nevertheless they suggest that the observation of the superconducting energy gap in the (**a-b**)-plane is difficult due to the presence of the uranium magnetic moments. So, accordingly more experiments seems necessary to make clear this point, perhaps the main reason of the absence of a gap in the **a-b** plane is due to the imperfect nesting in different parts of the Fermi surfaces.

In our point contact experiments we can observe clearly how the energy gap developes. In figure 7 it is shown a sequence of point contact characteristics of $dV/dI$ vs. $V$ curves at different temperatures, from 2.15 K to 0.3 K. At about 2.15 K it seems that only the wide feature exists (not shown in this figure). This is related to the SDW as we explained before. Upon decreasing the temperature the superconducting feature starts to develop around zero bias. This minimum becomes is well defined as temperature goes down. Naturally, implying the reinforcing of the superconducting process. This feature is related to the energy gap and it arises due to the well known process named Andreev reflections.[23] The superconducting critical temperature observed by point contact spectroscopy is close to about 2 K, which is high compared with 1.37 K by resistance measurements. The critical temperature enhancement is due to the applied pressure by the Au wire when the point contact is made. The pressure dependence of the antiferromagnetic SDW and superconducting transition was studied by Bakker et al.[24] They observed that $T_c$ increases when the pressure is applied in the c-direction of the crystallographic structure. The pressure applied by

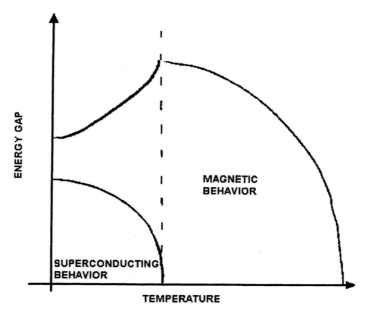

**Figure 6.** Schematic temperature behavior of the superconducting energy gap when magnetic order is present.

the Au wire according to the enhancement observed is high due to the small area of the point contact, this is less than 1 $\mu m^2$.

## IV.2 HoMo$_6$S$_8$

Resistance *versus* temperature measurements were performed to determine the transition temperatures to the superconducting state $T_{c1}$ as wel as to find the reentrant temperature $T_{c2}$ in a HoMo$_6$S$_8$ single crystal. Figure 2 shows the temperature behavior; the transition temperatures determined for $R = 0$ are $T_{c1} = 1.57$ K and $T_{c2} = 0.65$ K. The onset temperature near $T_{c1}$ is 1.78 K and the onset below $T_{c2}$ is 0.57 K. The resistance below $T_{c2}$ is 98% of the value of the resistance in the normal state above $T_{c1}$, suggesting that most of the compound has passed to the normal state when $T < T_{c2}$, but small portions still remain in the superconducting state. It is interesting to mention that the normal state below $T_{c2}$ depends on the magnetic history of the specimen, as was demonstrated by Giroud et al. They showed that in some cases, for $T < T_{c2}$, the electrical resistivity can have values which are of the same order or less than the normal value above $T_{c2}$.[25] This implies that in the sample there are still regions remaining superconducting, as we mentioned before.

The $dV/dI$ vs. $V$ curves obtained in contact made with HoMo$_6$S$_8$-Au(W) were measured between 0.3 K and 2 K. Figure 8 shows a set of complicated curves that have many interesting features (not so easy to interpret). For instances, the curve measured at a temperature of 2 K, above $T_{c1}$, presents a very smooth parabolic background, which is very characteristic of the normal state behavior found with point contact experiments. As temperature is decreased, below but close to $T_{c1}$, a very dim dip start to arise, tending to evolve first as an small depression. As temperature continues decreasing the dip is wider and deep. This dip is the signature of the feature originated

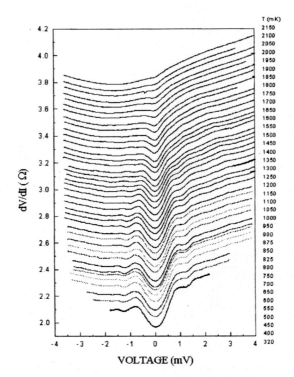

**Figure 7.** Differential resistance *vs* voltage of a URu$_2$Si$_2$Au(W) point contact measured between 0.3 K and 2.25 K. The column on the rigth is the temperature of each curve.

by the Andreev reflection due to the superconducting energy gap. As temperature follows decreasing and about $T_{M1} \simeq 0.75$ K an extra feature arises. This is a dim peak which is the feature that we relate of sinusoidal ferromagnetic order that settles at the ferromagnetic transition temperature. As temperatures decreases the peak grows in opposite direction to the superconducting feature. At temperature close to $T_{c2}$ the dip related to the superconducting energy gap is almost cancelled by the effect of the peak related to the ferromagnetic order, implying that the compound now is in the normal state. However, the superconducting gap feature is not completely destroyed. Two reasons are possible to explain this: The point contact was sensing a region in which a domain wall is located, thus sensing a small superconducting region trapped between domains. Second, the point contact is situated in a region where the local magnetic field is less than the upper critical magnetic field. In spite of this in other experiments we observe that the feature related to the superconducting energy gap disappear completely when a magnetic field about 0.6 Tesla is applied.[26]

A preliminary analysis of the behavior of the features observed in the curves shown in Fig. 8 is presented in Fig. 9. To analyze these we take three parameters: The amplitude of the dip $A_S$, related to the superconducting energy gap as measured from zero bias to the maximum observed around 2.5 mV; the amplitude of the structure related with the ferromagnetic ordering (peak) $A_F$, measured from the maximum around zero bias to near the minimum in positive bias, for temperatures below 0.65 K; and finally the voltage $\Delta V$ between the maxima at positive and negative bias. In

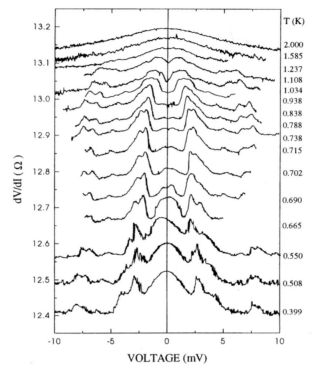

**Figure 8.** Differential resistance as function of bias voltage of $HoMo_6S_8$-Au(W) point contact.

this plots it is interesting to see that the amplitude of the ferromagnetic feature start to increase at about 0.8 K and presents a relative maximum at 0.65 K. Also it shows a minimum around 0.5 K which increases again. The behavior of $A_F$ between 0.65 K and around 0.8 K is in qualitative agreement with the results observed by neutron diffraction studies, that describe a sinusoidal magnetic ordering that increases when the temperature decreases, and brake down the superconducting order at about 0.65 K. This happens when the internal magnetic field exceeds the upper critical magnetic field and the ferromagnetic state sets in.[27,28]

The temperature behavior of the three parameters, between 0.65 K and 0.70 K, is somehow anomalous, if these are compared with the general trend of the data above and below this temperature. However, it is quite probable that this behavior produced by the competition between the two order parameters; the magnetization produced by the sinusoidal magnetic ordering and the energy gap of the superconducting state.

## V. Summary

In this work we studied two types of electronic systems; $URu_2Si_2$ and $HoMo_6S_8$. Both compounds presenting superconducting behavior with different magnetic ordering. Both can be catalogued as unconventional superconductors according to the Table I. The first one, a heavy fermion alloy, is superconducting at about 1.2 K

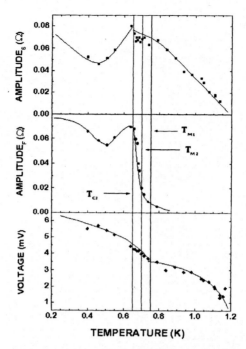

**Figure 9.** Temperature behavior of the amplitude of the peak related with the ferromagnetic order and energy gap. The lower graph represents the voltage between the peaks related with the energy gap. The vertical lines represent the transition temperatures of the sinusoidal magnetic order $T_{M1}$ and $T_{M2}$, and the reentrant transition $T_{c2}$.

and antiferromagnetic at about 17.5 K. The antiferromagnetic ordering is due to the formation of a SDW, whereas the pairing process that gives place to the superconductivity is believed to be of different symmetry that in normal BCS superconductors. The latter compound, is a Chevrel phase which is superconducting and ferromagnetic and presents a reentrant behavior to the normal state at about 0.65 K, due to the formation of two ferromagnetic transitions at 0.70 K and 0.75 K.

We observed that in the heavy fermion system exists a subtle balance between antiferromagnetism and superconductivity, giving as a results a coexistence of the two phenomena. In the second compound magnetism tends to destroy superconductivity given place to the reentrant behavior. Nevertheless, at still lower temperatures ($\sim$ 100 mK) superconductivity persists and is trapped between frontiers of magnetic domains.

Tunneling and point contact experiments show that the order parameter related to the SDW in the uranium compound, is anisotropic and that only some small portions of the Fermi surface are nested, given place to an imperfect nesting responsable of the superconducting behavior at low temperatures.

By examining other studies already published in the literature and with our observations we can reach the conclusion that in systems that present superconductivity and magnetism there are coexistence when the magnetic phase is antiferromagnetic, and in fact antiferromagnetism becomes a precursor of the superconducting behavior at lower temperatures (*i.e.*, also high $T_c$ cuprates). Whereas in systems where the magnetic ordering is of the ferromagnetic type there is a strong competition to break the superconducting process.

## Acknowledgments

This work was partially supported by the Consejo Nacional de Ciencia y Tecnología contracts No. G0017-E and 2295P-E, and by the Dirección General de Asuntos del Personal Académico contract No. IN105597. We thank F. Silva for technical assistance.

## References

1. V. L. Ginzburg, *Sov. Phys. JETP* **4**, 153 (1957).
2. B. T. Mathias, H. Suhl, and E. Corenzwit, *Phys. Rev. Lett.* **1**, 92 (1958).
3. A. A. Abrikosov and L. P. Gor'kov, *Sov. Phys. JETP* **12**, 1243 (1961).
4. R. Chevrel, M. Sergent, and J. Prigent, *J. Solid State Chem.* **3**, 515 (1971).
5. M. Giroud, Ph. D. Thesis, University of Grenoble, France (1987); A. Dinia, Ph. D. Thesis, University of Grenoble, France (1987).
6. W. A. Fertig et al., *Phys. Rev. Lett.* **38**, 387 (1977).
7. H. W. Meul et al., *Phys. Rev. Lett.* **53**, 497 (1984).
8. W. A. Little, *Phys. Rev.* **134**, A1416 (1964).
9. Y. G. Naidyuk, A. Nowack, I. K. Yanson, and P. N. Chubov, *Sov. J. Low Temp. Phys.* **17**, 614 (1991).
10. A. Nowack et al., *Z. Phys. B* **88**, 295 (1992).
11. K. Hasselbach, J. R. Kirtley, and P. Lejay, *Phys. Rev. B* **46**, 5826 (1992); *Physica B* **186-188**, 201 (1993).
12. A. Nowack et al., *Z. Phys. B* **97**, 77 (1995).
13. Y. G. Naidyuk et al., *Europhys Lett.* **33**, 557 (1996).
14. K. Hasselbach, P. Lejay, and J. Flouquet, *Phys. Lett. A* **156**, 313 (1991).
15. O. Peña and M. Sergent, *Progr. Solid State Chem.* **19**, 165 (1989).
16. U. Rauchschwalbe, *Physica B* **147**, 1 (1987).
17. G. Wexler, *Proc. Phys. Soc.* **89**, 927 (1966).
18. L. N. Bulaevskii, A. I. Buzdin, and M. Kulic, *Phys. Rev. B* **34**, 4928 (1986).
19. E. Fawcett, *Rev. Modern Phys.* **60**, 209 (1988).
20. R. Escudero, F. Morales, and P. Lejay, *Phys. Rev. B* **49**, 15271 (1994).
21. F. Morales, R. Escudero, and P. Lejay, *New Trends in Magnetism, Magnetic Materials, and Their Applications*, edited by J. L. Morán-López and J. M. Sanchez, (Plenum Press, New York, 1994).
22. M. B. Maple et al., *Phys. Rev. Lett.* **56**, 185 (1986).
23. G. E. Blonder, M. Tinkham, and T. M. Klapwijk, *Phys. Rev. B* **25**, 4515 (1982).
24. K. Bakker et al., *J. Magn. Magn. Mater.* **108**, 63 (1992).
25. M., Giroud et al., *J. Low Temp. Phys.* **69**, 419 (1987).
26. F. Morales et al., *Physica B* **218**, 193 (1996).
27. J. W. Lynn et al., *Phys. Rev. B* **24**, 3817 (1981).
28. J. W. Lynn et al., *Phys. Rev. B* **27**, 581 (1983).

# Intermediate Valence Model for $Tl_2Mn_2O_7$

C. I. Ventura and B. Alascio*

*Centro Atómico de Bariloche*
*8400 San Carlos de Bariloche*
*ARGENTINA*

## Abstract

There have been speculations about the need to find a new mechanism to explain the colossal magnetoresistance exhibited by this material, having pyrochlore structure and thus differing structurally and electronically from the manganites. We will report here our transport results based on a two band model, with conduction electrons and intermediate valence ions fluctuating between two magnetic configurations. The model has been previously employed to understand transport and thermodynamical properties of intermediate valence Tm compounds and, in its periodic version, to analyze the phase diagram. The results obtained with this model for the transport properties of $Tl_2Mn_2O_7$ are in good qualitative agreement with the experimental results.

## I. Introduction

The strong correlation between transport properties and magnetism in perovskite manganese oxides like $La_{1-x}M_xMnO_3$ (M = Ca, Sr, Ba) has been known for many years.[1] Shortly after the discovery of these materials, the double exchange mechanism was proposed to describe the interactions between the Mn ions which, due to the divalent substitution for La, are in a mixed valence state.[2] Electronic carrier hopping between heterovalent Mn pairs ($Mn^{3+}$–$Mn^{4+}$) is enhanced by the mutual alignment of the two magnetic moments. Thus the resistivity will depend on the spin disorder and is expected to display pronounced features at the ferromagnetic ordering transition temperature ($T_c$). On application of a magnetic field, which tends to align the local spins, the resistivity is expected to decrease.

Recently, colossal magnetoresistance (denoted MR hereafter) was observed near the ferromagnetic ordering temperature[3] and the interest in the study of these materials was renewed. At present, there is disagreement on whether theoretical models based only on the double exchange mechanism can account quantitatively for the observed transport and magnetic properties of the manganese perovskites. It has been proposed that other ingredients such as disorder,[4] Jahn Teller distortions,[5] etc. should be included and would play an important role.

In 1996 colossal magnetoresistance was observed for the non-perovskite $Tl_2Mn_2O_7$ compound.[6-8] This material has the pyrochlore $A_2B_2O_7$ structure,[9] consisting of $AO_8$ cubes and $BO_6$ octahedra linked to form a three-dimensional network of corner-sharing tetrahedra. This tetrahedral $Mn_2O_6$ (B = Mn) network differentiates the pyrochlores from the perovskite $ABO_3$ structure, with a cubic $MnO_6$ network. The pyrochlore structure is face-centered cubic with 8 formula units per unit cell. $Tl_2Mn_2O_7$ undergoes a ferromagnetic transition with $T_c \sim 140$ K (Refs. 5-7). Below $T_c$ the compound is ferromagnetic, whereas above $T_c$ it is paramagnetic. The magnetoresistance maximum around the ferromagnetic ordering temperature is similar to that obtained for the manganese perovskites. Mostly due to the absence of evidence for significant doping in the pyrochlore Mn-O sublattice (to produce the mixed valence responsible for double exchange in perovskite manganese oxides about 20–45% doping is needed), and due to the tendency of Tl to form 6$s$-conduction bands (unlike the perovskites where the rare-earth levels are electronically inactive), it has been put to question whether a double exchange mechanism similar to that of perovskites can account for the experimental results.[6-8] Hall data[6] show a very small number of $n$-type carriers ($\sim 0.005$ conduction electrons per formula unit) and this would seem to indicate a very small doping into the $Mn^{4+}$ state.[6,8] Among possible explanations, the authors of Refs. 6 and 8 mention that such Hall data could result from a small number of carriers in the Tl 6$s$-band, so that assuming $Tl^{3+}_{2-x}Tl^{2+}_{x}Mn^{4+}_{2-x}Mn^{5+}_{x}O_7$ with $x \sim 0.005$ the data could be accounted for.

Based on these facts we decided to explore the suitability of the intermediate valence (IV) model now to be introduced for $Tl_2Mn_2O_7$. The model was proposed originally for the study of Tm compounds.[10,11] The exactly solvable impurity model for valence fluctuations between two magnetic configurations was shown[10] to describe most of the peculiar features of the magnetic properties of paramagnetic intermediate valence Tm compounds. The impurity model resistivity exhibits an explicit quadratic dependence with the magnetization.[10] Such a behavior has also been found in transport experiments slightly above $T_c$ for colossal MR pyrochlores[6] and manganese perovskites.[12] With the periodic array of IV ions version of the model[11] the effect of the interactions between Tm ions was studied, obtaining the $T = 0$ phase diagram, specific heat and magnetic susceptibility for the paramagnetic phase in coherent potential approximation (CPA) and the neutron scattering spectrum. For manganese perovskites a similar model has been considered[13] to propose the possibility of a metal-insulator transition.

We will now present the model employed to describe a periodic lattice of intermediate valence (IV) ions, which fluctuate between two magnetic configurations associated to single ($S = \frac{1}{2}$) or double occupation ($S = 1$) of the ion, hybridized to a band of conduction states. The Hamiltonian considered is:[11]

$$H = H_L + H_c + H_H, \quad (1)$$

where

$$H_L = \sum_j (E_\uparrow |j\uparrow\rangle\langle j\uparrow| + E_\downarrow |j\downarrow\rangle\langle j\downarrow|) + (E_+|j+\rangle\langle j+| + E_-|j-\rangle\langle j-|),$$

$$H_c = \sum_{k,\sigma} \epsilon_{k,\sigma} c^+_{k,\sigma} c_{k,\sigma},$$

# Intermediate Valence Model for Tl$_2$Mn$_2$O$_7$

$$H_H = \sum_{i,j} V_{i,j} \left( |j+\rangle\langle j\uparrow| c_{i,\uparrow} + |j-\rangle\langle j\downarrow| c_{i,\downarrow} \right) + \text{h.c.}$$

The IV ions, which for Tl$_2$Mn$_2$O$_7$ we would identify with the Mn ions, are represented by $H_L$, which describes the $S = \frac{1}{2}$ magnetic configuration at site $j$ through states $|j\sigma\rangle$, ($\sigma = \uparrow, \downarrow$), with energies $E_\sigma$ split in the presence of a magnetic field $B$ according to:

$$E_{\uparrow(\downarrow)} = E - (+)\mu_0 B. \quad (2)$$

The $S = 1$ magnetic configuration is considered in the highly anisotropic limit, where the $S_z = 0$ state is projected out of the subspace of interest as in Refs. 10 and 11. The $S = 1$ states at site $j$ are represented by $|js\rangle$ ($s = +, -$) and energies $E_s$, split by the magnetic field as

$$E_\pm = E + \Delta \mp \mu_1 B. \quad (3)$$

Here, $H_c$ describes the conduction band, which for Tl$_2$Mn$_2$O$_7$ we would identify with the Tl $6s$-conduction band. Through hybridization with the conduction band, $H_H$ describes valence fluctuations between the two magnetic configurations at one site. For example, through promotion of a spin up electron into the conduction band at site $j$ the IV ion passes from state $|j+\rangle$ to state $|j\uparrow\rangle$. Notice that the highly anisotropic limit considered inhibits any spin flip scattering of conduction electrons, so that the direction of the local spin at each site is conserved. IV ions only hybridize with conduction electrons of parallel spin.

The Hamiltonian can be rewritten in terms of the following creation (and related annihilation) operators for the local orbitals (Mn $3d$ orbitals, in this case):[11]

$$d^+_{j,\uparrow} = |j+\rangle\langle j\uparrow|, \qquad d^+_{j,\downarrow} = |j-\rangle\langle j\downarrow|, \quad (4)$$

for which one has

$$[d_{i,\uparrow}, d^+_{j,\uparrow}]_+ = \delta_{i,j}(P_{i,\uparrow} + P_{i,+}),$$
$$[d_{i,\downarrow}, d^+_{j,\downarrow}]_+ = \delta_{i,j}(P_{i,\downarrow} + P_{i,-}),$$
$$P_{i,+} + P_{i,\uparrow} + P_{i,-} + P_{i,\downarrow} = 1, \quad (5)$$

where $P_{j,\alpha} = |j\alpha\rangle\langle j\alpha|$ are projection operators onto the local magnetic configuration states. The local Hamiltonian now reads:

$$H_L = (\Delta - \mu_D B)\sum_j d^+_{j,\uparrow} d_{j,\uparrow} + (\Delta + \mu_D B)\sum_j d^+_{j,\downarrow} d_{j,\downarrow}, \quad (5)$$

with

$$\mu_D = \mu_1 - \mu_0.$$

Due to the type of hybridization present, one can now consider the problem as described by two separate parts.[11] Given a certain configuration for the occupation of the local orbitals at all sites with spin up or down electrons, the spin up conduction electrons will only hybridize with those IV ions occupied by spin up electrons (*i.e.*, in $\uparrow$ or $+$ local states). One could simulate this by including a very high local correlation energy ($U \to \infty$) to be paid in the event of mixing with ions occupied by opposite spin electrons. Concretely, we can take:[9]

$$H = H_\uparrow + H_\downarrow,$$
$$H_\uparrow = (\Delta - \mu_D B)\sum_{j\in\uparrow} d^+_{j,\uparrow} d_{j,\uparrow} + U\sum_{j\in\downarrow} d^+_{j,\uparrow} d_{j,\uparrow}$$
$$+ \sum_k \epsilon_k c^+_{k,\uparrow} c_{k,\uparrow} + \sum_{ij}\left( V_{i,j} d^+_{j,\uparrow} c_{i,\uparrow} + \text{h.c.} \right). \quad (6)$$

$H_\downarrow$ is analogous to $H_\uparrow$, one having only to reverse the sign of $B$ and spin directions. We will ignore the split of the conduction band energies in the presence of the magnetic field $B$, and take for the hybridization: $V_{i,j} = V\delta_{i,j}$.

Given a certain configuration of up and down spin occupations of the sites, one can now solve two separate alloy problems, described by $H_\uparrow$ and $H_\downarrow$ respectively. These we solved in CPA approximation as in Ref. 11, introducing an effective diagonal self-energy for the local orbitals, $\Sigma_{(d)\sigma}(\omega)$ for the $H_\sigma$ alloy problem, through which on average translational symmetry is restored. The CPA equation obtained to determine self-consistently the self-energy relates it directly to the local Green function for the IV ions:

$$\Sigma_{(d)\uparrow}(\omega) = \frac{p-1}{\langle\langle d_{j,\uparrow}, d_{j,\uparrow}^+\rangle\rangle(\omega)}, \qquad \Sigma_{(d)\downarrow}(\omega) = \frac{-p}{\langle\langle d_{j,\downarrow}, d_{j,\downarrow}^+\rangle\rangle(\omega)}. \qquad (7)$$

Here $p$ denotes the concentration of spin up sites, and for the densities of states hold

$$\int_{-\infty}^{\infty} d\omega\, \rho_{c,\sigma}(\omega) = 1, \qquad \int_{-\infty}^{\infty} d\omega\, \rho_{d,\uparrow}(\omega) = p, \qquad \int_{-\infty}^{\infty} d\omega\, \rho_{d,\downarrow}(\omega) = 1-p. \qquad (8)$$

The CPA equation is solved self-consistently with the total number of particles equation, through which the chemical potential is determined.

To take into account the effect of temperature on the magnetization and describe qualitatively the experimental data in pyrochlores,[6,8] we will use a simple Weiss molecular field approximation to obtain the magnetization at each temperature[14] and through this the concentration $p$ of ions occupied by spin up electrons.

Considering now the determination of transport properties, using the Kubo formula it has been proved before[15,16] that no vertex corrections to the conductivity are obtained in a model such as this. Furthermore, in the absence of a direct hopping term between local orbitals only the conduction band will contribute to the conductivity.[16,17] As in Refs. 15–17, one can obtain the conductivity through Boltzmann equation in the relaxation time approximation limit as:

$$\sigma = \sigma_{c,\uparrow} + \sigma_{c,\downarrow}, \qquad \sigma_{c,\uparrow} = n_c e^2 \int d\omega \left(-\frac{\partial f(\omega)}{\partial \omega}\right) \tau_{c,\uparrow}^k(\omega)\Phi(\omega), \qquad (9)$$

where $f$ is the Fermi distribution, $n_c$ the total number of carriers per unit volume, and the relaxation time for spin up conduction electrons is:

$$\tau_{c,\uparrow}^k(\omega) = \frac{\hbar}{2|\text{Im}\{\Sigma_{c\uparrow}^k(\omega)\}|} \qquad (10)$$

and $\Phi = \frac{1}{N}\sum_k v_c^2(\epsilon_k)\delta(\omega - \epsilon_k)$, where $v_c(\epsilon_k)$ is the the conduction electron velocity. The extension of these formulas for spin down is straightforward.

Considering temperatures much lower than the Fermi temperature, one can approximate $\Phi(\omega) \sim v_F^2 \rho_c^{(0)}(\omega)$, being $v_F$ the Fermi velocity. The results presented here were obtained assuming for simplicity a semielliptic bare density of states for the conduction band $\rho_c^0(\omega)$, in which case the $k$-dependent self-energy for conduction electrons is related to the effective medium CPA self-energy through

$$\Sigma_{c\uparrow}^k(\omega) = \frac{V^2}{\omega - \sigma_{(d)\uparrow}(\omega)}, \qquad \sigma_{(d)\uparrow}(\omega) = \Delta - \mu_D B + \Sigma_{(d)\uparrow}(\omega). \qquad (11)$$

In figure 1 we show the results obtained for the resistivity employing this model with parameters: $W = 6$ eV for the semielliptic bare conduction half-bandwidth (centering the conduction band at the origin), $E = 0$, $\Delta = -4.8$ eV, $V = 0.6$ eV

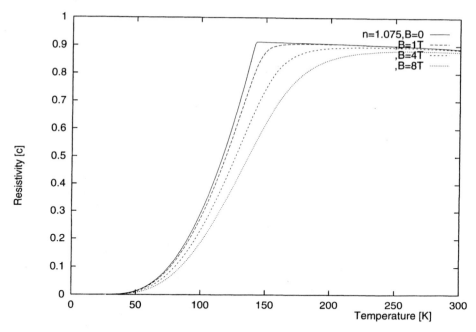

**Figure 1.** Periodic IV model. Resistivity as a function of temperature for magnetic fields: $B = 0$, 1, 4 and 8T. Parameters: $W = 6$ eV, $E = 0$, $\Delta = -4.8$ eV, $V = 0.6$ eV, $n = 1.075$, $T_c = 142$ K, $M_{\text{sat}} = 3\mu_B$, $\mu_D = 1\mu_B$. $c = (\text{eV})^2/[n_c e^2 v_F^2 \hbar]$, e.g., $c \sim 0.1$ $\Omega\,$cm for $n_c \sim 10^{21}/\text{cm}^3$ and $v_F \sim 10^7$ cm/s.

and total number of particles per site $n = 1.075$. To reproduce qualitatively the experimental magnetization data[6,8] we take $T_c$ as 142 K and a saturation value of 3 $\mu_B$ (like for free Mn$^{4+}$ ions) in the Weiss approximation, as well as $\mu_D = 1$ $\mu_B$. In figure 2 we show the gap in the CPA spin up densities of states obtained at $T = 0$ with those parameters. At temperatures above $T_c$ there still are gaps present (or at least pseudogaps, as is the case for the spin down bands at the higher magnetic fields shown $B = 4T$, $8T$). A filling of $n = 1.075$ corresponds to the Fermi level placed slightly above the bottom of the upper bands ($\mu \sim -4.51$ eV). The resistivity curves in Fig. 1 have a behavior around $T_c$ similar to that found in experiments[5–7]. An order of magnitude estimate for our resistivity results for $n = 1.075$ is compatible with the values experimentally found (see Fig. 1), in any case the absolute value of resistivity we obtain depends on the filling. In figure 3 we plot magnetoresistance results obtained with the same parameters. Here our results exhibit a difference in value between the MR maxima around $T_c$ for different magnetic fields and a crossing at higher temperatures of the MR curves obtained for those $B$, such as are present in the experimental data.[8] The description of the main features exhibited by transport measurements around $T_c$ is quite remarkable, considering the simplifications adopted here. Moreover, with the parameters employed (and due to the resulting presence of gaps or pseudogaps slightly below the Fermi level) we have a small effective number of electrons participating in the transport (mostly the few Tl conduction electrons above the gap) like Hall experiments indicate.[6,8] Nevertheless it is the total number of conduction electrons (which includes those below the gap and with our parameters

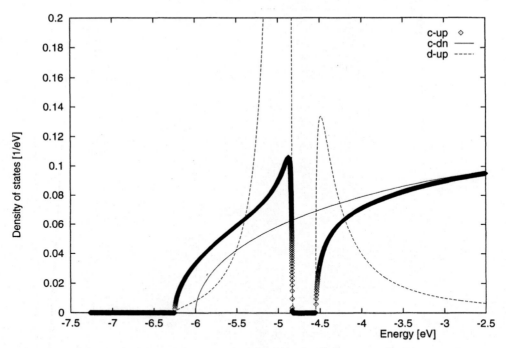

**Figure 2.** Periodic IV model. Densities of states as a function of energy at $T = 0$, $B = 1T$, parameters as in Fig. 1.

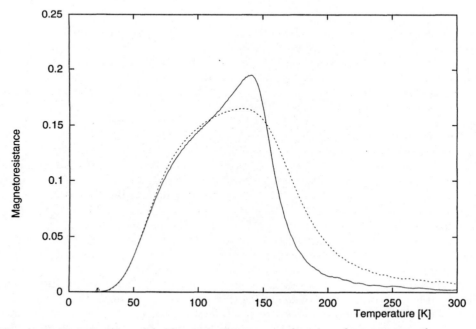

**Figure 3.** Periodic IV model. Magnetoresistance as a function of temperature, from curves of Fig. 1. Full line: $[\rho_{1T}(T) - \rho_{4T}(T)]/[\rho_{4T}(T)]$; dashed line: $[\rho_{4T}(T) - \rho_{8T}(T)]/[\rho_{4T}(T)]$.

is about 0.15, ten times the number of carriers above the gap) which would represent the real doping ($x$) into the $Mn^{4+}$ state to take into consideration. This could solve the difficulties with Hall data which are mentioned in Refs. 6 and 8.

We will now briefly comment on the transport results obtained with the impurity version of the model discussed above.[10] Using parameters for the impurity in accordance with those employed for the lattice, and a filling such that the Fermi level falls slightly above the peak of the impurity density of states, one can obtain transport results which are very similar to those presented above. Nevertheless, it would be hard to reconcile the impurity picture with the Hall data indicating a very small number of carriers effective in transport.[6,8]

To conclude we will indicate that the agreement between our results and the experimental data available on colossal MR pyrochlore $Tl_2Mn_2O_7$ is indeed very reasonable, and we believe that it can certainly be improved by fine tuning of the parameters of the model. The presence of gaps or pseudogaps in the electronic structure, such as those considered here in the periodic model, not only could solve the problems posed by the Hall data which are mentioned by the authors of Refs. 6 and 8. But their presence should cause observable effects in other experiments, such as spin polarized tunneling and optical properties which would be interesting to investigate.

## References

* Members of the Carrera del Investigador Científico of Consejo Nacional de Investigaciones Científicas y Tecnológicas (CONICET), Argentina.
1. G. H. Jonker and J. H. Van Santen, *Physica* **16**, 337 (1950).
2. C. Zener, *Phys. Rev.* **82**, 403 (1951); P. W. Anderson and H. Hasegawa, *Phys. Rev.* **100**, 675 (1955); P. G. De Gennes, *Phys. Rev.* **181**, 141 (1960).
3. R. Von Helmholt *et al.*, *Phys. Rev. Lett.* **71**, 2331 (1993); S. Jin byline *et al.*, *Science* **264**, 413 (1994).
4. R. Allub and B. Alascio, *Solid State Commun.* **99**, 613 (1996); R. Allub and B. Alascio, *Phys. Rev. B* **55**, 14113 (1997); Q. Li, J. Zang, A. R. Bishop, and C. M. Soukoulis, cond-mat/9612046 preprint.
5. A. J. Millis, P. B. Littlewood, and B. I. Shraiman, *Phys. Rev. Lett.* **74**, 5144 (1995); A. J. Millis, B. I. Shraiman, and R. Mueller, *Phys. Rev. Lett.* **77**, 175 (1996).
6. Y. Shimakawa, Y. Kubo, and T. Manako, *Nature* **379**, 55 (1996).
7. S. W. Cheong, H. Y. Hwang, B. Batlogg, and L. W. Rupp Jr., *Solid State Commun.* **98**, 163 (1996).
8. M. A. Subramanian, *Science* **273**, 81 (1996).
9. H. Fujinaka, N. Kinomura, and M. Korizumi, *Mater. Res. Bull.* **14**, 1133 (1979).
10. C. A. Balseiro and B. Alascio, *Phys. Rev. B* **26**, 2615 (1982); J. Mazzaferro, C. A. Balseiro, and B. Alascio, *Phys. Rev. Lett.* **47**, 274 (1981).
11. J. Mazzaferro, Thesis, Inst. Balseiro - Univ. Cuyo, (1982); A. A. Aligia, J. Mazzaferro, C. A. Balseiro, B. Alascio, *J. Magn. Magn. Materials* **40**, 61 (1983); A. A. Aligia, Thesis, Inst. Balseiro - Univ. Cuyo, (1984).
12. J. Fontcuberta *et al.*, *Phys. Rev. Lett.* **76**, 1123 (1996).
13. J. Mazzaferro, C. A. Balseiro, and B. Alascio, *J. Phys. Chem. Sol.* **46**, 1339 (1985).
14. Y. Lassailly, A. K. Bhattacharjee, and B. Coqblin, *Phys. Rev. B* **31**, 7424 (1985).
15. B. Velický, *Phys. Rev.* **184**, 614 (1969).

16. F. Brouers and A. V. Vedyayev, *Phys. Rev. B* **5**, 348 (1972); F. Brouers, A. D. Vedyayev, and M. Giorgino, *Phys. Rev. B* **7**, 380 (1973).
17. O. Sakai, S. Seki, and M. Tachiki, *J. Phys. Soc. Jpn* **45**, 1465 (1978).

# Spin Density Waves in Dimerized Systems

S. Caprara,[1] M. Avignon,[1] and O. Navarro[2]

[1] Laboratoire d'Etudes des Propriétés Electroniques des Solides
Centre National de la Recherche Scientifique
B. P. 166, 38042 Grenoble Cedex 9
FRANCE

[2] Instituto de Investigaciones en Materiales
Universidad Nacional Autónoma de México
Apartado postal 70-360, 04510, México D. F.
MEXICO

## Abstract

We study the effect of strong electronic correlations in the presence of chemically-induced charge transfer. We consider the Hubbard model at half-filling with an external staggered field as the simplest model to investigate the condition for the existence of a spin density wave competing with a charge density wave. The slave-boson approach is used to treat the electronic correlations in the strong coupling regime. We also consider the effect of a Brillouin-zone boundary phonon via a dimerization of the hopping term, thus introducing a competition with a paramagnetic bond-alternating phase and determine the phase diagram.

## I. Introduction

Dimerized systems with an induced charge transfer between neighboring sites may be produced by a modulation of the chemical environment of the system or by a deformation of the lattice coupled to the electronic density. In both cases to produce a true charge-density-wave (CDW) state a suitable screening of the long-range coulombic forces is needed. This may be provided, for instance, by a corresponding modulation of the substrate which induces the local potential or the lattice deformation.

In the presence of a strong intra-atomic correlation the possibility for a spin-density-wave (SDW) state arises. However, since spin and charge ordering are generally competitive, two cases are possible: Either the existence of a spin-ordered state excludes any charge ordering and vice versa, or charge and spin order coexist. Indeed if the charge transfer between neighboring sites is produced by an external potential (*e.g.*, a modulation of the chemical environment) a coexistence is possible if the intra-atomic energy scale responsible for spin ordering is sufficiently large. If the charge transfer is produced by a lattice deformation associated to an elastic potential, the presence of spin order excludes charge ordering through an increase of the self-consistent elastic potential. In this case the SDW phase stays undeformed, whereas the deformed CDW phase is paramagnetic.[1] There is still a lattice deformation which competes with spin ordering, without introducing any charge modulation, and arises when long and short bonds are formed, with a corresponding modification of the hopping term. This dimerization produces the opening of a gap at the boundary of the Brillouin zone thus stabilizing the paramagnetic (PM) phase with respect to the SDW phase when the coupling to the lattice deformation is stronger than the intra-atomic energy responsible for spin ordering.

When some or all of the above mechanisms are present different physical behaviors are possible and a correspondingly rich phase diagram arises. In this paper we devote our analysis to the stability of the SDW phase in the presence of a chemically-induced CDW and of a bond dimerization.

## II. Model

The physical scenario discussed in Sect. I is analyzed in this Section by means of the one-dimensional model defined by the Hamiltonian

$$\widetilde{\mathcal{H}} = -t \sum_{n,\sigma} (1 + Y_n) \left( \widetilde{f}^+_{n,\sigma} \widetilde{f}_{n+1,\sigma} + \text{h.c.} \right) + I \sum_{n,\sigma} (-1)^n \widetilde{f}^+_{n,\sigma} \widetilde{f}_{n,\sigma}$$
$$+ U \sum_n \widetilde{f}^+_{n\uparrow} \widetilde{f}_{n\uparrow} \widetilde{f}^+_{n\downarrow} \widetilde{f}_{n\downarrow} + \frac{t}{\pi \lambda} \sum_n Y_n^2, \quad (1)$$

where the intersite distance is unity, the fermionic operators $\widetilde{f}^+_{n,\sigma}$, $\widetilde{f}_{n,\sigma}$ act in the Wannier representation, $t$ is the nearest-neighbors hopping parameter, $I$ is the amplitude of a staggered local potential, which is produced by some external (crystal) field, $U$ is the onsite coulombic repulsion, and $\lambda$ is the dimensionless coupling to the deformation of the bond $Y_n$. This is the simplest model to investigate the possibility of coexistence of a SDW, which may result from the $U$ term, and a CDW, produced by the $I$ term. The possibility of a dimerization of the hopping term due to the coupling to a bond deformation is also taken into account.

Let us first consider the case in which the bonds are not deformed.[2] In the non-interacting (*i.e.*, for $U = 0$) half-filled system, indeed, a CDW exists induced by the external field $I$. If we let $n_A$ and $n_B$ to be the number of particles on the two inequivalent sites of the bipartite chain (with $n_A + n_B = 2$ in the half-filled case), then the charge order parameter $m_e \equiv \frac{1}{2}(n_B - n_A)$ is given by

$$m_e = \frac{I}{\pi} \int_{-\pi/2}^{\pi/2} \frac{dk}{\sqrt{I^2 + 4t^2 \cos^2 k}}, \quad (2)$$

which saturates towards 1 as $I$ is increased. When the electron-electron interaction $U$ is present the possibility for a magnetic state arises. This state can be characterized by a SDW amplitude $m_s \equiv n_{A\uparrow} - n_{A\downarrow} = n_{B\downarrow} - n_{B\uparrow}$. At $I = 0$, $m_s$ is found to be an increasing function of the ratio $U/t$ both in the Hartree-Fock and slave-boson approximations. As $U/t \to \infty$, $m_s$ saturates towards 1.

The simultaneous presence of the staggered external field $I$ and the on-site electron-electron interaction $U$ implies an interplay of charge and spin degrees of freedom which we want to clarify both in the weak-coupling and strong-coupling regimes by means of the Hartree-Fock (HF) and slave-boson (SB) approximations respectively.

On a general ground the parameter $I$ opposes to SDW formation whereas the parameter $U$ can substantially reduce the CDW order parameter even in the presence of a sizeable external field $I$. However self-consistency imposes a complicate interdependence of $m_s$ and $m_e$ which can give rise to discontinuous (first-order) phase transitions between magnetic and nonmagnetic phases.

The weak-coupling regime ($U < t$) can be treated within the HF approximation. In the following we make use of the parametrization:

$$n_{A\sigma} = \tfrac{1}{2}(1 + \sigma m_s - m_e), \qquad n_{B\sigma} = \tfrac{1}{2}(1 - \sigma m_s + m_e).$$

The HF spin-dependent atomic levels are then given by

$$E_{A\sigma} \equiv I + U n_{A,-\sigma} = \tfrac{1}{4}U(1 - \sigma m_s - m_e),$$

and

$$E_{B\sigma} \equiv -I + U n_{B,-\sigma} = \tfrac{1}{4}U(1 + \sigma m_s + m_e),$$

so that the gap in the energy spectrum is given by

$$\Delta_{-\sigma} \equiv \tfrac{1}{2}(E_{A\sigma} - E_{B\sigma}) = I - \tfrac{1}{2}U(m_e + \sigma m_s).$$

The self-consistency equations for $m_s$ and $m_e$ are then

$$m_s = \frac{1}{2\pi t}\sum_\sigma \sigma \kappa_\sigma \Delta_\sigma \mathcal{K}(\kappa_\sigma^2), \qquad m_e = \frac{1}{2\pi t}\sum_\sigma \kappa_\sigma \Delta_\sigma \mathcal{K}(\kappa_\sigma^2), \tag{3}$$

where $\kappa_\sigma = 2t/\sqrt{4t^2 + \Delta_\sigma^2}$ and $\mathcal{K}(x)$ is the complete elliptic integral of the first kind. Introducing the auxiliary parameters $m_\sigma = m_e - \sigma m_s$ so that $\Delta_\sigma = I - \tfrac{1}{2}U m_\sigma$ one can reduce the above equations to $m_\uparrow = (1/\pi t)\kappa_\downarrow \Delta_\downarrow \mathcal{K}(\kappa_\downarrow^2)$ and $m_\downarrow = (1/\pi t)\kappa_\uparrow \Delta_\uparrow \mathcal{K}(\kappa_\uparrow^2)$, where it is evident that $m_\uparrow$ is determined via $m_\downarrow$ and vice versa. From the numerical point of view the last two equations can be reduced to an equation for a single variable (for instance $m_\uparrow$) the other being univoquely determined by the knowledge of the first.

The energy per lattice site corresponding to each self-consistent solution of Eq. (3) is

$$E = -\tfrac{1}{4}U m_\uparrow m_\downarrow - \frac{2t}{\pi}\sum_\sigma \frac{\mathcal{E}(\kappa_\sigma^2)}{\kappa_\sigma},$$

where $\mathcal{E}(x)$ is the complete elliptic integral of the second kind. We found that the paramagnetic-antiferromagnetic transition predicted by the HF approximation is of first order in the weak-coupling regime, but becomes of second order in the strong-coupling regime (see Fig. 1). However, when $U > t$ the HF predictions are unreliable and the SB approach[3] is more appropriate.

Since we are working on a bipartite lattice we introduce a set of Kotliar-Ruckenstein SB operators,[4] with the corresponding Lagrange multipliers, on each sublattice. At mean-field level, in the case of coexisting charge and spin density waves, we introduce the parametrization

$$\begin{aligned}
\langle p_{i,\sigma}\rangle &= p_{A\sigma}, & \langle e_i\rangle &= e_A, & \langle d_i\rangle &= d_A, & \langle \lambda_i\rangle &= \lambda_A, & \langle \Lambda_{i,\sigma}\rangle &= \Lambda_{A\sigma}, & \text{for } i \in A; \\
\langle p_{i,\sigma}\rangle &= p_{B\sigma}, & \langle e_i\rangle &= e_B, & \langle d_i\rangle &= d_B, & \langle \lambda_i\rangle &= \lambda_B, & \langle \Lambda_{i,\sigma}\rangle &= \Lambda_{B\sigma}, & \text{for } i \in B;
\end{aligned} \tag{4}$$

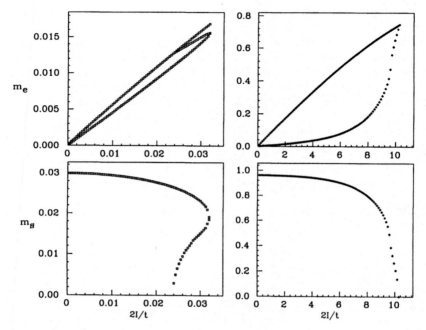

**Figure 1.** Charge- (top) and spin- (bottom) density-wave amplitudes within the HF approximation as a function of $2I/t$, for $U/t = 1$ (left) and $U/t = 10$ (right).

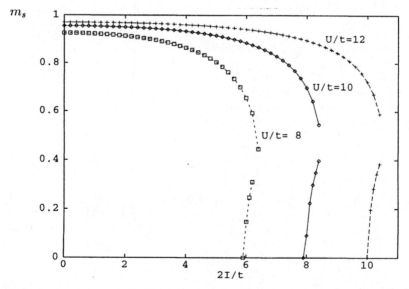

**Figure 2.** Spin-density-wave amplitude within the SB approximation as a function of $2I/t$, for $U/t = 8$, 10, and 12.

where, as in the previous Sections, $A$ and $B$ indicate the two different sublattices respectively, $p_{i,\sigma}$, $e_i$, and $d_i$ are the SB operators to label singly occupied, empty and doubly occupied sites respectively, $\lambda_i$ is the Lagrange multiplier to enforce the completeness relation on each site and $\Lambda_{i,\sigma}$ are the Lagrange multipliers to enforce correct fermion counting. To connect the above parameters to the parameters of the previous Section observe that

$$p_{A(B)\sigma} \equiv p_0 + (-)\sigma m_s/4p_0, \qquad d_{A(B)} \equiv d_0 - (+)m_e/4d_0,$$
$$e_{A(B)} \equiv d_0 + (-)m_e/4d_0, \qquad \Lambda_{A(B)\sigma} \equiv \tfrac{1}{2}U + (-)[\Lambda_c - \sigma\Lambda_s],$$

where $p_0$ and $d_0$ are new parameters, and $\Lambda_c$, $\Lambda_s$ play the role of the charge and spin effective fields respectively. The average particle-density per spin on each sublattice is given by $n_{A\sigma} \equiv \langle f^+_{i,\sigma} f_{i,\sigma}\rangle = d_A^2 + p_{A\sigma}^2 =$ for $i \in A$ and $n_{B\sigma} \equiv \langle f^+_{i,\sigma} f_{i,\sigma}\rangle = d_B^2 + p_{B\sigma}^2$ for $i \in B$, where $f_{i,\sigma}$ are the pseudofermionic operators in the Wannier representation. After introducing the effective hopping parameter $\tilde{t}_\sigma = t z_\sigma^2$, where

$$z_\sigma^2 \equiv z_{A\sigma} z_{B\sigma} = \frac{(p^+_{A\sigma} e_A + d^+_A p_{A,-\sigma})(p^+_{B\sigma} e_B + d^+_B p_{B,-\sigma})}{\sqrt{(1 - e_A^2 - p_{A,-\sigma}^2)(1 - d_A^2 - p_{A\sigma}^2)(1 - e_B^2 - p_{B,-\sigma}^2)(1 - d_B^2 - p_{B\sigma}^2)}}$$

is the hopping-renormalization factor, we performed a Fourier transform in the reduced-Brillouin-zone scheme to obtain the mean-field Hamiltonian in the form

$$\mathcal{H}_{mf} = \sum_{k,\sigma} \left(f^+_{k,\sigma}\ f^+_{k+Q,\sigma}\right) \begin{pmatrix} E_\sigma - 2\tilde{t}_\sigma \cos k & \Delta_\sigma \\ \Delta_\sigma & E_\sigma + 2\tilde{t}_\sigma \cos k \end{pmatrix} \begin{pmatrix} f_{k,\sigma} \\ f_{k+Q,\sigma} \end{pmatrix}$$
$$+ \tfrac{1}{2} N_s U (d_A^2 + d_B^2) - \tfrac{1}{2} N_s \sum_\sigma \left[\Lambda_{A\sigma}(p_{A\sigma}^2 + d_A^2) + \Lambda_{B\sigma}(p_{B\sigma}^2 + d_B^2)\right]$$
$$+ \tfrac{1}{2} N_s \left[\lambda_A \left(\sum_\sigma p_{A\sigma}^2 + e_A^2 + d_A^2 - 1\right) + \lambda_B \left(\sum_\sigma p_{B\sigma}^2 + e_B^2 + d_B^2 - 1\right)\right]. \quad (5)$$

In the above formulas $f^+_{k,\sigma}$, $f_{k,\sigma}$ are the pseudo-fermionic operators in the Bloch representation, the sum over $k$ runs on the reduced Brillouin zone, $Q = \pi$, $E_\sigma = \tfrac{1}{2}(\Lambda_{A\sigma} + \Lambda_{B\sigma}) = \tfrac{1}{2}U$ is independent of $\sigma$, $\Delta_\sigma = \tfrac{1}{2}(2I + \Lambda_{A\sigma} - \Lambda_{B\sigma}) = I + \Lambda_c - \sigma\Lambda_s$, and $N_s$ is the number of lattice sites. The eigenvalues of the matrix Hamiltonian in (5) give the quasi-particle bands $E^\pm_{k\sigma} = E_\sigma \pm \sqrt{\Delta_\sigma^2 + 4\tilde{t}_\sigma^2 \cos^2 k}$. At half-filling and zero temperature only the two lower bands are occupied and the mean-field energy per lattice site is

$$E = E_0 + \frac{1}{N_s} \sum_{k,\sigma} E^-_{k\sigma}, \qquad (6)$$

where $E_0$ is the energy per lattice site associated to the last two lines in (5). The self-consistency equations are obtained by requiring (6) to be stationary with respect to the parameters (4), and have the general form

$$\frac{\partial E_0}{\partial P} + \frac{1}{N_s} \sum_{k,\sigma} \frac{\partial E^-_{k\sigma}}{\partial P} = 0, \qquad (7)$$

where $P$ represents generically one of the parameters (4).

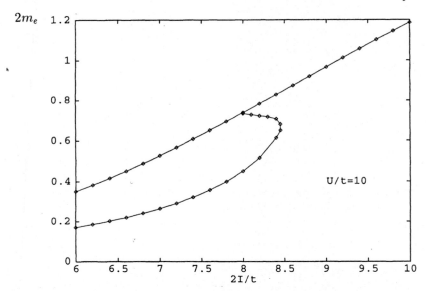

**Figure 3.** Charge-density-wave amplitude within the SB approximation as a function of $2I/t$, for the representative value $U/t = 10$.

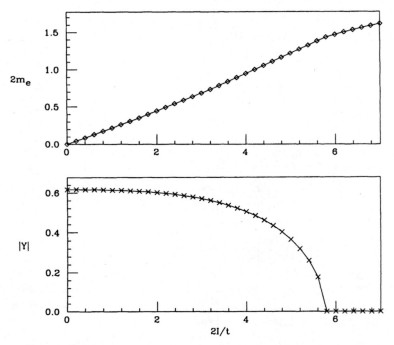

**Figure 4.** Charge-density-wave amplitude (top) and bond dimerization (bottom) in the case $U/t = 4$, $\lambda = 0.5$, when the SDW phase has a higher energy than the dimerized PM phase at $I/t = 0$. The dimerized PM phase evolves into a dimerized CDW phase which in turn undergoes a second-order phase transition towards the undimerized CDW phase as $I/t$ is increased.

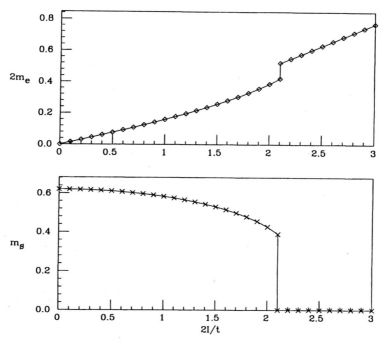

**Figure 5.** Charge-density-wave amplitude (top) and spin-density-wave amplitude (bottom) in the case $U/t = 4$, $\lambda = 0.1$, when the SDW phase has a lower energy than the dimerized PM phase at $I/t = 0$. The SDW phase undergoes a first-order phase transition towards the undimerized CDW phase as $I/t$ is increased.

We found that, contrary to the HF approximation, the SB approximation gives a first-order antiferromagnetic-paramagnetic phase transition even in the strong-coupling regime ($U \gg t$). The region of hysteresis is generally narrow when $U \gg t$, and the jump of the SDW amplitude at the transition is sizeable (see Fig. 2). The PM state is always characterized by a larger CDW amplitude (see Fig. 3). The corresponding phase diagram is found in the $2I/t$ vs. $U/t$ plane in Fig. 8.

Since the SB formalism provides a description which agrees with HF results in the weak-coupling regime, in the following we apply the SB approach down to $U/t = 0$. Let us now consider the effect of a dimerization of the hopping term in Eq. (1) due to the coupling to a phonon, by taking $\lambda > 0$ and $Y_A = -Y_B$. When $I, U > 0$ a PM phase is stabilized with respect to the SDW phase in the region of small $U/t$ by the opening of a gap at the boundary of the Brillouin zone. The SDW phase is always undeformed (*i.e.*, spin ordering and bond deformation do not coexist) and at sufficiently large $U/t$ it has a lower energy than the PM dimerized phase. A first-order phase transition between a PM and a SDW antiferromagnetic phase is then produced by increasing $U/t$ at a fixed $\lambda$. The value of $U/t$ at the phase transition increases with increasing $\lambda$. On the other hand if $U = 0$ and $I > 0$, there are two possibilities: If $I$ is small the resulting CDW phase has dimerized bonds, *i.e.*, $Y_A, Y_B \neq 0$. As $I$ is increased the CDW phase with dimerized bonds undergoes a second-order phase transition towards a CDW phase with undimerized bonds, where $Y_A = Y_B = 0$. When $U, I, \lambda > 0$ the phase diagram is divided into three regions. For large $U/t$ a SDW amplitude is present, together with a CDW amplitude as soon as $I > 0$. The bonds in the spin-

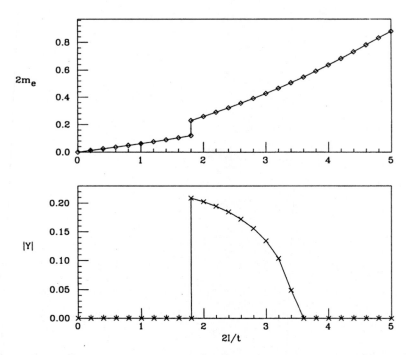

**Figure 6.** Charge-density-wave amplitude (top) and bond dimerization (bottom) in the case $U/t = 4$, $\lambda = 0.3$, when the SDW phase has a lower energy than the dimerized PM phase at $I/t = 0$. Contrary to the the case of Fig. 5, however, the SDW phase undergoes a first-order phase transition towards the dimerized CDW phase which in turn undergoes a second-order phase transition towards the undimerized CDW phase as $I/t$ is increased. The spin-density-wave amplitude is the same as in Fig. 5, but the jump to the phase with $m_s = 0$ is located at $2I/t = 1.8$.

ordered phase are never dimerized. For large $I/t$ the pure CDW phase exists, and the bonds are not dimerized. Finally for small $I/t$ and $U/t$ a PM phase exists, which is charcterized by dimerized bonds, and a finite CDW amplitude as soon as $I > 0$. The transition between the CDW phases with dimerized and undimerized bonds, which is produced for instance as $I/t$ is increased at fixed $\lambda, U/t$, is always of second order (see Fig. 4). The SDW phase may either undergo a direct phase transition to the CDW phse with undimerized bonds (see Fig. 5), or to the CDW phase with dimerized bonds, which in turn evolves towards the CDW with undimerized bonds (see Fig. 6). It is interesting to observe that, for $U/t$ sufficiently large the bond dimerization in the CDW phase is first increasing with increasing $I/t$, and then decreasing until it vanishes at the phase transition to the CDW phase with undimerized bonds (see Fig. 7). This behavior may, however, be overshadowed by the presence of the SDW phase, since for large $U/t$ it has generally a lower energy than the dimerized CDW phase. A schematic idea of the full phase diagram is given in Fig. 8.

## III. Conclusions

We investigated the competition of spin ordering and dimerization by means of a generalized Hubbard model, both in the case when dimerization is associated with

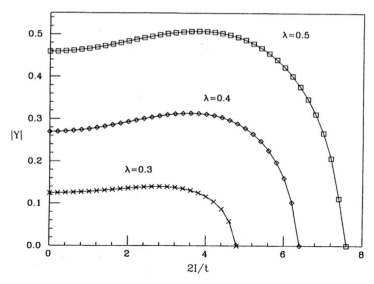

**Figure 7.** Mean-field value of the bond dimerization $Y \equiv Y_A = -Y_B$ within the SB approximation as a function of $2I/t$ for $U/t = 6$ and different values of the adimensional electron-phonon coupling $\lambda$. For this value of the parameters, however, the dimerized CDW phase is a local minimum of the energy, but the SDW phase has a lower energy.

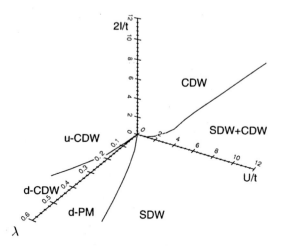

**Figure 8.** Phase diagram of the Hubbard model in the presence of chemical and bond dimerization, within the SB approximation. The prefix d- (u-) stand for dimerized- (undimerized-) bonds.

the appearence of a charge modulation and when the dimerization is produced by a deformation of the bonds. We found that when the dimerization is produced by an elastic deformation of the bonds it is incompatible with spin ordering, *i.e.*, the system is either dimerized or antiferromagnetic. When the dimerization is imposed by some external field (provided for instance by the chemical environment of the system) a coexistence of spin order and dimerization becomes possible. As the strength of the intra-atomic potential responsible for spin ordering is reduced the SDW phase undergoes a first-order phase transition towards a PM dimerized phase, characterized by a the presence of a CDW amplitude and/or a bond dimerization.

## References

1. S. Caprara, M. Avignon, and O. Navarro, unpublished.
2. A similar model, where however the direct effect of $U$ on the charge-density-wave amplitude was prevented by subtracting the Hartree contribution in the $U$ term, was studied in V. Tugushev, S. Caprara, and M. Avignon, *Phys. Rev. B* **54**, 5466 (1996).
3. S. E. Barnes, *J. Phys. F* **6**, 1375 (1976); P. Coleman, *Phys. Rev. B* **29**, 3035 (1984).
4. G. Kotliar and A. R. Ruckenstein, *Phys. Rev. Lett.* **57**, 1362 (1986); we refer to this paper for any detail about the SB technique on a bipartite lattice.

# Bond-Order-Wave *versus* Spin-Density-Wave Dimerization in Polyacetylene

G. M. Pastor and M. B. Lepetit

Laboratoire de Physique Quantique
Unité Mixte de Recherche 5626 du CNRS
Université Paul Sabatier
118 route de Narbonne, F-31062 Toulouse Cedex
FRANCE

## Abstract

The dimerization of polyacetylene is studied in the framework of a distance dependent Hubbard Hamiltonian for the valence $\pi$ electrons. The underlying one-dimensional many-body problem is solved numerically using the density matrix renormalization group (DMRG) method and the resulting ground-state energy is optimized in order to determine the dimerization $\delta$ and the average bond-length $\overline{R}$. The strength of the Coulomb repulsion $U/t$ is varied from the uncorrelated or Hückel limit all over to the strongly correlated or Heisenberg case. While $\overline{R}$ is not significantly affected by the value of $U/t$, $\delta$ shows a remarkable non-monotonic behavior. The differences between the bond-order-wave (small $U/t$) and the spin-Peierls (large $U/t$) regimes are discussed.

## I. Introduction

The low-energy electronic properties of polyacetylene are dominated by a single, half-filled band involving the $2p_z$ orbitals of the C atoms which are perpendicular to the chain ($\pi$ orbitals). The other 3 valence electrons of the C's, as well as those of the H's, form stronger C-C and C-H $\sigma$ bonds ($sp^2$ hybridization). Within a single-particle band-structure picture, it is easy to see that the periodic structure with equal C-C distances should be unstable and that a bond-length alternation or dimerization should set in for such a polymer chain. However, the situation is far from simple when electron interactions come into play. Indeed, the role of electron correlations on the dimerization of polyacetylene is a long standing problem which has motivated a considerable research activity over many years.[1] Simple models as well as *ab initio* methods have been applied, together with a variety of approximations, in order to describe the fundamental interplay between the delocalization of the valence $\pi$ electrons and the associated local charge fluctuations. Depending on the values of the intra-atomic Coulomb repulsion $U$ and of the nearest neighbor (NN) hopping integral $t$ of the $\pi$ system, two qualitatively different regimes may be distinguished. For small $U/t$ the dimerization can be regarded as a bond-order wave which results from the opening of a gap at the Fermi surface of the one-dimensional (1D) single-particle band structure ($U/t \to 0$, Peierls distortion). In contrast, for large $U/t$, charge fluctuations are severely reduced and the low-energy properties of the $\pi$ system are dominated by the spin degrees of freedom. In this case the dimerization enhances the antiferromagnetic coupling in one every two pairs of NN spins at the expense of reducing the coupling between different pairs. The resulting spin-Peierls state can be regarded as an alternation of the strength of the AF correlations along the chain.

Besides the importance of elucidating how the dimerization $\delta$ and the average bond-length $\overline{R}$ depend on the strength of the Coulomb interactions, the dimerization of polyacetylene is a very interesting problem from a purely methodological point of view. On the one hand, a few exact results are available to compare with: (i) the Bethe-Ansatz solution for infinite Hubbard chains and arbitrary values of $U/t$,[2] (ii) the noncorrelated limit ($U = 0$) for any value of $\delta$ and $\overline{R}$, and (iii) finite-cluster Lanczos calculations in the strongly correlated limit (Heisenberg model) including extrapolations to infinite chains.[3] On the other, determining $\delta$ is a very difficult problem, since it requires an extremely precise calculation of the ground-state energy $E$. In fact, the energy differences involved are $10^{-4}$–$10^{-5}$ times smaller than the total binding energy per site. This explains the quantitative disagreements between the different approaches tried so far[1] and sets a serious challenge to any method attempting to determine the ground-state energy of correlated one-dimensional systems.

The present study of the dimerization of trans-polyacetylene is based upon the many-body model Hamiltonian presented in Sect. II. The ground-state energy and other derived physical properties are calculated using the real-space density matrix renormalization group approach (DMRG) as described in Sect. III. Representative results are discussed in Sect. IV. Finally, Sect. V summarizes the main conclusions. A detailed account of this work may be found elsewhere.[4]

## II. Model Hamiltonian

The binding in trans-polyacetylene is given by two main contributions. First, a skeleton of $\sigma$ bonds which is built from $sp^2$ hybridized orbitals of the C atoms and which favors a nondimerized state with rather long bonds. Second, a weaker $\pi$ system based upon the $2p_z$ orbitals of C, which is responsible for the dimerization. In order to describe these polymer chains we consider a distance-dependent Hamiltonian of the form $H = H_\sigma + H_\pi$, where $H_\sigma$ approximates the $\sigma$ binding as a sum of effective pair potentials $E_\sigma(r_{ij})$ and $H_\pi$ describes explicitly the $\pi$ electrons in the framework of a Hubbard Hamiltonian:

$$H = \sum_i E_\sigma(r_{i,i+1}) + \sum_{i,s} t(r_{i,i+1})\left(a^\dagger_{i,s} a_{i+1,s} + a^\dagger_{i+1,s} a_{i,s}\right) + U \sum_i n_{i\uparrow} n_{i\downarrow}. \quad (1)$$

As usual, $a^\dagger_{i,s}$ ($a_{i,s}$) refers to the creation (annihilation) operator of the $p_z$ electron of spin $s$ at atom $i$ ($n_{is} = a^\dagger_{i,s} a_{i,s}$ is the corresponding number operator). The parameters

$E_\sigma(r_{ij})$ and $t_{ij}(r_{ij})$ entering Eq. (1) are derived from accurate *ab initio* configuration interaction (CI) calculations on the ground state ($^1A_g$) and on the lowest excited state ($^3B_u$) of the ethylene molecule (*i.e.*, the dimer). The CI results are available as a function of the inter-atomic distance $r_{ij}$ (Ref. 5). Let us recall that both states have valence character and that their representation in the minimal basis which spans the Hubbard Hamiltonian is

$$|^1A_g\rangle = \cos\theta \left(|a\bar{b}\rangle + |b\bar{a}\rangle\right)/\sqrt{2} + \sin\theta \left(|a\bar{a}\rangle + |b\bar{b}\rangle\right)/\sqrt{2}, \qquad (2)$$

and

$$|^3B_{u0}\rangle = \left(|a\bar{b}\rangle - |b\bar{a}\rangle\right)/\sqrt{2}. \qquad (3)$$

Here, $a$ and $b$ refer to the $p_z$ orbitals of the two C atoms. From Eqs. (1), (2), and (3) it is straightforward to obtain the relations between the eigen-energies and the model parameters:

$$E\left(^1A_g\right) = E_\sigma(r_{ab}) + \tfrac{1}{2}\left(U - \sqrt{U^2 + [4t(r_{ab})]^2}\right), \qquad (4)$$

$$E\left(^3B_u\right) = E_\sigma(r_{ab}). \qquad (5)$$

Notice that the $^3B_u$ state is purely neutral (no $U$ contribution) and that it has completely localized $\pi$ electrons (no $t$ contribution). Within this model, the scalar term $E_\sigma(r_{ij})$ corresponds to the energy of the first excited (triplet) valence state of the ethylene molecule and is independent of the value of the parameters describing the $\pi$ binding (*i.e.*, $t$ and $U$). The intra-atomic Coulomb repulsion $U$ is taken to be independent of the inter-atomic distance. Its value is varied from $U = 0$ to $U = +\infty$ in order to analyze how the dimerization changes as the electronic correlations increase. For a given value of $U$, the hopping integral $t(r_{ab})$ is obtained from

$$t(r_{ab}) = -\tfrac{1}{2}\sqrt{\Delta E(r_{ab})\left[\Delta E(r_{ab}) + U\right]}, \qquad (6)$$

where $\Delta E(r_{ab}) = E(^3B_u) - E(^1A_g)$ is the first excitation energy at the distance $r_{ab}$.

The strength of the Coulomb interaction among $\pi$ electrons can be quantified by the ratio $U/|t_0|$, where $t_0 = t(r_0)$ is the hopping integral at the equilibrium distance of the ethylene molecule ($r_0 = 2.65$ a.u.). In the following several physical properties are investigated as a function of $U/|t_0|$. Notice that as $U$ increases $|t_{ij}(r_{ij})|$ also increases in order to yield always the same ground-state energy surface for the dimer. In particular for large $U/t$, the binding energy scales with $J \propto t^2/U$ and therefore $t_{ij} \propto \sqrt{J(r_{ij})}\sqrt{U}$. Our approach differs from most previous studies where $U$ is varied independently of $t$, thus implying a weakening of the elementary C-C $\pi$ bond as $U$ increases.

## III. Calculation Method

We consider the real-space density matrix renormalization group (DMRG) algorithm proposed in Ref. 6 for the study of infinite open chains. In this Section we just intend to recall very briefly the essentials of the method and to state some aspects which are specific to the present study of dimerized chains. For further details, as well as for alternative implementations on finite chains, the reader should refers to the original papers by White.[6]

In the present DMRG approach the properties of the infinite system are derived by extrapolating the results of a succession of calculations on finite systems. Each one of these finite-system calculations is considered as a renormalization group (RG)

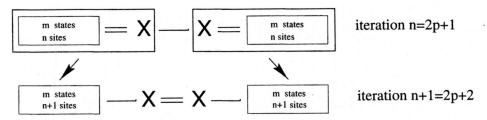

**Figure 1.** Illustration of the superblock renormalization procedure. A cross represents a single C atom or Hubbard site. Single (double) lines represent long (short) bonds.

iteration. The length $n$ of the chain increases very slowly at each iteration. For example, in the present case, $n$ is increased by 2 sites (see Fig. 1). In spite of the fact that $n$ increases, the dimension of the many-body Hilbert space is kept constant by means of the following approximation. At each RG iteration the Hilbert space of the $n$-site chain is projected onto a reduced space spanned by a limited number of states which allow to take into account the main part of the eigenstates one attempts to calculate (typically, the ground-state wave function). These states are constructed as a product of many-body states corresponding to different regions in real space, also referred to as blocks (see Fig. 1). The error introduced by the truncation of the Hilbert space is minimized when the projection subspace is generated by the eigenvectors with the largest eigenvalues of the density matrix reduced to each block. In other words, only the states having the largest occupations are kept.

Figure 1 illustrates the renormalization procedure. At each RG iteration the system is divided in two side blocks spanned by $m$ states (typically, $m = 35\text{--}80$ in the present calculations) and two central sites which require $m_1$ states each to be represented exactly (e.g., $m_1 = 4$ for a Hubbard site). The Hilbert space at the iteration $N$ is obtained as the direct product of the four blocks and thus has the size $(m \times m_1)^2$. Four main steps are involved in going from iteration $N$ to iteration $N + 1$: (i) The ground state corresponding to the Hilbert space of the iteration $N$ is obtained. (ii) The reduced ground-state density matrix of the superblock formed by one of the side blocks and its neighboring site is calculated (see Fig. 1). (iii) The reduced density matrix is diagonalized and the eigenvectors yielding the $m$ largest eigenvalues (i.e., occupations) are obtained. (iv) The superblock Hilbert space of size $m \times m_1$ is projected onto the $m$ most populated states derived in the previous step; the renormalized interactions within the blocks, and between the blocks and the central sites, are obtained by performing the corresponding unitary transformations. The resulting Hamiltonian is then used for calculating the ground-state at the following iteration, i.e., we return to step (i) until convergence is achieved.

Notice that in the present renormalization scheme, the boundary between superblocks corresponds alternatively to a long bond (single line, even iterations) or to a short bond (double line, odd iterations). In this way, short and long bonds are treated on the same footing (see Fig. 1). Nevertheless, an asymmetry remains which is unavoidable in the case of open chains: for all iterations the bonds at the ends of the chain are short ones ($\delta > 0$). This corresponds to the physically correct situation of larger contractions at the ends of a finite chain.[7] Note that the ground-state energy $E$ is not a pair function of the dimerization $\delta$ [i.e., $E(\delta) \neq E(-\delta)$] as one should expect for a periodic infinite chain. This is a drawback which has consequences for the extrapolated properties of infinite chains.

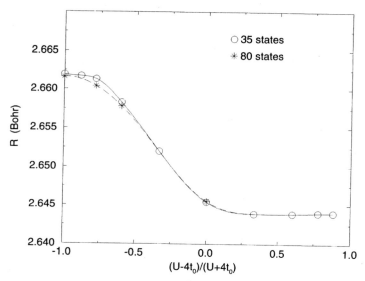

**Figure 2.** Average bond-length $\overline{R} = \frac{1}{2}(r_{i,i+1} + r_{i-1,i})$ of polyacetylene (in a.u.) as a function of the Coulomb interaction strength $U/|t_0|$. $t_0 = t(r_0)$, where $r_0 = 2.65$ a.u. is the equilibrium distance of the ethylene molecule.

The ground-state wave function is calculated using the Lanczos method, which takes advantage of the fact that the many-body states are obtained as a direct product of the states of different blocks and that the interblock interactions are sparse. The extrapolation to the infinite length limit is performed safely by considering chains having up to about 300 atoms. Once the $\pi$ electron energy of the infinite chain is obtained, we minimize $E_\pi + E_\sigma$ with respect to the dimerization $\delta = \frac{1}{2}(r_{i,i+1} - r_{i-1,i})$ and the average bond length $\overline{R} = \frac{1}{2}(r_{i,i+1} + r_{i-1,i})$ in order to obtain their equilibrium values. The accuracy of the results is controlled by considering different numbers $m$ of states kept in each superblock (e.g., $m = 35$, 80 and 150).

## IV. Results

We have performed geometry optimizations under the bond alternation constraint $r_{2i,2i\pm1} = \overline{R} \pm \delta$. Results for the optimized average inter-atomic distance $\overline{R}$ of polyacetylene are given in Fig. 2 as a function of $(U - 4t_0)/(U + 4t_0)$. For small $U/t_0$, $\overline{R} \simeq 2.662 a_0$ while in the strongly correlated limit $\overline{R} \simeq 2.644 a_0$. The crossover between the weakly and strongly correlated regimes is monotonous and occurs within a limited range of $U/t_0$, namely, from $U/t_0 \simeq 1/2$ to $U/t_0 \simeq 4$. Although the changes in $\overline{R}$ are quantitatively very small, they are of the same order of magnitude as the corresponding variations of the dimerization $\delta$. Results for $\delta$ are discussed later on.

Taking into account that the singlet and triplet energy surfaces of the dimer are always the same, independently of the value of $U/t_0$, the decrease of $\overline{R}$ with increasing $U/t_0$ indicates that electron correlations within the $\pi$ band are more efficient in the infinite chain than in the dimer as the Coulomb interactions become stronger. This tendency is confirmed by our results on the binding energy per atom given in Fig. 3. Indeed, the ground-state energy $E$ decreases with increasing $U/t_0$ in a

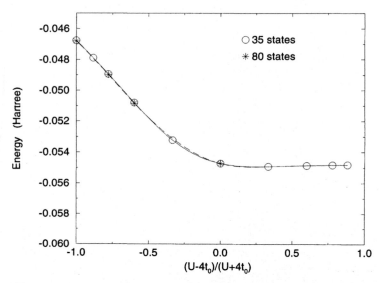

**Figure 3.** Ground-state energy of polyacetylene (in a.u.) corresponding to the optimal $\overline{R}$ and $\delta$ as a function of the Coulomb interaction strength $U/|t_0|$ [$t_0 = t(2.65$ a.u.$)$].

similar way as $\overline{R}$. In other words, the difference between the $\pi$-electron binding energy in the dimer and in the infinite polymer chain becomes larger as the strength of Coulomb interaction increases. However notice that in the spin-Peierls regime, i.e., for $U/t_0 > 4$, both $E$ and $\overline{R}$ are nearly independent of $U/t_0$. In this limit, the low-energy properties are dominated by the spin degrees of freedom (localized-electrons) and can be approximated quite accurately using a Heisenberg Hamiltonian with an exchange constant $J \sim t^2/U$. In our approach the distance dependent $J$ corresponds to the singlet-triplet excitation energy of the dimer,[8] which is the same for all values of $U/t_0$. Therefore, the decrease of $\overline{R}$ and $E$ are intrinsically related to the itinerant electron behavior.

In figure 4 results are given for the dimerization $\delta$ as a function of $(U - 4t_0)/(U + 4t_0)$. For $U = 0$, the DMRG calculations keeping $m = 80$ states per superblock are in reasonable agreement with the exact tight-binding solution. As expected, $\delta$ is larger in the strongly correlated limit than in the weakly correlated case. For large $U/t_0$ our calculations agree with the extrapolations from finite-cluster exact diagonalizations performed using the Heisenberg model.[3] However, it is remarkable that $\delta$ is a non-monotonic function of $U/t_0$. Indeed, it presents a minimum for $U/t_0 \simeq 0.5$ and a slight maximum for $U/t_0 \simeq 6.4$. This clearly illustrates the subtle competition between attractive and repulsive interactions in polyacetylene.

Concerning the convergence of the DMRG method we may notice that neither $\overline{R}$ nor $E$ are very much affected by changing the number of states $m$ kept in a superblock from $m = 35$ to $m = 80$ (see Figs. 2 and 3). This is in agreement with our calculations on the ground-state energy of nondimerized chains which compare extremely well with the Bethe-Ansatz results even if as few as $m = 35$ states are kept per block at each RG iteration.[4] For the dimerization, however, the behavior is more complicated. In the strongly correlated domain ($U/t > 4$), increasing the number of states in a superblock from $m = 35$ to $m = 80$ does not change significantly the dimerization.

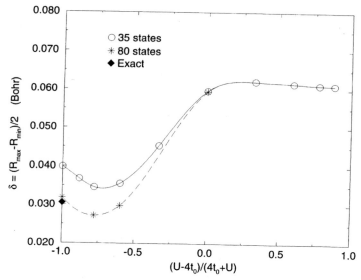

**Figure 4.** Dimerization $\delta = \frac{1}{2}(r_{i,i+1} - r_{i-1,i})$ of polyacetylene (in a.u.) as a function of the Coulomb interaction strength $U/|t_0|$ [$t_0 = t(2.65 \text{ a.u.})$].

One may conclude safely that the DMRG method has converged satisfactorily already for $m = 35$. However, in the delocalized domain ($U/t < 4$) there are large differences between the $m = 35$ and the $m = 80$ results (see Fig. 4). Although $m = 35$ yields the correct qualitative behavior concerning the minimum for $U/t_0 \simeq 0.5$, it largely overestimates the absolute value of the dimerization. The error increases as $U/t_0$ decreases and reaches up to 30% for $U = 0$. Even the $m = 80$ calculations are not fully converged for very small $U$, since $\delta$ is still overestimated by about 5%. A more detailed analysis of our results, in particular concerning the $U = 0$ case, shows that the performance of the DMRG method is very sensitive to the Coulomb interaction strength and to the degree of distortion. Strongly correlated or strongly dimerized systems, i.e., situations presenting a large tendency to electronic localization, are much easier to handle than the fully delocalized regime.[4]

## V. Summary and Discussion

The dimerization of polyacetylene has been determined in the framework of a Hubbard Hamiltonian for the valence $\pi$ electrons. The model parameters —i.e., distance dependence of the hopping integrals and of the $\sigma$-electron contribution to the binding energy—have been extracted from accurate *ab initio* calculations on the ethylene molecule. The strength of the Coulomb repulsion among $\pi$ electrons, $U/t$, has been varied from the uncorrelated limit all over to the strongly correlated or Heisenberg case. The underlying one-dimensional (1D) many-body problem has been solved numerically using the density matrix renormalization group (DMRG) method and the resulting ground-state energy has been optimized in order to determine the dimerization $\delta$ and the average bond-length $\overline{R}$ as a function of $U/t$. While $\overline{R}$ is rather insensitive to $U/t$, $\delta$ shows a remarkable non-monotonic behavior. The differences between the bond-order-wave (small $U/t$) and the spin-Peierls (large $U/t$) regimes have been quantified.

The strong sensitivity of $\delta$ to correlation effects may be exploited to analyze the convergence properties and accuracy of the DMRG calculations as well as the reliability of this approach for studying subtle electronic properties involving electron correlations, distortions and small energy differences. Besides the dimerization of polyacetylene, there are many other interesting problems where the physics is dominated by a similar subtle competition between opposite tendencies (*e.g.*, dimerization *vs.* delocalization, metal-insulator transitions, etc.). In order to be able to study such problems, it is necessary to treat the limits of short and long correlation length with comparable accuracies. Therefore, one is interested in finding criteria to quantify the accuracy of the DMRG procedure and to predict the appropriate number of block states to be kept in different practical situations. From this point of view the study of the dimerization of polyacetylene chains may also serve as a reference, which opens a variety interesting applications in the field of one-dimensional strongly correlated systems.

## Acknowledgments

Computer resources provided by IDRIS (Orsay) are gratefully acknowledged.

## References

1. See, for instance, H. C. Longuet-Higgins and L. Salem, *Proc. R. Soc. London, Ser. A* **25**, 172 (1959); W. P. su, J. R. Schrieffer, and A. J. Heeger, *Phys. Rev. Lett.* **42**, 1698 (1979); A. Karpfen and J. Petkov, *Solid State Commun.* **29**, 251 (1979); S. Suhai, *Chem. Phys. Lett.* **96**, 619 (1983); M. Takahashi and J. Paldus, *Int. J. Quantum Chem.* **28**, 459 (1985); C. M. Liegener, *J. Chem. Phys.* **88**, 6999 (1988); S. R. Chubb, *Phys. Rev. Lett.* **62**, 2016 (1989); G. König and G. Stollhoff, *Phys. Rev. Lett.* **65**, 1239 (1990); J. Ashkenazi et al., *J. Chem. Phys.* **104**, 8553 (1996).
2. E. H. Lieb and F. Y. Wu, *Phys. Rev. Lett.* **20**, 1445 (1968).
3. S. Capponi et al., *Chem. Phys. Lett.* **255**, 238 (1996).
4. M. B. Lepetit and G. M. Pastor, *Phys. Rev. B* **56**, 4447 (1997).
5. M. Said, D. Maynau, J. P. Malrieu, and M. A. Garcia Bach, *J. Am. Chem. Soc.* **106**, 571 (1984); *J. Am. Chem. Soc.* **106**, 580 (1984).
6. S. R. White, *Phys. Rev. Lett.* **69**, 2863 (1992); *Phys. Rev. B* **48**, 10 345 (1993).
7. C. S. Yannoni and T. C. Clarke, *Phys. Rev. Lett.* **51**, 1191 (1983); H. Kahlert, O. Leitner, and G. Leising, *Synthetic Metals* **17**, 467 (1987).
8. For finite $U$ minor differences result from the fact that we fit $t(r_{ij})$ by using the Hubbard Hamiltonian.

# Impurity States in Kondo Insulators

P. Schlottmann

*Department of Physics*
*Florida State University*
*Tallahassee, FL 32306*
*USA*

## Abstract

Kondo insulators are compounds with small-gap semiconductor properties. A Kondo hole is the charge neutral substitution of a rare earth or actinide atom by a nonmagnetic analog. Kondo holes break the translational invariance and give rise to boundstates in the gap, which pin the Fermi level and determine the magnetic, thermal and transport properties. A finite concentration of Kondo holes generates an impurity band inside the gap of the semiconductor. For small concentration, $c$, the height and width of the impurity band in the density of states are proportional to $\sqrt{c}$. We consider arbitrary clusters of Kondo holes embedded into the symmetric nondegenerate Anderson lattice with nearest-neighbor tight-binding conduction band on a simple cubic lattice. Properties of boundstates at the Fermi level are related to the connectivity of the cluster. The impurity band undergoes an insulator-metal transition, which is reduced to the classical site percolation of Kondo holes with first, second and fourth nearest-neighbor bonds. The critical percolation concentration is estimated at $c_{cr} = 0.10$. Of great interest are also the consequences of doping, ligand defects and the effect of Kondo holes on magnetic instabilities. Within a Kotliar-Ruckenstein mean-field approximation the Kondo insulator is unstable to long-range antiferromagnetism for $U > U_c$ and to ferromagnetism in sufficiently large fields. The paramagnetic-antiferromagnetic phase boundary is re-entrant as a function of the concentration of Kondo holes.

## I. Introduction

Kondo insulators are stoichiometric compounds with small-gap semiconductor properties. As a consequence of coherence a hybridization gap opens at the Fermi level. The Kondo insulators SmS, SmB$_6$, and TmSe were already an exciting topic more than fifteen years ago.[1] The more recent discovery of several Ce, Yb, and U based Kondo insulators, *e.g.*, CeNiSn (Ref. 2), $Ce_3Bi_4Pt_3$ (Ref. 3), $CeFe_4P_{12}$ (Ref. 4), $YbB_{12}$ (Ref. 5), $UFe_4P_{12}$ (Ref. 4), UNiSn (Ref. 6), and FeSi (Ref. 7), has renewed and enhanced the interest in this subject. Most Kondo insulators are nonmagnetic and have a van Vleck dominated low $T$ susceptibility. Exceptions are TmSe and UNiSn for which antiferromagnetic long-range order has been reported and $UFe_4P_{12}$ orders ferromagnetically at low temperatures. SmS and UNiSn undergo a metal-insulator transition as a function of pressure and temperature, respectively. Finally, a Bose-Einstein condensation of excitons is believed to take place in $TmSe_{1-x}Te_x$ (Ref. 8 and 9) and $Sm_{1-x}La_xS$ (Ref. 10).

Transport and thermodynamic properties of Kondo insulators at low temperature frequently differ from the ones expected for a standard semiconductor (for a review see Ref. 11). Deviations from an exponential activation are found and different measurements lead to different gap energies. In particular, it has been argued that the spin and charge gaps are different in these compounds.[12] Moreover, in view of the small energy gaps involved the properties strongly depend on strains in the crystal and impurities. For instance, optical reflectance and transmission measurements and the low-frequency Raman response of $SmB_6$ (Refs. 13 and 14) revealed the existance of states in the gap at low temperatures, which in this case probably are intrinsic to the semiconductor. At high temperatures the gap is smeared and the properties are those of a metal. Alternatively, the suppression of low-frequency optical conductivity in $Ce_3Bi_4Pt_3$ correlates with the quenching of the $4f$ moment.[15]

The formation of a coherent state (in this case the gap) in the Kondo lattice can be studied introducing disorder into the system,[16] *i.e.*, by alloying nonmagnetic impurities substituting for the rare earth or actinide ions. A charge neutral substitution, *i.e.*, a missing $f$ electron without changing the total number of electrons at a given site (for instance La for Ce or Th for U) is known as a Kondo hole. A Kondo hole in a Kondo insulator gives rise to a ($\delta$-function-like) boundstate in the energy gap.[17,18] These states only appear in the coherent state and disappear in the continuum at higher temperatures. With increasing concentration of Kondo holes an impurity band develops,[19] which gradually fills the hybridization gap. For a low concentration of Kondo holes the width and height of this band depend nonanalytically on the impurity concentration (proportional to $\sqrt{c}$) and the Fermi level is pinned within this band. As a consequence of this finite bandwidth there is a small low temperature regime in which the specific heat is proportional to $T$ and the susceptibility is finite as $T \to 0$.[19,20] For sufficiently low concentrations of Kondo holes the states within the impurity band are localized and have no mobility. The system is then an insulator. As a function of Kondo hole concentration there is a insulator-metal transition (not yet verified experimentally), which for electron-hole symmetry and on a simple cubic lattice with nearest neighbor hopping for the conduction band occurs at a critical concentration $c_{cr}$ of about 10%.[21] Electron-hole asymmetry and hopping beyond nearest-neighbors reduce $c_{cr}$.

Theoretical studies of pure (undoped) Kondo insulators are usually based on the periodic Anderson model or Anderson lattice. Numerous slave-boson mean-field approaches have been applied to obtain thermodynamic and transport properties.[22–26] Gaussian fluctuations about the mean-field saddle-point were incorporated by Karbowski,[27] who obtained that magnetic interactions among the sites are short-ranged and weak, in agreement with the experimental findings for some of the compounds. Varma,[28] on the other hand, discusses exchange and double exchange mechanisms in Kondo insulators and concludes from their weakness in many compounds that Kondo insulators have a substantial valence admixture. Ueda et al.[29] studied the phase diagram of the one-dimensional Kondo lattice combining numerical and exact results. The formation of a boundstate inside the hybridization gap of low-carrier-density systems is attributed to the screening of $f$ holes (exciton-like state) by Kasuya.[30] The spin and charge gaps of the Kondo lattice at half-filling were studied perturbatively in $t/J$ in Ref. 31.

Magnetic instabilities in pure Kondo insulators have been studied within Kotliar and Ruckenstein's slave-boson mean-field approach[32] involving four auxiliary bosons per site[33] (see also Refs. 34 and 35). On a bipartite lattice the paramagnetic phase is unstable (second-order phase transition) to long-range antiferromagnetic order above a critical value of the Coulomb repulsion $U$. The paramagnetic and antiferromagnetic phases are both insulating. An external magnetic field interferes destructively with the antiferromagnetic long-range order, but favors ferromagnetism. For sufficiently large fields the system jumps into a metallic ferromagnetic state (first-order transition).

Since the Kondo hole band pins the Fermi level, there are low-energy impurity states not present in the pure Kondo insulator, which affect the correlations as well as the magnetic instabilities.[36] The probability of finding a doubly-occupied or empty site does not significantly change with the Kondo hole concentration in the paramagnetic phase. This is the consequence of the interplay of two cancelling effects, namely, on the one hand, the impurity band widens the gap, but, on the other hand, reduces the effective Coulomb repulsion among $f$ electrons. The density of impurity states in the gap gives rise to a Stoner criterion for a ferromagnetic instability (second-order transition). This critical $U$ is, however, much larger than the one required for the antiferromagnetic instability. The low-energy states in the impurity band favor antiferromagnetic order overcoming the reduced (by the Kondo holes) effective $U$. As a function of Kondo hole concentration the critical $U$ for the antiferromagnetic instability first decreases with $c$, goes through a minimum and then reverses increasing for larger $c$. Hence, as a function of $c$ the system may display re-entrant behavior.

The results for the antiferromagnetic phase are similar to those anticipated by Doniach and Fazekas,[37] who however consider a different mechanism. Doping shifts the Fermi level out of the gap and into the band of heavy particles giving rise to a dilute gas of heavy electrons (holes). The exchange interaction between heavy particles leads to an antiferromagnetic groundstate at relatively low doping levels, but for larger doping (dirty metal) the system may revert to a nonmagnetic state.

The rest of the paper is organized as follows. In Section II we introduce the model, i.e., the Anderson lattice and the Kondo holes. The scattering off a cluster of Kondo holes is first solved exactly for $U = 0$. The correlations are then introduced via the $f$-electron self-energy within the local approximation. Results for isolated Kondo holes, a pair of Kondo holes and finite clusters of impurities are summarized in Sect. III. The connectivity of the boundstates may lead to an insulator-metal transition at a finite critical concentration. The self-consistent calculation leading to the impurity band is presented in Sect. IV. Some properties of the impurity band are discussed. Other possible perturbations, like doping and ligand defects are analyzed in Sect. V. The magnetic instabilities of the Kondo insulator and their interplay with the impurity band are reviewed in Sect. VI within the Kotliar-Ruckenstein mean-field saddle-point approximation.[32] Concluding remarks follow in Sect. VII.

## II. Formation of Boundstates in Kondo Insulators

### II.1 Model

We consider the Anderson lattice without orbital degeneracy:

$$H_0 = \sum_{\mathbf{k}\sigma} \epsilon_\mathbf{k} c^\dagger_{\mathbf{k}\sigma} c_{\mathbf{k}\sigma} + \epsilon_f \sum_{i\sigma} f^\dagger_{i\sigma} f_{i\sigma} + U \sum_i n_{i\uparrow} n_{i\downarrow} + V \sum_{\mathbf{k}\sigma} \left( c^\dagger_{\mathbf{k}\sigma} f_{\mathbf{k}\sigma} + f^\dagger_{\mathbf{k}\sigma} c_{\mathbf{k}\sigma} \right), \quad (1)$$

where $\epsilon_f$ is the $f$-level energy, $U$ is the Coulomb repulsion in the $f$-shell, $n_{i\sigma} = f^\dagger_{i\sigma} f_{i\sigma}$, $c^\dagger_{i\sigma}$ ($f^\dagger_{i\sigma}$) creates a conduction electron ($f$ electron) with spin $\sigma$ at the site $\mathbf{R}_i$, and $c^\dagger_{\mathbf{k}\sigma}$ ($f^\dagger_{\mathbf{k}\sigma}$) is a Bloch state with momentum $\mathbf{k}$. Here $\epsilon_\mathbf{k}$ is the dispersion, later assumed to be nearest-neighbor tight-binding on a simple cubic lattice.

Kondo holes at the sites $\mathbf{R}_j$, $j = 1, \ldots, N_i$, are introduced by the following scattering potential, $H_i$,

$$H_i = \left(\frac{\Delta \epsilon_f}{N}\right) \sum_{j,\mathbf{k},\mathbf{k}',\sigma} e^{i(\mathbf{k}'-\mathbf{k})\mathbf{R}_j} f^\dagger_{\mathbf{k}\sigma} f_{\mathbf{k}'\sigma}, \quad (2)$$

where $N$ is the number of sites. Here $\Delta\epsilon$ locally raises the $f$ electron energy and prevents the occupation of the $f$ state. The limit $\Delta\epsilon \to \infty$ corresponds then to a missing $f$ electron at the sites $\mathbf{R}_j$, and the impurities are nonmagnetic.[17,18]

### II.2 Single Impurity Scattering Problem for $U = 0$

Since the impurity potential is factorizable, the $U = 0$ scattering problem can be solved exactly.[17] It is convenient to introduce the matrix Green's function

$$\widehat{G}_{\mathbf{k},\mathbf{k}'}(z) = \begin{pmatrix} \langle\langle f_\mathbf{k}; f^\dagger_{\mathbf{k}'} \rangle\rangle_z & \langle\langle f_\mathbf{k}; c^\dagger_{\mathbf{k}'} \rangle\rangle_z \\ \langle\langle c_\mathbf{k}; f^\dagger_{\mathbf{k}'} \rangle\rangle_z & \langle\langle c_\mathbf{k}; c^\dagger_{\mathbf{k}'} \rangle\rangle_z \end{pmatrix}, \quad (3)$$

where we dropped the spin index, since up and down spins are decoupled. Using the equation of motion for $\widehat{G}$ we obtain after some algebra

$$\widehat{G}_{\mathbf{k},\mathbf{k}'}(z) = \widehat{G}^0_\mathbf{k}(z)\delta_{\mathbf{k},\mathbf{k}'} + \widehat{G}^0_\mathbf{k}(z)\widehat{T}_{\mathbf{k},\mathbf{k}'}(z)\widehat{G}^0_{\mathbf{k}'}(z), \quad (4)$$

with the $T$ matrix given by

$$\widehat{T}_{\mathbf{k},\mathbf{k}'}(z) = \left(\frac{1}{N}\right) e^{-i\mathbf{k}\cdot\mathbf{R}} \widehat{M} \left[\hat{I} - \widehat{G}^0(z)\widehat{M}\right]^{-1} e^{i\mathbf{k}'\cdot\mathbf{R}}, \quad (5)$$

where $\hat{I}$ is the identity and

$$\widehat{M} = \Delta\epsilon_f \begin{pmatrix} 1 & 0 \\ 0 & 0 \end{pmatrix}. \quad (6)$$

Here $\widehat{G}^0_\mathbf{k}(z)$ is the Green's function without impurity and

$$\widehat{G}^0(z) = \left(\frac{1}{N}\right) \sum_\mathbf{k} \widehat{G}^0_\mathbf{k}(z). \quad (7)$$

These equations considerably simplify in the limit $\Delta\epsilon_f \to \infty$ yielding

$$\widehat{T}_{\mathbf{k},\mathbf{k}'}(z) = -\frac{1}{N} e^{-i(\mathbf{k}-\mathbf{k}')\cdot\mathbf{R}} \left[G^0_{ff}(z)\right]^{-1} \begin{pmatrix} 1 & 0 \\ 0 & 0 \end{pmatrix}, \quad (8)$$

where $G^0_{ff}(z)$ is the $f$ component of Eq. (7). Since $\Delta\epsilon_f$ suppresses the $f$ electron occupation at the Kondo hole, this result does not depend on the actual value of the hybridization at the impurity site, which may be different from V.

## II.3 Many Impurity Scattering Problem for $U = 0$

Following similar steps as for the isolated Kondo hole[9] we obtain for the one $f$-electron Green's function for $\Delta\epsilon_f \to \infty$ (we suppress the $ff$ subindex)

$$G_{\mathbf{kk'}\sigma}(z) = \delta_{\mathbf{k},\mathbf{k'}}G^0_{\mathbf{k}\sigma}(z) + G^0_{\mathbf{k}\sigma}(z)T_{\mathbf{kk'}}(z,\{\mathbf{R}_j\})G^0_{\mathbf{k'}\sigma}(z),$$
$$G^0_{\mathbf{k}\sigma}(z) = (z - \epsilon_f)^{-1} + [V/(z-\epsilon_f)]^2[z - V^2/(z-\epsilon_f) - \epsilon_{\mathbf{k}}]^{-1}, \quad (9)$$

where the $T$-matrix explicitly depends on the space configuration of the Kondo holes. We introduce a transform of the $f$-electron Green's function of the pure insulator

$$\mathcal{G}^0_\sigma(z,\mathbf{R}) = \frac{1}{N}\sum_{\mathbf{k}} e^{i\mathbf{k}\cdot\mathbf{R}} G^0_{\mathbf{k}\sigma}(z) = \frac{\delta_{\mathbf{R},0}}{z-\epsilon_f} + \left(\frac{V}{z-\epsilon_f}\right)^2 F\left[z - \frac{V^2}{z-\epsilon_f}, \mathbf{R}\right], \quad (10)$$

where $\mathbf{R} = 0$ is the zero vector and

$$F[z',\mathbf{R}] = \frac{1}{N}\sum_{\mathbf{k}} \exp(i\mathbf{k}\cdot\mathbf{R})/(z' - \epsilon_{\mathbf{k}}), \quad (11)$$

is the transform of the spectral function of the conduction band. Denoting $\mathbf{R}_{lj} = \mathbf{R}_l - \mathbf{R}_j$ we define the $N_i \times N_i$ matrix $\tilde{\mathcal{G}}^0_\sigma(z,\mathbf{R}_{jl})$ with entries $l$ and $j$ that correlates the impurity sites. In the limit $\Delta\epsilon_f \to \infty$ (no $f$-electron at Kondo hole sites) the $T$-matrix is the Fourier transform of the inverse of the matrix $\tilde{\mathcal{G}}^0_\sigma(z,\mathbf{R}_{jl})$, i.e.

$$T_{\mathbf{kk'}}(z,\{\mathbf{R}_j\}) = -\frac{1}{N}\sum_{jl} e^{i\mathbf{k'}\cdot\mathbf{R}_j - i\mathbf{k}\cdot\mathbf{R}_l}[\tilde{\mathcal{G}}^0_\sigma(z,\mathbf{R}_{jl})]^{-1}, \quad (12)$$

and the local $f$-DOS at a site $\mathbf{R}_0$ is[38,39]

$$\rho_{f\sigma}(\omega,\mathbf{R}_0) = -\left(\frac{1}{\pi N}\right)\text{Im}\left\{\sum_{\mathbf{k},\mathbf{k'}} e^{i(\mathbf{k}-\mathbf{k'})\cdot\mathbf{R}_0} G_{\mathbf{kk'}\sigma}(z)\right\}$$
$$= -\pi^{-1}\text{Im}\{\mathcal{G}^0_\sigma(z,0)\}$$
$$+ \pi^{-1}\sum_{jl}\text{Im}\left\{\mathcal{G}^0_\sigma(z,\mathbf{R}_{0j})[\tilde{\mathcal{G}}^0_\sigma(z,\mathbf{R}_{jl})]^{-1}\mathcal{G}^0_\sigma(z,\mathbf{R}_{l0})\right\}, \quad (13)$$

where $z = \omega + i0$ and Im denotes *imaginary part*. The situation of an isolated Kondo hole is contained as a special limit.

## II.4 Correlations in the $f$-band

In the absence of impurities the DOS of the semiconductor consists of a full valence and an empty conduction band, separated by the hybridization gap. The gap edges are approximately at $\epsilon_f \pm V^2/6t$ (assuming a simple cubic lattice with nearest-neighbor hopping $t$) and the chemical potential $\mu$ lies in this gap. Inside the gap the imaginary part of the Green's function vanishes, so that $\mathcal{G}^0_\sigma(\omega+i0,\mathbf{R})$ is purely real. Correlations are incorporated into the $f$-band via the $f$-self-energy, which we consider within the local approximation,[40] i.e., the self-energy does not depend on $\mathbf{k}$ and just enters the Green's function as a renormalization of $\epsilon_f$. The $\mathbf{k}$-dependence of the correlations is believed to be less relevant in heavy-fermion systems than the energy dependence which reduces the hybridization gap to the size of the Kondo temperature. The imaginary part of $\Sigma_U(z)$ is zero inside the gap of the DOS. In the range of interest the real part of $\Sigma_U(z)$ is approximately given by $\Sigma_U(\omega) = Un_f/2 - (\gamma - 1)(\omega - \mu)$. Below we incorporate the Hartree-Fock shift $Un_f/2$ into $\epsilon_f$. The Kondo effect is contained in the parameter $\gamma$, which reduces the gap to $2V^2/\gamma 6t$. Note that the local approximation neglects the **k**-dependence arising from the interaction $U$, as well as the one induced into $\Sigma_U(\omega)$ via the broken translational invariance by the impurities (*i.e.*, $\Sigma_U$ is not calculated self-consistently).

## III. Properties of Kondo Hole Boundstates

### III.1 Spectral Properties

We first show that the local $f$-DOS vanishes identically at any impurity site.[39] If $\mathbf{R}_0$ is the site of a Kondo hole then $\mathcal{G}_\sigma(z, \mathbf{R}_{0j})$ is a row of the matrix $\tilde{\mathcal{G}}_\sigma(z, \mathbf{R}_{jl})$ and the second term of Eq. (13) reduces to $(1/\pi)\,\mathrm{Im}\{\mathcal{G}_\sigma(z, 0)\}$. Hence, the first and second terms cancel each other, so that the $f$-DOS vanishes identically for all frequencies. This result is not surprising, since $\Delta\epsilon_f \to \infty$ should suppress the $f$-DOS to zero. Here the superindex 0 in $\tilde{\mathcal{G}}^0_\sigma(z, \mathbf{R}_{jl})$ is dropped to indicate that it refers to the matrix dressed with $\gamma$ (self-energy due to correlations).

For a Kondo insulator with $N_i$ Kondo holes we expect $N_i$ boundstates in the hybridization gap with energies given by the poles in Eq. (13), *i.e.*, by the solutions of [see Eq. (12)]

$$\det\!\left[\tilde{\mathcal{G}}_\sigma(\omega + i0, \mathbf{R}_{jl})\right] = 0. \tag{14}$$

For the electron-hole symmetric Anderson lattice (the Hartree-Fock shifted $\epsilon_f$ is zero) the boundstate energies are symmetrically distributed with respect to the Fermi level at $\mu = 0$. Hence, if $N_i$ is odd there is always a boundstate at the Fermi level. In general, if $N_i$ is large we expect an accumulation of boundstates with energy close to zero. These low-energy boundstates are the relevant ones for the low-temperature thermodynamic properties of an impure Kondo insulator[19,20] and their mobility, *i.e.* connectivity, determines whether the system is a metal or an insulator.

In order to study the spectral weights of the boundstates we first recall[17,18,38,39] some properties of the transform function $\mathcal{G}_\sigma(z, \mathbf{R})$. To be specific we consider a simple cubic lattice with nearest-neighbor tight-binding dispersion for the conduction electrons, $\epsilon_\mathbf{k} = -2t\cos(k_x) - 2t\cos(k_y) - 2t\cos(k_z)$ with $|k_i| \leq \pi$, where the lattice parameter is set equal to one. Denoting $|R_x| = n_x$, $|R_y| = n_y$ and $|R_z| = n_z$, we have

$$F[z', \mathbf{R}] = (-i)\int_0^\infty d\lambda\, \exp(iz'\lambda)\, i^{n_x + n_y + n_z}\, J_{n_x}(2t\lambda)\, J_{n_y}(2t\lambda)\, J_{n_z}(2t\lambda), \tag{15}$$

where $J_n(2t\lambda)$ are Bessel functions of integer order. The argument $z' = z - V^2/(\gamma z)$ of the function $F$ for the symmetric Anderson lattice becomes large as $z \to 0$, *i.e.*, it diverges at the Fermi level. To leading order for $|z'| \to \infty$ we obtain

$$F[z', \mathbf{R}] = t^n\, (z')^{-n-1}/n_x!\, n_y!\, n_z!, \tag{16}$$

where $n = n_x + n_y + n_z$, and inserting into Eq. (10) we have

$$\mathcal{G}_\sigma(z, 0) = \frac{z}{z^2\gamma - V^2} + \frac{z\gamma\, 6t^2 V^2}{(z^2\gamma - V^2)^3},$$

$$\mathcal{G}_\sigma(z, \mathbf{R} \neq 0) = \frac{t^n}{n_x! \, n_y! \, n_z!} \frac{V^2 \, (z\gamma)^{n-1}}{(z^2\gamma - V^2)^{n+1}}. \tag{17}$$

For $\mathbf{R} \neq 0$ we have to distinguish the cases: (i) $n = 1$ for which $\mathcal{G}_\sigma^0(z, \mathbf{R})$ remains finite as $z \to 0$, and (ii) $n > 1$ for which the quantity has a zero of order $(n-1)$ at $z = 0$.

Consider now the ratio of $\mathcal{G}_\sigma(z, \mathbf{R})$ for two $\mathbf{R}$ differing by one unit in $n$

$$\mathcal{G}_\sigma(z, \mathbf{R}_{n+1})/\mathcal{G}_\sigma(z, \mathbf{R}_n) \approx -tz\gamma/V^2, \tag{18}$$

where we used that $V \gg V^2/6t$. For low-energy boundstates, $|z|$ is small since the states are close to the Fermi level, the spectral weight of a site then decreases dramatically with the distance to the closest Kondo hole. In particular, for states at the Fermi level all the spectral weight is located on nearest-neighbor sites to Kondo holes. All other sites have zero spectral weight if $z = 0$. Hence, the spectral weights for boundstates at the Fermi level of clusters of Kondo holes separated by more than two hoppings are disconnected. For a different dispersion (hopping beyond nearest-neighbors) or a different lattice, the extension of the above arguments leads to disconnected boundstates if two clusters of Kondo holes cannot be joined by two hoppings.

## III.2 Properties of Isolated Kondo Holes

In this Subsection we consider for instance the charge neutral substitution of a single Ce atom by its nonmagnetic analog La. Since the total number of electrons is constant, the Kondo hole boundstate pins the Fermi level. This determines the magnetic and thermal properties of the impurity state. The boundstate appears in both, the $f$ and conduction electron densities of states.

Since the the $\delta$-function is at the Fermi level, a Kondo hole in a Kondo insulator has magnetic properties. Its zero-field susceptibility follows a Curie law. Even a small magnetic field polarizes the boundstate at $T = 0$ and induces a Schottky anomaly in the specific heat, in analogy to a free spin $\frac{1}{2}$. This spin will precess in a homogeneous field if a small oscillating transversal field is applied and should in principle be observable by electron paramagnetic resonance. In practice, however, the linewidth of the resonance has to be smaller than a fraction of one-tenth of a meV for the effect to be observable. A linewidth may originate from relaxation processes and/or from inhomogeneous broadening. Defects and strains in the crystal introduce inhomogeneities which may give rise to a distribution of $g$ factors broad enough so that the line is not observable. On the other hand, inhomogeneities also introduce a distribution of boundstate energies, reducing in this way its magnetic character. Very large magnetic fields, however, would still polarize Kondo holes with a distribution of boundstate energies, but the field also modifies the bulk properties of the Kondo insulator, in particular, it reduces the gap.[22] Kondo holes also affect the infrared absorption spectrum.

For the spectral properties we may consider the Kondo hole at the origin. As shown above the $f$-DOS vanishes identically at the impurity site. For the electron-hole symmetric Anderson lattice the boundstate is at $z = \mu = 0$. Hence, according to Eq. (18) all the spectral weight of the boundstate is located on the nearest-neighbor sites and the corresponding $f$ electron residue for one nearest-neighbor site (the lattice coordination is 6) is given by[17]

$$t^2/(V^2 + 6t^2\gamma). \tag{19}$$

Note that as the Coulomb correlations increase, also $\gamma$ increases and the $f$ spectral weight decreases. Electron-hole asymmetry gives rise to nonzero spectral weight everywhere except at the impurity site, but the residue falls off rapidly with the distance from the impurity, so that the boundstate is still almost completely localized on the nearest-neighbor sites. According to Eq. (17) the spectral weight falls off faster than exponential with the distance from the impurity (factorials in the denominator) and with the $n$th power of the position of the pole with respect to the center of the gap over the size of the gap.

### III.3 Properties of a Pair of Kondo Holes

In this Subsection we limit ourselves to electron-hole symmetry and a tight-binding band with nearest-neighbor hopping on a simple cubic lattice. In this case Eq. (14) refers to a $2 \times 2$ determinant and the two boundstates are given by its zeroes[38]

$$\mathcal{G}_\sigma(z, \mathbf{0}) = \pm \mathcal{G}_\sigma(z, \mathbf{R}_{12}). \qquad (20)$$

We must distinguish the situation where the two Kondo holes are on nearest-neighbor sites, $n_{12} = 1$, from the case where they are not nearest-neighbors, $n_{12} > 1$.

For two impurities on nearest-neighbor sites, $n_{12} = 1$, Eq. (20) has two symmetric solutions at finite $z$, which have to be determined numerically. These solutions correspond to the bonding and antibonding states and for reasonable parameter values the energies are very close to the band edges. The boundstates are predominantly localized within the unit cells neighboring the impurities. One electron with spin-up and one with down-spin occupy the lower-energy boundstate, so that the groundstate when the two impurities are nearest-neighbors is a relatively strong singlet state. The energy difference between the bonding and antibonding states is the exchange coupling or binding energy. Since $\mu = 0$ for electron-hole symmetry, the boundstate is nonmagnetic and it requires an irrealistically large magnetic field to break up this singlet.

For $n_{12} > 1$, on the other hand, Eq. (20) has a twofold solution at $z = 0$. Hence, there is no binding energy and the impurities are independent and non-interacting. The singlet and triplet states are degenerate, so that the boundstates are magnetic (in a magnetic field two electrons with upspin occupy the states). The spectral weight is nonzero only on sites neighboring at least one of the impurities. Hence, if $n_{12} > 2$ the residues are given by Eq. (19), the two boundstates non-interfering and said to be disconnected. However, if $n_{12} = 2$ the two impurities have at least one common nearest-neighbor, so that the boundstates are shared among the two Kondo holes. This situation is refered to as a connected cluster. The spectral weights of the neighboring sites depend on the relative position of the two Kondo holes.

### III.4 Arbitrary Clusters and Insulator-Metal Transition

The above arguments for a pair of Kondo holes can be extended to arbitrary configurations of Kondo holes, but multiple scattering and the interference between the different scattering sites considerably complicate the problem and give rise to interactions among the impurities.[39] As discussed above the $N_i$ energies of the boundstates are given by the zeroes of the determinant, Eq. (14). These energies depend on the spatial distribution of the $N_i$ Kondo holes. The electron-hole symmetry requires that the energy distribution be symmetric with respect to the center of the gap. In particular, if $N_i$ is large there is an accumulation of boundstates with energy close to zero (this eventually leads to the impurity band). These states are the most relevant ones, because they determine the low temperature thermodynamics and transport properties.

For all boundstates the spectral weight is predominantly localized on the sites neighboring the Kondo holes, and the residue decreases fast with the distance from the closest Kondo hole. In particular, it follows from Eq. (18) that for zero-energy boundstates only sites neighboring a Kondo hole can have nonzero spectral weight. Hence, clusters of Kondo holes separated by more than one lattice site are disconnected for boundstates at the Fermi level and wavefunctions of disconnected clusters have no overlap.

The connectivity increases with increasing $N_i$ until eventually we obtain a percolating cluster. This percolation threshold corresponds to the insulator-metal transition. In other words, we have traced out all the electron degrees of freedom and are left with a simple geometrical problem, namely to find the percolation threshold. On a simple cubic lattice with nearest-neighbor hopping the insulator-metal transition in the impurity band then reduces to the site percolation of Kondo holes with first, second and fourth nearest-neighbor bonds.[21] We have estimated the critical concentration of Kondo holes using (i) the low density mean cluster size expansion and (ii) a small cell renormalization which yields $c_{\text{cr}}$ as the scaling fixed point. The results of both methods are consistent with $c_{\text{cr}} = 0.099$.

The mean size of finite clusters $S(p)$, where $p$ is the probability that a site is occupied, can be expanded in a power series in $p$, $S(p) = 1 + \sum_{n=1}^{\infty} a_n p^n$, by summing over the probabilities of occurrence of all possible clusters up to a desired order. The convergence of the series expansion in $p$ is then analyzed with Padé approximants or with the ratio method. The assumption that $S(p)$ diverges as $A(1 - p/p_c)^{-(1+g)}$ when $p \to p_c^-$, leads to the large $n$ asymptotic form $a_n \approx A n^g / \Gamma(1 + g) p_c^n$ (Ref. 41). $p_c$ is then determined from the ratio of consecutive coefficients, $p_c = (1 + g/n)(a_{n-1}/a_n)$. Since critical properties do not depend on the range of the bonds (as long as they are short-ranged), we may use $g = \gamma - 1 \approx 0.8$.[42]

The small cell renormalization (like the block spin renormalization frequently used for thermal critical phenomena) is based on self-similarity under scaling. Scaling assumes that the linear dimension $b$ of the cell is much smaller than the correlation length $\xi$. The $M = b^3$ sites of the cell are replaced by a single supersite. The criterion for such replacement in thermal critical phenomena is the majority rule. For a percolation problem the connectivity is the essential ingredient[43] and we consider a supersite occupied if the bottom and top faces of the cell are connected, and empty otherwise. The percolation threshold is then given by the critical fixed point of the renormalization transformation.

Assuming a constant mean free path in the vicinity of $c_{\text{cr}}$ the electrical conductivity tends to zero proportional to $(c - c_{\text{cr}})^2$ as $c \to c_{\text{cr}}^+$ and in the critical region the correlation length diverges as $\xi \propto (c - c_{\text{cr}})^{-\nu}$ with $\nu \approx 0.9$ (Ref. 21).

If the tight-binding dispersion in the conduction band extends beyond nearest-neighbors, also the physical extension of the boundstates increases accordingly. The spectral weight for boundstates at the Fermi level is nonzero on all sites that can be reached by simple hopping from a Kondo hole. This substantially increases the connectivity of the clusters, lowering in this way the percolation threshold. Hence, 10% of Kondo holes is an upper bound for the insulator to become a metal.

For a small electron-hole asymmetry or for states away from the Fermi level also sites beyond nearest-neighbors to a Kondo hole have a small spectral weight. The spectral weight decreases very fast with distance from the impurities. Hence, electron-hole asymmetry, an a.c. electric field, a constant magnetic field or a finite temperature increase the connectivity. Two types of connections have to be distinguished: The strong ones we considered in Ref. 21, and weak ones induced by the above parameters. Below the $c_{cr}$ for the strong connections the impure Kondo lattice is expected to behave like a very poor metal or a dirty semiconductor.

## IV. Impurity Band

For a finite concentration of impurities, $c$, a finite density of boundstates develops at the Fermi energy, which has to be treated self-consistently. The perturbed Green's function, Eq. (4), is written as a Dyson equation

$$[\widehat{G}^0_{\mathbf{k}}(z)^{-1} - \widehat{\Sigma}_{\mathrm{imp}}(z, \epsilon_f)]\widehat{G}_{\mathbf{k}}(z) = \hat{1}, \tag{21}$$

with an impurity self-energy, which to lowest order in $c$ is given by $cN\widehat{T}_{\mathbf{kk}}(z)$. The Dyson self-energy effectively renormalizes $\epsilon_f$, since according to Eq. (8) only the $f$-$f$ component of $\widehat{T}$-matrix is nonzero and momentum independent for $\mathbf{k} = \mathbf{k}'$. Denoting with $\tilde{\epsilon}_f(z)$ the renormalized quantity we arrive at the following self-consistency condition

$$\tilde{\epsilon}_f(z) = \epsilon_f + \Sigma_{\mathrm{imp}}(z, \tilde{\epsilon}_f(z)) + \Sigma_U(\omega), \tag{22}$$

where inside the gap we may use again $\Sigma_U(z) = Un_f/2 - (\gamma - 1)(z - \mu)$. Note that $\Sigma_U(z)$ was computed within the local approximation, i.e. neglecting the $\mathbf{k}$-dependence arising from fluctuations due to the Coulomb interaction (see Subsection II.4). In the same spirit we do not self-consistently incorporate the effects of the Kondo hole band into the self-energy. The impurities break the translational invariance; this affects the $f$-electron propagator, giving rise to an additional $\mathbf{k}$-dependence in the self-energy, which is also neglected here.

The self-consistency Eq. (22) is equivalent to summing over all noncrossing impurity diagrams[44] (crossing diagrams give rise to a $\mathbf{k}$-dependence) and corresponds to the lowest order correction expansion in the coherent potential approximation (CPA) (for a similar calculation within the CPA see Ref. 45). The solution of Eq. (22) depends on the conduction density of states. For the symmetric Anderson model we obtain for $\omega = 0$ (center of the band)

$$\tilde{\epsilon} = -i\sqrt{c}\, V^2/\langle \epsilon_{\mathbf{k}}^2 \rangle^{1/2}, \tag{23}$$

where $\langle \epsilon_{\mathbf{k}}^2 \rangle$ is the second order moment of the conduction band and proportional to $D^2$. Hence, the renormalized $f$-level energy acquires an imaginary part for small frequencies, which depends nonanalytically on the concentration of impurities. A similar analysis for $\omega \neq 0$ yields an impurity bandwidth proportional and of the order of $\tilde{\epsilon}_f$. The impurity band is approximately semielliptic with both height and width proportional to $\sqrt{c}$ for small $c$; hence, as expected, the number of states in the impurity band is proportional to $c$. Note that because of the nonanalytic dependence on $c$, the self-consistency of Eq. (22) and the limit $c \to 0$ cannot be interchanged. Similar results are obtained for the asymmetric Anderson model.

In order to discuss the $f$-density of states over the entire energy range, the complete self-energy $\Sigma_U(z)$ is needed. The physical situation is usually qualitatively well-described with $\Sigma_U(z)$ calculated to second order in $U$. This restricts the calculation to small $U$, but effects and trends are not expected to change qualitatively for larger

$U$, so that conclusions are valid quite generally. Also the density of states of the conduction band, i.e., the spectral function $F(z)$, has to be specified. Since the results do not critically depend on the form of $F(z)$, we may assume an elliptic density of states of halfwidth $D$, i.e.

$$F(z) = \frac{1}{N} \sum_{\mathbf{k}} \frac{1}{z - \epsilon_{\mathbf{k}}} = \frac{2}{D^2} \left( z - \sqrt{z^2 - D^2} \right). \quad (24)$$

The function (24) corresponds to Eq. (11) evaluated at $\mathbf{R} = \mathbf{0}$.

The $f$-density of states has in principle a five peak structure. The center peak is at the Fermi level and represents the Kondo hole impurity band. This band has no coherent character. Most of the $f$-DOS is located in the two coherent Kondo peaks close to the Kondo gap edges. Finally, the self-energy $\Sigma_U(z)$ induces two additional incoherent structures at energies of the order of $U$. These broad peaks correspond to charge promotions in and out of the $f$-band, i.e., to the "atomic" transitions $f^0 \to f^1$ and $f^1 \to f^2$.

The magnetic and thermal properties of the impure Kondo insulator are determined by the Kondo hole impurity band, which is the one pinning the Fermi level. The low-temperature specific heat has a component proportional to $T$, arising from the finite density of states in the impurity band. The mass-enhancement associated with the $\gamma$-coefficient depends on the concentration of Kondo holes and is roughly proportional to $\sqrt{c}$ for small $c$. Such a dependence of $\gamma$ has been verified for La-substituted $Ce_3Bi_4Pt_3$ (Ref. 3). The temperature range of this linear $T$ dependence is given by the width of the impurity band. The susceptibility is Pauli like at low $T$ and becomes a Curie law at higher $T$. As expected, both the zero-temperature susceptibility and the $\gamma$-coefficient, track the density of states at the Fermi level.[19,20]

## V. Doping and Ligand Defects

The above results are easily extended to more general defects and to doping. Consider an isolated Kondo hole impurity at the site $\mathbf{R}$ described by the following Hamiltonian

$$H_i = \frac{1}{N} \sum_{\mathbf{k}\mathbf{k}'\sigma} e^{i(\mathbf{k}'-\mathbf{k})\cdot\mathbf{R}} \left[ \Delta\epsilon_f f^\dagger_{\mathbf{k}\sigma} f_{\mathbf{k}'\sigma} + W c^\dagger_{\mathbf{k}\sigma} c_{\mathbf{k}'\sigma} + \Delta V (f^\dagger_{\mathbf{k}\sigma} c_{\mathbf{k}'\sigma} + c^\dagger_{\mathbf{k}\sigma} f_{\mathbf{k}'\sigma}) \right], \quad (25)$$

where in addition to $\Delta\epsilon_f$, we now have $W$, the potential scattering of the conduction electrons at the impurity, and $\Delta V$, a local change in the hybridization, e.g., due to a ligand defect.

Again for $U = 0$ the scattering problem is solved exactly in terms of a $T$-matrix, as in Eq. (5), but with the matrix $\widehat{M}$ given by

$$\widehat{M} = \begin{pmatrix} \Delta\epsilon_f & \Delta V \\ \Delta V & W \end{pmatrix}. \quad (26)$$

The boundstates in the gap of the Kondo insulator are given by the poles of the $T$-matrix. The components of $\widehat{G}^0(z)$ are related to the density of states of the conduction electrons, which we assume semi-elliptic, Eq. (15). Denoting $g(z) = 1/(z - \epsilon_f)$ we obtain

$$\begin{aligned} G^0_{ff}(z) &= g(z) + V^2 g(z)^2 F[z - V^2 g(z)], \\ G^0_{cf}(z) &= G^0_{fc}(z) = V g(z) F[z - V^2 g(z)], \\ G^0_{cc}(z) &= F[z - V^2 g(z)]. \end{aligned} \quad (27)$$

The imaginary part of these quantities vanishes inside the gap and the Coulomb correlations in the $f$-band are introduced via a self-energy. The above equations completely determine the $f$ electron Green's function for one Kondo hole.[46]

For the isolated impurity we limit ourselves to discuss two interesting situations. In the limit $\Delta\epsilon_f \to \infty$ the local change in hybridization is irrelevant (the $f$-level is not occupied) and $W$ introduces a phase shift, which affects the position of the boundstate in the gap. If the Kondo hole is a charge neutral substitution (e.g., La for Ce) the boundstate pins the Fermi level and the results are similar to the ones discussed in Subsect. III.2. On the other hand, if the impurity is a dopand (e.g., tetravalent Th for trivalent Ce), there is an additional conduction electron and the boundstate will be filled for both spin-components. The Fermi level is not pinned and lies in the gap between the boundstate and the empty band edge, so that the properties of the impurity are nonmagnetic (in contrast to the charge neutral substitution, where the boundstate pins the Fermi level and is magnetic.

The other interesting case is the ligand defect. In principle, all three scattering amplitudes in Eq. (25) could be nonzero, but the most relevant one is now $\Delta V$. For simplicity we then consider $W = \Delta\epsilon_f = 0$. Carrying out the matrix products in Eq. (5) we obtain a denominator that is quadratic in $\Delta V$. The zeroes of this denominator determine the positions of the $\delta$-function-like boundstates. There are then, in principle, two poles, both corresponding to negative values of $\Delta V$ (local reduction of the hybridization). In general, it requires a threshold value of $\Delta V$ for a boundstate to develop and $\Delta V$ should not exceed a critical value.

The properties of the ligand defect again depend on the characteristics of the impurity. If the defect is a donor or acceptor the Fermi level will be pinned at one of the boundstates and the impurity is magnetic as discussed above. If the defect is charge neutral, on the other hand, it does not pin the Fermi level and its properties are nonmagnetic (note that the properties are inverted with respect to the $\Delta\epsilon_f \to \infty$ case).

A finite concentration of impurities, $c$, again gives rise to impurity bands. We procede as in Sect. IV and rewrite the perturbed Green's function as a Dyson equation with an impurity self-energy $\hat{\Sigma}_{\text{imp}}(z,\epsilon_f,V) = cN\hat{T}_{\mathbf{kk}}(z)$. This matrix self-energy renormalizes the frequency, the $f$-level energy and the hybridization; a self-consistent treatment of this self-energy leads to[46]

$$\tilde{z} = z - \Sigma_{cc}(\tilde{z},\tilde{\epsilon}_f,\tilde{V}),$$
$$\tilde{\epsilon}_f = \epsilon_f + \Sigma_{ff}(\tilde{z},\tilde{\epsilon}_f,\tilde{V}) - \Sigma_{cc}(\tilde{z},\tilde{\epsilon}_f,\tilde{V}) + \Sigma_U(z),$$
$$\tilde{V} = |V + \Sigma_{fc}(\tilde{z},\tilde{\epsilon}_f,\tilde{V})|. \qquad (28)$$

Here we take the absolute value of the hybridization to keep it real; this corresponds to a gauge transformation, which has no effect on the $f$-DOS. Eqs. (28) are equivalent to the sum of all diagrams with noncrossing impurity lines. They lead to a complex $f$-level energy and finite imaginary part of $\tilde{z}$, which broaden the $\delta$ function of the Kondo hole boundstate giving rise to an impurity band of finite width. Note that again $\Sigma_U$ is not calculated self-consistently, in the sense that the effect of the impurity band on the self-energy (as well as the $\mathbf{k}$-dependence) is neglected here.

For $\Delta\epsilon_f \to \infty$ the results are very similar to those discussed in Sect. IV. The $f$-DOS for the situation $\Delta V \neq 0$ and $\Delta\epsilon_f = W = 0$ (ligand defects) has a reduced height of the main Kondo peaks close to the gap edges and impurity tails protude into the gap. If the ligand defects are charge neutral the Fermi level remains within the gap and the system is a nonmagnetic insulator with a reduced gap. If, on the other hand, the ligand impurity dopes the crystal the Fermi level lies in the continuum; hence, the low-temperature specific heat is proportional to $T$, the susceptibility is Pauli-like and depending on the position of the mobility edge the system will be a metal or an insulator.

## VI. Magnetic Instabilities of a Kondo Insulator

### VI.1 Formulation

The Kondo insulators TmSe and UNiSn are antiferromagnetic at low $T$, while UFe$_4$P$_{12}$ has long-range ferromagnetic order. To study magnetic instabilities it is convenient to reformulate the many-body problem in terms of "auxiliary bosons." The slave-boson method has been extensively used for the $U \to \infty$ limit in terms of one "slave boson" per site.[47-49] Kotliar and Ruckenstein[32] extended the method to the finite $U$ situation by introducing four "slave bosons" per site to study the Hubbard model. This slave-boson technique was later applied to a model for highly correlated bands of hybridized Cu $3d$ and O $2p$ orbitals[50,51] and the Anderson lattice.[33] The slave-boson approach has been formulated with spin-rotational invariance,[52] but this does not affect the mean-field results. In this Section we briefly review the interplay of the Kondo hole impurity band with magnetic order.

We introduce four Bose creation and annihilation operators for each site,[32,33] i.e., $e^\dagger$, $e$ and $d^\dagger$, $d$ for the empty and doubly occupied states, and $p_\sigma^\dagger$, $p_\sigma$ for the single occupied states, which act as projectors onto the corresponding electronic states. They satisfy the completeness relation and the projector condition

$$e_i^\dagger e_i + p_{i\uparrow}^\dagger p_{i\uparrow} + p_{i\downarrow}^\dagger p_{i\downarrow} + d_i^\dagger d_i = 1, \qquad f_{i\sigma}^\dagger f_{i\sigma} = p_{i\sigma}^\dagger p_{i\sigma} + d_i^\dagger d_i. \qquad (29)$$

In the physical subspace defined by Eq. (1) the operators $f_{i\sigma}^\dagger$ and $f_{i\sigma}$ are replaced by $Z_{i\sigma}^\dagger f_{i\sigma}^\dagger$ and $f_{i\sigma} Z_{i\sigma}$, so that the matrix elements are invariant in the combined fermion-boson Hilbert-space. The definition of the operators $Z_{i\sigma}$ is not unique and we choose the same expression as in Refs. 32 and 33, which yields the correct matrix elements and the correct expectation value of $\langle Z_{i\sigma}^\dagger Z_{i\sigma} \rangle$ within the mean-field approximation as $U \to 0$. The constraints, Eqs. (29), are incorporated via Lagrange multipliers, $\lambda_i^{(1)}$ and $\lambda_{i\sigma}^{(2)}$, respectively.

In the mean-field (saddle-point) approximation we replace all boson operators by their expectation values. For a half-filled band (two electrons per site) with electron-hole symmetry, $\epsilon_f = -\frac{1}{2}U$, we have for the paramagnetic and ferromagnetic phases

$$\sum_\sigma \langle f_{i\sigma}^\dagger f_{i\sigma} \rangle = 1,$$

$$\langle p_{i\sigma}^\dagger \rangle = \langle p_{i\sigma} \rangle = p_\sigma,$$

$$\langle e_i^\dagger \rangle = \langle e_i \rangle = \langle d_i^\dagger \rangle = \langle d_i \rangle = d,$$

$$\langle Z_{i\sigma}^\dagger \rangle = \langle Z_{i\sigma} \rangle = \langle Z_{i-\sigma}^\dagger \rangle = \langle Z_{i-\sigma} \rangle = Z,$$

$$Z = \frac{d(p_\uparrow + p_\downarrow)}{[(p_\uparrow^2 + d^2)(p_\downarrow^2 + d^2)]^{1/2}}, \qquad (30)$$

and the mean-field expression of Hamiltonian (1) is

$$H_{mf} = \sum_{\mathbf{k}\sigma} \epsilon_{\mathbf{k}} c_{\mathbf{k}\sigma}^\dagger c_{\mathbf{k}\sigma} + \sum_{i\sigma} \epsilon_{f\sigma} f_{i\sigma}^\dagger f_{i\sigma} + VZ \sum_{i\sigma} (c_{i\sigma}^\dagger f_{i\sigma} + f_{i\sigma}^\dagger c_{i\sigma})$$
$$+ N_s d^2 (U + 2\lambda^{(1)} - \lambda_\uparrow^{(2)} - \lambda_\downarrow^{(2)}) - N_s \lambda^{(1)}$$
$$+ N_s p_\uparrow^2 (\lambda^{(1)} - \lambda_\uparrow^{(2)}) + N_s p_\downarrow^2 (\lambda^{(1)} - \lambda_\downarrow^{(2)}), \tag{31}$$

where $\epsilon_{f\sigma} = \epsilon_f - \sigma B + \lambda_\sigma^{(2)} = -\sigma \epsilon_m^*$ is the renormalized $f$ level energy. The constraints (29) only apply to sites occupied by a rare earth (actinide) atom, but not to those with Kondo holes, so that $N_s = N(1-c)$ with $c$ being the Kondo hole concentration.

The minimization of the groundstate energy of Hamiltonian (29) with respect to $\epsilon_m^*$, $\lambda^{(1)}$, $p_\uparrow$, $p_\downarrow$ and $d$ yields

$$\langle f_{i\sigma}^\dagger f_{i\sigma} \rangle = p_\sigma^2 + d^2, \qquad V \frac{\partial Z}{\partial p_\sigma} \sum_{i\sigma'} \langle f_{i\sigma'}^\dagger c_{i\sigma'} \rangle + N_s p_\sigma (\lambda^{(1)} - \lambda_\sigma^{(2)}) = 0,$$
$$p_\uparrow^2 + p_\downarrow^2 + 2d^2 = 1, \qquad V \frac{\partial Z}{\partial d} \sum_{i\sigma'} \langle f_{i\sigma'}^\dagger c_{i\sigma'} \rangle + 2 N_s \lambda^{(1)} d = 0. \tag{32}$$

It is straightforward to eliminate the parameter $\lambda^{(1)}$ from the last two equations. The expectation values $\langle f_{i\sigma}^\dagger f_{i\sigma} \rangle$ and $\langle f_{i\sigma}^\dagger c_{i\sigma} \rangle$ are obtained from the one-particle Green's functions evaluated self-consistently in the presence of the Kondo holes.

Kondo holes are again introduced via the scattering potential, Eq. (2), using the procedure already described in Subsection II.2, but now for the Hamiltonian $H = H_{mf} + H_i$. The only difference is that in $H_{mf}$ the hybridization is renormalized by the constraints imposed by the correlations, i.e., $V$ is to be replaced by $VZ$ and $\epsilon_{f\sigma}$ by $-\sigma \epsilon_m^*$. The correlations in the $f$-band are then no longer introduced through the self-energy $\Sigma_U(z)$, but via the self-consistently determined mean-field slave-bosons. The results for isolated Kondo holes, clusters of Kondo holes, and the metal-insulator transition in the impure Kondo insulator derived in Section III without slave bosons remain unchanged and can be taken over. The fact that these results are independent of the formulation is a strong indication of their robustness.[36]

To generalize the results to a finite concentration of impurities we again self-consistently sum over all the non-crossing diagrams. This leads to a Dyson equation with an impurity self-energy as discussed in Section IV. For $\Delta \epsilon_f \to \infty$ only the $f$-level energy is renormalized and determined by the self-consistency[19,20,44,46]

$$\tilde{\epsilon}_f = \epsilon_f + \Sigma_{ff}(z, \tilde{\epsilon}_f). \tag{33}$$

Equation (33) leads to a complex $f$ level energy with finite imaginary part, which broadens the $\delta$ function of the Kondo hole boundstate giving rise to an impurity band of finite width.

For given $\epsilon_m^*$, $d$, $p_\uparrow$, and $p_\downarrow$ the self-consistent solution of Eq. (33) yields the one-particle Green's functions, which are then used to calculate the expectation values $\langle f_{i\sigma}^\dagger f_{i\sigma} \rangle$ and $\langle f_{i\sigma}^\dagger c_{i\sigma} \rangle$. These expectation values are needed in Eq. (32) to determine $\epsilon_m^*$, $d$, $p_\uparrow$, and $p_\downarrow$ self-consistently. To preserve electron-hole symmetry in the presence of Kondo holes, we assume that we have as many doubly occupied $f$-sites as empty $f$-sites. Both give rise to a singlet state in the $f$-shell and, hence, to the same effects. Below we treat the paramagnetic, ferromagnetic and antiferromagnetic states separately.

## VI.2 Paramagnetic Groundstate

In the paramagnetic phase and in the absence of an external magnetic field $\langle f_{i\sigma}^\dagger f_{i\sigma}\rangle = \frac{1}{2}$, $\epsilon_m^* = 0$, and $p_\uparrow = p_\downarrow = p$, such that

$$4V(p^2 - d^2)\frac{1}{N}\sum_{\mathbf{k}\sigma}\langle f_{\mathbf{k}\sigma}^\dagger c_{\mathbf{k}\sigma}\rangle + (1-c)U dp = 0, \qquad p^2 + d^2 = \tfrac{1}{2}. \tag{34}$$

The expectation value is evaluated by integrating over the off-diagonal Green's function, calculated self-consistently including the impurity band. When the total number of $f$-electrons is calculated by integrating over $\text{Im}\{G_{ff}(\omega + i0)\}$, the result is not $\frac{1}{2}$ unless $c/2$ is added in to take into account the Kondo holes with double $f$-electron occupation (required to preserve the electron-hole symmetry). For $c = 0$ we recover the pure Kondo insulator discussed in Ref. 33. The parameter $d^2$, representing the probability of having two $f$-electrons at one site (or an empty shell) in the Kondo lattice, depends strongly on the Coulomb repulsion $U$ and weakly on the Kondo hole concentration $c$.

For the pure insulator $d^2$ determines the renormalization of the Kondo gap, $32d^2(1/2 - d^2)V^2/D$. As a function of $U$, $d^2$ has the value $1/4$ for $U = 0$ and decreases to zero as $U \to \infty$. Hence, also the gap monotonically decreases as a function of $U$ and for large $U$ we obtain asymptotically

$$d^2 = a\frac{D^2}{8V^2}\exp\left(-\frac{UD}{8V^2}\right), \tag{35}$$

which has the characteristic exponential Kondo dependence, but the exponent differs by a factor of 2 from the usual Kondo impurity exponential. This is known as the "lattice enhancement of the Kondo effect,"[37,53] which is an artifact of Gutzwiller-type approximations. In Eq. (35) $a$ is a constant that depends on the chosen density of states for the conduction band, e.g., $a = 1$ for a square density of states.

The Kondo hole impurity band introduces three competing effects. First, the impurity band shifts the band edges of the semiconductor, increasing the magnitude of the gap. Second, the integration over the impurity band itself increases the magnitude of the expectation value $d^2$. Finally, the factor $(1 - c)U$ in Eq. (34) reflects the reduction of correlations due to the Kondo holes. The quantity $d^2(U,c)/d^2(U,c=0)$ decreases with $c$ approximately linearly, although there is a small sublinear term proportional to $c^{1/2}$. In general, the decrease of $d^2$ is small (a few per cent, increasing with $U$), because of the compensation of the several effects.

The impurity band arises from the imaginary part of $\tilde{\epsilon}_f$, which for the center of the band is given by[19,36]

$$\tilde{\epsilon}_f(\omega = 0) = -i\frac{32d^2(1/2 - d^2)V^2}{D}\sqrt{c}. \tag{36}$$

This again explicitly shows the nonanalytic behavior as $c \to 0$ of the height and width of the band (see also Sect. IV).

## VI.3 Ferromagnetic Groundstate

The pure Kondo insulator can undergo a first order phase transition from the paramagnetic into a ferromagnetic state in two ways.[33] For sufficiently large $U$ a metallic ferromagnetic state has lower energy than the paramagnetic one. The ferromagnetic state has a strongly reduced $d^2$ and hence a small effective hybridization. However, this state is energetically less favorable than the antiferromagnetic state. As a function of magnetic field the up- and down-spin bands are shifted with respect to each other. This again reduces the hybridization gap, leading to a stable metallic ferromagnetic state. The magnetic field quenches long-range antiferromagnetic order and favors ferromagnetism. The critical field for the transition is of the order of 0.1 V (Ref. 33).

The existence of an impurity band opens a third possibility for a transition into a ferromagnetic state. Due to the finite density of states at the Fermi level, $\rho_{\rm imp}(0)$ per spin component, the self-consistent solution of Eqs. (32) and (33) leads to an instability criterion of the Stoner type for a second-order transition. In zero external magnetic field we obtain

$$U_c = \frac{4p^2}{\rho_{\rm imp}(0)} - \frac{4p}{d(1-c)} V \frac{1}{N} \sum_{\mathbf{k}\sigma} \langle f^\dagger_{\mathbf{k}\sigma} c_{\mathbf{k}\sigma} \rangle. \qquad (37)$$

$U_c$ is a decreasing function of $c$, which tends to infinity as $c \to 0$. Since $U_c$ is larger than the critical $U$ required to get an antiferromagnetic groundstate, this instability has no consequences on the phase diagram of a Kondo insulator.

## VI.4 Antiferromagnetic Groundstate

To study antiferromagnetic order we introduce two interpenetrating sublattices, denoted with $a$ and $b$, and nearest-neighbor hopping from one sublattice to the other. The expectation values of the $p_{i\sigma}$ slave-boson operators and the $\lambda_i^{(2)}$ Lagrange multipliers are now different on the two sublattices,[32,33] e.g. for electron-hole symmetry,

$$p_{a\uparrow} = p_{b\downarrow} = p_\uparrow, \quad p_{a\downarrow} = p_{b\uparrow} = p_\downarrow,$$
$$-\epsilon_m^* = \lambda_{a\uparrow}^{(2)} - \tfrac{1}{2}U = \lambda_{b\downarrow}^{(2)} - \tfrac{1}{2}U,$$
$$= -\lambda_{a\downarrow}^{(2)} + \tfrac{1}{2}U = -\lambda_{b\uparrow}^{(2)} + \tfrac{1}{2}U. \qquad (38)$$

Antiferromagnetic order reduces the Brillouin zone to one-half of its original size and increases the number of bands from two to four, so that the Green's function matrix is now a $4 \times 4$ matrix. The $4 \times 4$ matrix separates into two $2 \times 2$ matrices coupled only through the parameter $\epsilon_m^*$, which vanishes for the paramagnet but is nonzero in the antiferromagnetic phase,[33] and the Kondo hole scattering Eq. (2). The one-particle Green's function can now be expressed in terms of a $4 \times 4$ $T$-matrix, which in the limit $\Delta\epsilon_f \to \infty$ has nonzero elements only in the $f$-electron subspace [see Eq. (8)] spanned by the $2 \times 2$ subspace of two $f$-electron bands labeled 1 and 2. For an isolated Kondo hole the results depend on whether the sublattice $a$ or $b$ contains the impurity.

For a finite concentration of Kondo holes $c$, assumed equally distributed between the two sublattices (each has $c/2$ Kondo holes), the self-consistent sum over all non-crossing impurity diagrams gives rise to a $2 \times 2$ $f$-electron self-energy matrix

$$\Sigma_{11}(z) = \Sigma_{22}(z) = -c\frac{G_{11}(z) + G_{22}(z)}{\det(z)},$$
$$\Sigma_{12}(z) = \Sigma_{21}(z) = c\frac{G_{12}(z) + G_{21}(z)}{\det(z)}, \qquad (39)$$

where $G_{ij}(z) = (1/N)\sum_{\mathbf{k}}[\hat{G}_{\mathbf{k}\sigma}(z)]_{ij}$ summed over the reduced Brillouin zone, is to be determined self-consistently with the impurity self-energy and $\det(z) = [G_{11}(z) + G_{22}(z)]^2 - [G_{12}(z) + G_{21}(z)]^2$.

In the pure system the transition from paramagnetism to antiferromagnetism is second order. To study the onset of antiferromagnetism, it is then sufficient to linearize in the sublattice magnetization. This gives rise to two self-consistent equations, one for the diagonal part and one for the off-diagonal part of the $f$-electron Green's function matrix. As to be expected, the diagonal one is identical to Eq. (33) for the paramagnetic state (if quadratic corrections in the sublattice magnetization are neglected), while the off-diagonal one renormalizes $\epsilon_m^*$ and determines the onset long-range order.

For a semielliptic density of states of the conduction electrons we obtain $U_c = 0.5920$ for $c = 0$. This result differs slightly from the value obtained in Ref. 33, $U_c = 0.54$, where a square density of states was used. As a function of $c$ the critical $U$ first decreases, reaches a minimum and then increases again. Hence, $U_c$ as a function of $c$ shows a slight re-entrant behavior from paramagnetic to antiferromagnetic to paramagnetic groundstates. This result is similar to the finding in Ref. 37.

## VII. Concluding Remarks

We briefly reviewed the effects of nonmagnetic impurities embedded into a Kondo insulator. The Kondo lattice is described by the symmetric non-degenerate periodic Anderson model. Correlations in the $f$-shell were introduced (i) within a self-energy evaluated in the local approximation (*i.e.*, neglecting the **k**-dependence in the spirit of the $d \to \infty$ limit) and (ii) in terms of four slave bosons per site[32,33] within a standard mean-field approximation. A finite concentration of Kondo holes forms an impurity band in the gap of the Kondo insulator. The results for clusters of Kondo holes and the impurity band are essentially the same within both approaches. This is a strong indication of the robustness of the results.

A Kondo hole introduces a boundstate in the gap, whose spectral weight is localized in the neighboring unit cells to the impurity. The boundstate, if, caused by a charge neutral substitution, pins the Fermi level and has magnetic properties. The properties of arbitrary clusters of Kondo holes are summarized in Sect. III. Boundstates with energy at the Fermi level can be classified as connected and disconnected. Tracing out the electron degrees of freedom the problem of an insulator-metal transition within the impurity band is reduced to the site percolation of Kondo holes. On a simple cubic lattice with nearest-neighbor hopping and electron-hole symmetry the percolation corresponds to first, second and fourth nearest-neighbor sites. The critical concentration of Kondo holes has been estimated at 10%, which is believed to be an upper limit for the insulator to become a metal, since electron-hole asymmetry and a tight-binding dispersion beyond nearest-neighbors enhances the physical extension of the boundstates and hence increases the connectivity accordingly. The threshold could be further reduced if the **k**-dependence of the self-energy is considered.

With a finite concentration of Kondo holes the boundstates accumulate at the Fermi level giving rise to an impurity band. For small $c$ the height and width of the impurity band in the $f$-electron density of states are both proportional to $\sqrt{c}$. The impurity band has consequences on the optical properties in the infrared range of Kondo insulators, and at very low temperatures it gives rise to a finite magnetic susceptibility (Pauli-like in the paramagnetic phase) and a term proportional to $T$ in the specific heat. The mass enhancement depends on the concentration of Kondo holes and is roughly proportional to $\sqrt{c}$ for small $c$. The temperature range of this linear term is determined by the width of the impurity band.

There are several experimental studies on impure Kondo insulators. The predicted $\sqrt{c}$-dependence of $\gamma$ has been verified for La-substituted $Ce_3Bi_4Pt_3$ (Ref. 3). The valence instability and electrical properties of Yb and La substituted $SmB_6$ have been studied long ago.[54] The gap of CeNiSn was found to close with 15% La substitution[55] and is also smeared with increasing $x$ in $CeNi_{1-x}Pt_xSn$ (Ref. 56).

In Section V, we briefly discussed the effects of other impurities, namely dopands and ligand defects. Doping changes the number of carriers in the system and reverses magnetic and nonmagnetic properties of impurities, as a consequence of a pinning/depinning of the Fermi level. Boundstates due to ligand defects appear in pairs and are usually located at the edges of the Kondo gap. A finite concentration of ligand defects reduce the height of the main Kondo peaks and give rise to tails protuding into the gap.

Magnetic properties of Kondo insulators have been investigated within the framework of the Kotliar-Ruckenstein formulation in terms of four auxiliary bosons per site. Within a standard mean-field approximation the auxiliary boson operators are replaced by their expectation values. We studied the interplay between the Kondo holes and the magnetic long-range order in the system.[36]

The finite density of states at the Fermi level due to the impurity band opens the possibility of a second order phase transition from the paramagnetic state to the ferromagnetic one (only first order transitions are allowed in the pure Kondo insulator). This instability arises in analogy to the Stoner criterion. This critical $U$, however, is quite large so that the system is already an antiferromagnet. The only possibility to induce a ferromagnetic groundstate is then with a very large external magnetic field.

The instability of the pure Kondo insulator to antiferromagnetic long-range order has been studied in Ref. 33. The transition from the paramagnetic groundstate is continuous (second order). The impurity band reduces the critical $U$ for this transition. As a function of $c$ the $U_c$ first decreases, reaches a minimum at about $c = 0.06$ and then it increases again. Hence, as a function of $c$ a re-entrant paramagnetic-antiferromagnetic-paramagnetic boundary is obtained for a small range of $U$. Although caused by a different mechanism, such re-entrant behavior has been predicted by Doniach and Fazekas[37] for a doped Kondo semiconductor.

If taken literally our calculations indicate that Kondo insulators should always be antiferromagnets in contrast to the experimental observations. Several approximations went into the model and the Kotliar-Ruckenstein mean-field approximation. (i) The orbital degeneracy, which is believed to increase the threshold for long-range order,[34,35,53] has been neglected. (ii) Gaussian fluctuations about the mean-field saddle-point also reduce the magnetic correlations[27] and hence raise the threshold for the antiferromagnetic instability. (iii) On the other hand, the two-loop approximation in the Anderson lattice generates a Ruderman-Kittel-Kasuya-Yosida (RKKY) interaction between local $f$ moments via the polarization of the conduction electrons,[57] favoring magnetic order. (iv) The normalization of the $Z_\sigma$-factors has been chosen to reproduce the weak-coupling limit for which the approach is intended for.

# Acknowledgments

The support of the Department of Energy under grant DE-FG05-91ER45443 is acknowledged.

# References

1. See *e.g.*, *Valence Instabilities*, edited by P. Wachter and H. Boppart (North-Holland, Amsterdam, 1982).
2. T. Takabatake, Y. Nakazawa, and M. Ishikawa, *Jpn. J. Appl. Phys. Suppl.* **26**, 547 (1987).
3. M. F. Hundley *et al.*, *Physica B* **171**, 254 (1991).
4. G. P. Meisner *et al.*, *J. Appl. Phys.* **57**, 3075 (1985).
5. M. Kasaya, F. Iga, M. Takigawa, and T. Kasuya, *J. Magn. Magn. Mater.* **47-48**, 429 (1985).
6. N. Bykowetz *et al.*, *J. Appl. Phys.* **63**, 4127 (1988).
7. Z. Schlesinger *et al.*, *Phys. Rev. Lett.* **71**, 1748 (1993).
8. J. Neuenschwander and P. Wachter, *Phys. Rev. B* **41**, 12693 (1990).
9. B. Bucher, P. Steiner, and P. Wachter, *Phys. Rev. Lett.* **67**, 2717 (1991).
10. P. Wachter, A. Jung, and P. Steiner, *Phys. Rev. B* **51**, 5542 (1995).
11. G. Aeppli and Z. Fisk, *Comments Cond. Matt. Phys.* **16**, 155 (1992).
12. A. Severing *et al.*, *Phys. Rev. B* **44**, 6832 (1991).
13. T. Nanba *et al.*, *Physica B* **186-188**, 440 (1993).
14. P. Nyhus, S. L. Cooper, Z. Fisk, and J. Sarrao, *Phys. Rev. B* **52**, 14308 (1995).
15. B. Bucher, Z. Schlesinger, P. C. Canfield, and Z. Fisk, *Phys. Rev. Lett.* **72**, 522 (1994).
16. N. B. Brandt and V. V. Moschalkov, *Adv. Phys.* **33**, 373 (1984).
17. R. Sollie and P. Schlottmann, *J. Appl. Phys.* **69**, 5478 (1991).
18. R. Sollie and P. Schlottmann, *J. Appl. Phys.* **70**, 5803 (1991).
19. P. Schlottmann, *Phys. Rev. B* **46**, 998 (1992).
20. P. Schlottmann, *Physica B* **186-188**, 375 (1993).
21. P. Schlottmann and C. S. Hellberg, *J. Appl. Phys.* **79**, 6414 (1996).
22. A. Millis, in *Physical Phenomena at High Magnetic Fields*, edited by E. Manousakis *et al.*, (Addison-Wesley, 1992), p.146.
23. P. S. Riseborough, *Phys. Rev. B* **45**, 13984 (1992).
24. C. Sanchez-Castro, K. S. Bedell, and B. R. Cooper, *Phys. Rev. B* **47**, 6879 (1993).
25. P. S. Riseborough, *Physica B* **199-200**, 466 (1994).
26. S. Doniach, C. Fu, and S. A. Trugman, *Physica B* **199-200**, 450 (1994).
27. J. Karbowski, *Phys. Rev. B* **54**, 728 (1996).
28. C. M. Varma, *Phys. Rev. B* **50**, 9952 (1994).
29. K. Ueda, M. Sigrist, H. Tsunetsugu, and T. Nishino, *Physica B* **194-196**, 255 (1994).
30. T. Kasuya, *Europhys. Lett.* **26**, 277 (1994).
31. J. Pérez-Conde and P. Pfeuty, *Physica B* **223-224**, 438 (1996).
32. G. Kotliar and A. Ruckenstein, *Phys. Rev. Lett.* **57**, 1362 (1986).
33. V. Dorin and P. Schlottmann, *Phys. Rev. B* **46**, 10800 (1992).
34. V. Dorin and P. Schlottmann, *J. Appl. Phys.* **73**, 5400 (1993).
35. V. Dorin and P. Schlottmann, *Phys. Rev. B* **47**, 5095 (1993).
36. P. Schlottmann, *Phys. Rev. B* **54**, 12324 (1996).
37. S. Doniach and P. Fazekas, *Phil. Mag. B* **65**, 1171 (1992).
38. P. Schlottmann, *Physica B* **206-207**, 816 (1995).
39. P. Schlottmann, *Physica B* **223-224**, 435 (1996).
40. W. Metzner and D. Vollhardt, *Phys. Rev. Lett.* **62**, 324 (1989); W. Metzner, *Z. Phys. B* **77**, 253 (1989); E. Müller-Hartmann, *ibid.* **76**, 211 (1989); H. Schweitzer and G. Czycholl, *Solid State Commun.* **69**, 171 (1989).

41. D. S. Gaunt and A. J. Guttmann, in *Phase Transitions and Critical Phenomena*, edited by C. Domb and M. S. Green, (Academic Press, London, 1974),VOL. 3 p. 181.
42. D. Stauffer, *Introduction to Percolation Theory* (Taylor and Francis, London, 1985).
43. P. J. Reynolds, H. E. Stanley and W. Klein, *Phys. Rev. B* **21**, 1223 (1980).
44. R. Freytag and J. Keller, *Z. Phys. B* **80**, 241 (1990).
45. S. Wermbter, K. Sabel and G. Czycholl, *Phys. Rev. B* **53**, 2528 (1996).
46. P. Schlottmann, *J. Appl. Phys.* **75**, 7044 (1994).
47. S. E. Barnes, *J. Phys. F* **6**, 1375 (1976); **7**, 2637 (1977).
48. P. Coleman, *Phys. Rev. B* **29**, 3035 (1984).
49. N. Read and D. M. Newns, *J. Phys. C* **16**, 3273 (1983).
50. C. A. Balseiro, M. Avignon, A. G. Rojo, and B. Alascio, *Phys. Rev. Lett.* **62**, 2624 (1989).
51. A. Sudbo and A. Houghton, *Phys. Rev. B* **42**, 4105 (1990).
52. T. Li, P. Wölfle, and P. Hirschfeld, *Phys. Rev. B* **40**, 6817 (1989).
53. T. M. Rice and K. Ueda, *Phys. Rev. Lett.* **55**, 995 (1985).
54. M. Kasaya et al., in *Valence Fluctuations in Solids*, edited by L. M. Falicov, W. Hanke and M. B. Maple (North-Holland, Amsterdam, 1981), p.251.
55. F. G. Aliev et al., *Physica B* **163**, 358 (1990).
56. S. Nishigori et al., *Physica B* **186-188**, 406 (1993).
57. A. Houghton, N. Read, and H. Won, *Phys. Rev. B* **35**, 5123 (1987).

# Real-Space Ground-State of a Generalized Hubbard Model

O. Navarro,[1] E. Flores,[1] and M. Avignon[2]

[1] Instituto de Investigaciones en Materiales
Universidad Nacional Autónoma de México
Apartado postal 70-360, 04510, México D. F.
MEXICO

[2] Laboratoire d'Etudes des Propriétés Electroniques des Solides–CNRS
Associated with Université Joseph Fourier
B. P. 166, 38042 Grenoble Cedex 9
FRANCE

## Abstract

We study a generalized Hubbard model with onsite interaction $U$, intersite interaction $V$, and a general correlated hopping subject to the condition that the number of doubly occupied sites is conserved. This study is carried out in the real space using a new mapping method. The method is based on mapping the original many-body problem onto an equivalent tight-binding one with impurities in a higher dimensional space. For a linear chain, we have obtained an exact solution of the problem of one hole and one doubly-occupied site moving in a ferromagnetic spin background, and a metal-insulator transition for some values of the Hubbard parameters. Preliminary results are also given for the two-dimensional case.

## I. Introduction

The Hubbard model is one of the simplest models used to describe the electron-electron interaction,[1] where only the onsite electronic correlation is taken into account. This model also assigns the same hopping rate $t$ to three different hopping processes regardless of the occupation of the two sites involved. Besides the onsite interaction $U$, other contributions of the electron-electron interaction are required,[2] such as the nearest-neighbor interactions and the bond-charge interaction term. The Hamiltonian which includes these interactions is often called the generalized Hubbard Hamiltonian, and can be written as

$$H = \sum_{\langle i,j \rangle, \sigma} t_{i,j}^{\sigma}(c_{i,\sigma}^{+}c_{j,\sigma} + \text{h.c.}) + U\sum_{i} n_{i,\uparrow}n_{i,\downarrow} + \tfrac{1}{2}V\sum_{\langle i,j \rangle} n_{i}n_{j}, \quad (1)$$

where $\langle i,j \rangle$ denotes nearest-neighbor sites, $c_{i,\sigma}^{+}$ ($c_{i,\sigma}$) is the creation (annihilation) operator with spin $\sigma = \downarrow$ or $\uparrow$ at site $i$, and $n_i = n_{i,\uparrow} + n_{i,\downarrow}$ where $n_{i,\sigma} = c_{i,\sigma}^{+}c_{i,\sigma}$. The occupation-dependent hopping amplitude, $t_{i,j}^{\sigma}$, is given by

$$\begin{aligned} t_{i,j}^{\sigma} &= t_{AA}(1 - n_{i,-\sigma})(1 - n_{j,-\sigma}) + t_{BB}n_{i,-\sigma}n_{j,-\sigma} \\ &+ t_{AB}[n_{j,-\sigma}(1 - n_{i,-\sigma}) + n_{i,-\sigma}(1 - n_{j,-\sigma})]. \end{aligned} \quad (2)$$

The three parameters $t_{AA}$, $t_{BB}$, and $t_{AB}$ are the hopping amplitudes from a singly occupied to an empty site, from a doubly occupied to a singly site and from a doubly occupied to an empty site, respectively. The special case $t_{AA} = t_{BB} = t_{AB} = t$ corresponds to the $t$-$U$-$V$ extended Hubbard model, which has been studied intensively by analytical and numerical methods.[1] When $t_{AA} + t_{BB} - 2t_{AB} = 0$ and $t_{AA} \neq t_{BB}$, the generalized model given by Eq. (1) reduces to the Hirsch and Marsiglio model of hole superconductivity.[2] The hopping $t_{AB}$ favors nearest-neighbor antiferromagnetic correlations while $t_{AA}$ and $t_{BB}$ move particles without disturbing the underlying magnetic background. However in the case $t_{AB} = 0$, the antiferromagnetic correlations are suppressed, and a Mott-transition exist in two and three dimensions.[3] The generalized Hamiltonian [see Eq. (1)] has been studied previously by several authors.[4,5]

In this paper we apply a new mapping method[6] to obtain an exact solution of the generalized Hubbard model under the restriction $t_{AB} = 0$ and for the case of one hole and one doubly-occupied site moving in a ferromagnetic spin background, like a "Wannier exciton."

## II. Method

For simplicity, we introduce first the mapping method for a one-dimensional system. Let us consider two electrons with opposite spins in an $N$-site chain ($N = 2, 3, 4, \ldots$). For a system of four sites, the states are: $|1\rangle = |\pm 000\rangle$, $|2\rangle = |+-00\rangle$, $|3\rangle = |+0-0\rangle$, $|4\rangle = |+00-\rangle$, $|5\rangle = |0+0-\rangle$, $|6\rangle = |0+-0\rangle$, $|7\rangle = |0\pm 00\rangle$, $|8\rangle = |-+00\rangle$, $|9\rangle = |-0+0\rangle$, $|10\rangle = |0-+0\rangle$, $|11\rangle = |00\pm 0\rangle$, $|12\rangle = |00+-\rangle$, $|13\rangle = |000\pm\rangle$, $|14\rangle = |00-+\rangle$, $|15\rangle = |0-0+\rangle$, $|16\rangle = |-00+\rangle$. Where $+$ and $-$ represent an electron with spin up and spin down, respectively, and 0 represents an empty site. Furthermore, $\pm$ indicates a doubly occupied site. In general, the number of states is given by $N^2$.

In figure 1, a geometric representation of these 16 states is given, where the circles represent the states. In the extended Hubbard Hamiltonian, a state with a site occupied by two electrons requires a energy $U$, such as the states $|1\rangle$, $|7\rangle$, $|11\rangle$, and $|13\rangle$. Likewise, a state in which two electrons are situated in nearest-neighbor sites, requires an energy $V$, for instance, the states $|2\rangle$, $|6\rangle$, $|8\rangle$, $|10\rangle$, $|12\rangle$, and $|14\rangle$ have a self-energy $V$. Finally, the remaining states with two electrons placed at distant sites do not require any energy [see Eq. (1)]. Moreover, the amplitude of the transition probability for nearest-neighbor states in Fig. 1 is precisely $t$, since the difference between these two states is only the hopping of one electron.

For the case of two electrons in an infinite periodic chain, the network of states has the shape of a two-dimensional square lattice with an infinite number of impurities localized onsites along the principal diagonal and two next diagonal chains (see Fig. 1).

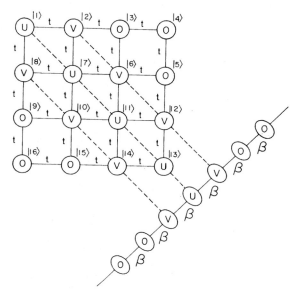

**Figure 1.** Geometric representation of the two-electron states for a chain of four sites. The states are represented by circles with site-energy indicated inside. The direction of the projection procedure is shown by dashed lines. The final chain is formed by effective states represented by ellipses.

This network of states has an exact solution, since the Hamiltonian is of a tight-binding type. A simple way to obtain the solution is taking advantage of the translation symmetry of the impurities and mapping the two-dimensional lattice onto a linear chain with impurities, as it is shown in Fig. 1, where $\beta = 2t\cos(Ka/\sqrt{2})$, the lattice parameter $a = 1$, and $K$ is the wave vector in the projection direction. In other words, the lattice is dealt within two separated spaces, in one direction it is real space and the other it is reciprocal space. For each $K$, we have a one-dimensional lattice that can be solved by means of the transfer matrix approach firstly introduced by Falicov and Yndurain.[7] The one-dimensional results must be integrated with respect to $K$ within the first Brillouin zone.

For the case of $n$ electrons in a $d$-dimensional crystal, the geometric structure of the network of states belongs to a $nd$-dimensional lattice, where the Pauli exclusion principle must be taken into account. For $n = 2$ and $d = 2$, i.e., a two-dimensional square lattice with two electrons and an infinite number of sites, the states form a hypercubic network in a four-dimensional space. The way to get the solution is similar to the one-dimensional case, except that the mapping is made from a four-dimensional network of states onto a two-dimensional one. This mapped square lattice of effective states has a center impurity with a self-energy $U$, surrounded by four states with a self-energy given by $V$ and other states with self-energies equal to zero. Furthermore, there are two effective hopping parameters $\beta_x = 2t\cos(K_x a/\sqrt{2})$ and $\beta_y = 2t\cos(K_y a/\sqrt{2})$, where $K_x$ and $K_y$ are wave vectors on the $x$-$y$ projection plane. The two-dimensional results must be integrated with respect to $K_x$ and $K_y$ within the first Brillouin zone.

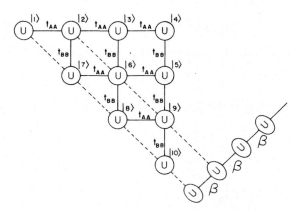

**Figure 2.** Geometric representation of the states of one hole and one doublon (doubly occupied site) in a ferromagnetic spin background for a chain of five sites, the direction of the projection procedure is shown by dashed lines and the effective states are represented by ellipses.

## III. Results

For a system of one hole and one doubly-occupied site moving in a ferromagnetic spin background with the hopping parameter $t_{AB} = 0$ [see Eq. (1)], the mapping method can be applied as is described in the following lines. For example, the case of a five-site chain and $V = 0$, has ten possible states: $|1\rangle = |\pm0+++\rangle$, $|2\rangle = |\pm+0++\rangle$, $|3\rangle = |\pm++0+\rangle$, $|4\rangle = |\pm+++0\rangle$, $|5\rangle = |+\pm++0\rangle$, $|6\rangle = |+\pm+0+\rangle$, $|7\rangle = |+\pm0++\rangle$, $|8\rangle = |++\pm0+\rangle$, $|9\rangle = |++\pm+0\rangle$, $|10\rangle = |+++\pm0\rangle$. Spin up and down are denoted by $+$ and $-$, respectively, a doubly-occupied site by $\pm$, and a hole by $0$. A site occupied by two electrons requires an energy $U$, and the amplitude of the transition probability for nearest-neighbor states will be $t_{AA}$ and $t_{BB}$ for the hopping from a singly-occupied to an empty site and from a doubly-occupied to a singly-occupied site, respectively. In general, for an $N$-site chain the number of states is given by $N(N-1)/2$. In figure 2 these ten states are represented geometrically, where circles denote states with self-energy as indicated. This network of states has an exact solution since it can be described by a *single-body* tight-binding Hamiltonian. A simple way to obtain the solution is by taking advantage of the translational symmetry and projecting the two-dimensional state lattice onto a linear chain of effective states (similar to the procedure given in Ref. 7), as is shown in Fig. 2, where $\beta = (t_{AA} + t_{BB})\cos(Ka/\sqrt{2}) + i(t_{AA} - t_{BB})\sin(Ka/\sqrt{2})$, and $K$ is the wave vector in the projection direction, we take the lattice parameter $a = 1$.

For the case of a two-dimensional square lattice, with one hole and one doubly-occupied site moving in a ferromagnetic spin background of infinite sites, the states form a hypercubic network in a four-dimensional space. The way to get the solution is similar to the one-dimensional case, except that the mapping must be made from a four-dimensional network of states onto a two-dimensional one. This mapped square lattice of effective states has a center forbidden state and the other states with site energies given by $U$. Furthermore, there are two effective hopping parameters $\beta_x = (t_{AA} + t_{BB})\cos(K_x a/\sqrt{2}) + i(t_{AA} - t_{BB})\sin(K_x a/\sqrt{2})$ and

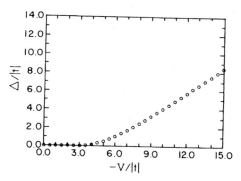

**Figure 3.** Pairing energy ($\Delta$) of two electron with antiparallel spin in the limit of $U \to \infty$ for a square lattice.

$\beta_y = (t_{AA} + t_{BB})\cos(K_y a/\sqrt{2}) + i(t_{AA} - t_{BB})\sin(K_y a/\sqrt{2})$, where $K_x$ and $K_y$ are wave vectors on the $x$-$y$ projection plane.

The ground-state energy in the one-dimensional system is given by $U - z(|t_{AA}| + |t_{BB}|)$, with $z$ the coordination number. The critical value of $U$ is then $U_c = z(|t_{AA}| + |t_{BB}|)$; this critical value marks the metal-insulator transition. For $U \geq U_c$ the ground state is a Mott insulator state with one particle per site, and for $0 < U < U_c$ the ground state has metallic behavior and contains a finite number of doubly-occupied sites (doublons), and equal number of empty sites (holes).

Our problem of one hole and one doubly-occupied site can be transformed into one of two-particles with opposite spin, infinite onsite repulsion and an attractive intersite interaction $-V$ (Ref. 5). The solution of this two-particle problem can be found using the above mapping method and the hopping parameter given by $t = t_{AA} = t_{BB}$. In this case, the projected lattice is a semi-infinite linear chain of effective states with an impurity located at the surface with self-energy $V$, and the other states with self-energy equal to zero. Furthermore, the effective hopping parameter is given by $\beta = 2t\cos(Ka/\sqrt{2})$. For $K = 0$ the ground state energy of the localized states is given by $E_l = V + (2t)^2/V$, valid only for $|V| > |2t|$, which leads to $E_l > 4|t|$. In figure 3 we exhibit the pairing energy as a function of the parameter $V$ for two electrons with opposite spin on a square lattice. Note that the pairing energy for two holes with opposite spin is the same as that for two electrons since we have a bipartite lattice.[8] The pairing energy has been calculated from the difference of energies between the lowest pairing state ($K = 0$) and the original lower band edge when there is no electron-electron interaction.

In conclusion, we have studied a generalized Hubbard model with hopping depending on the occupation, using a mapping method. For $t_{AB} = 0$, the problem of one hole and one doubly-occupied site in a ferromagnetic background is solved exactly. The model displays a metal-insulator transition for $V = 0$ at $U_c = z(|t_{AA}| + |t_{BB}|)$.

## Acknowledgments

This work was partially supported by DGAPA-UNAM Grant IN102196 and by CONA-CyT Grant 2661P-A9507.

# References

1. R. Micnas, J. Ranninger, and S. Robaszkiewicz, *Rev. Mod. Phys.* **62**, 113 (1990); *The Hubbard Model: Recent Results*, edited by M. Rosetti, Series on Advances in Statistical Mechanics (World Scientific, Singapore, 1992), Vol. 7.
2. J. E. Hirsch and F. Marsiglio, *Phys. Rev. B* **41**, 2049 (1990).
3. A. A. Aligia, L. Arrachea, and E. R. Gagliano, *Phys. Rev. B* **51**, 13 774 (1995); E. R. Gagliano, A. A. Aligia, L. Arrache, and M. Avignon, *Physica B* **223-224**, 605 (1996).
4. R. Strack and D. Vollhardt, *Phys. Rev. Lett.* **70**, 2637 (1993); A. A. Ovchinikov, *Mod. Phys. Lett. B* **7**, 21 (1993).
5. L. Arrachea and A. A. Aligia, *Phys. Rev. Lett.* **73**, 2240 (1994); Gagliano, A. A. Aligia, L. Arrache, and M. Avignon, *Phys. Rev. B* **51**, 14 012 (1995).
6. O. Navarro and C. Wang, *Solid State Commun.* **83**, 473 (1992).
7. L. M. Falicov and F. Yndurain, *J. Phys. C* **8** 147 (1975).
8. L. A. Pérez, O. Navarro, and C. Wang, *Phys. Rev. B* **53**, 15 389 (1996).

# Coherent and Ultracoherent States in Hubbard and Related Models

K. A. Penson[1,*] and A. I. Solomon[2,†]

[1] Laboratoire de Physique Théorique des Liquides
Université Pierre et Marie Curie–CNRS
URA 765, Tour 16, 5ème étage, boite 121
4, place Jussieu, 75252 Paris Cedex 05
FRANCE

[2] Laboratoire de Gravitation et Cosmologie Relativistes
Université Pierre et Marie Curie–CNRS
URA 769, Tour 22, 4ème étage, boite 142
4, place Jussieu, 75252 Paris Cedex 05
FRANCE

## Abstract

We consider the Hubbard model and extensions on bipartite lattices. We define a dynamical group based on the $\eta$-pairing operators introduced by Yang, and define coherent pairing states, which are combinations of eigenfunctions of $\eta$ operators. The coherent states are defined through the exponentiation of $\eta$-operators. In addition we introduce the so-called *ultracoherent* states through exponentiation of certain functions of $\eta$-operators. The coherent states permit exact calculation of numerous physical properties of the system, including energy, various fluctuation and correlation functions, as well as pairing off-diagonal long-range order (ODLRO) to all orders. This approach is complementary to that of BCS, in that these are superconducting coherent states associated with the exact model, while not eigenstates of the Hamiltonian.

## I. Introduction

The Hubbard model is the most widely used interacting fermion model in condensed matter physics.[1-3] Applications range from the theoretical description of insulators to that of conductors and superconductors. It is believed that high-$T_c$ superconductivity may be described by a form of Hubbard model. The model can be solved exactly only in one spatial dimension (1D); at present no exact solution is available in higher dimensions. The success of the Hubbard model lies in its simplicity: It represents electrons on a lattice with a Coulomb interaction so strongly screened that all that remains is a purely local term, coupling electrons in opposite spin directions.

Although a complete solution is lacking for dimension > 1, one might nevertheless imagine that a partial solution (a portion of the exact spectrum, some eigenfunctions, etc.) would also be of interest. But even this reduced objective has proved elusive; it was only in 1989 that Yang[4] obtained such a partial solution, treating only a subspace generated by fermion pairs.

In this note we first briefly describe Yang's method. We then propose three new kinds of collective wave function based on Yang's eigenfunctions. The first is an analogue of spin coherent states (SCS)[5] obtained through exponentiation of a certain operator. The advantage of SCS is that practically all the physical quantities, such as energy, arbitrary moments of the Hamiltonian, fluctuations, correlation functions, etc., can be exactly calculated by purely algebraic means. The other collective wave functions, which we call *ultracoherent* states (UCS) are obtained through further exponentiation. The UCS are closely related to some combinatorial generating functions. This observation permits exact calculation, as for the SCS.

The functions which we introduce are parametrized by complex numbers related to the average values of the pair number operator, analogous to the BCS superconducting order parameter. Finally, we discuss the variational meaning of coherent state parameters.

## II. The Model and $\eta$-Pairing

We adopt the definitions and notation of Ref. 4. Let $a_\mathbf{r}^+$ and $b_\mathbf{r}^+$ be real-space creation operators for spin-up and spin-down electrons respectively and $a_\mathbf{k}^+$ and $b_\mathbf{k}^+$ their Fourier transforms. The Hubbard Hamiltonian on a $L \times L \times L = M$ cube ($L$ even) with periodic boundary conditions is ($\varepsilon > 0$):

$$H = T_0 + T_1 + V, \tag{1}$$

$$T_0 = 6\varepsilon \sum_\mathbf{k} (a_\mathbf{k}^+ a_\mathbf{k} + b_\mathbf{k}^+ b_\mathbf{k}), \tag{2}$$

$$T_1 = -2\varepsilon \sum_\mathbf{k} (\cos k_x + \cos k_y + \cos k_z)(a_\mathbf{k}^+ a_\mathbf{k} + b_\mathbf{k}^+ b_\mathbf{k}), \tag{3}$$

$$V = 2W \sum_\mathbf{r} a_\mathbf{r}^+ a_\mathbf{r} b_\mathbf{r}^+ b_\mathbf{r}, \tag{4}$$

where $2W$ is the on-site Hubbard interaction of arbitrary sign. We introduce the $\eta$-operator which creates an electron pair with momentum $\mathbf{\Pi}$:

$$\eta^+ = \sum_\mathbf{r} e^{i\mathbf{\Pi r}} a_\mathbf{r}^+ b_\mathbf{r}^+ = \sum_\mathbf{k} a_\mathbf{k}^+ b_{\mathbf{\Pi}-\mathbf{k}}^+. \tag{5}$$

It follows that $[T_1, \eta^+] = 0$ and

$$[H, \eta^+] = E\eta^+, \tag{6}$$

with $E = 12\varepsilon + 2W$. The operator $\eta$ satisfies:

$$[\eta^+, \eta] = 2\eta_3, \tag{7}$$

where $\eta_3 = \frac{1}{2}\sum_\mathbf{r} n_\mathbf{r} - \frac{1}{2}M$, with local occupation numbers $n_\mathbf{r} = a_\mathbf{r}^+ a_\mathbf{r} + b_\mathbf{r}^+ b_\mathbf{r}$. The $\eta$'s satisfy the angular momentum commutation relations of SU(2), with $(\eta)^{M+1} = 0$. Introducing $J_0 \equiv H/E - \eta_3$, we see that $[J_0, \eta] = 0$. We may therefore consider the resulting u(2) algebra:

$$(\eta^+, \eta, \eta_3, J_0) \tag{8}$$

as the dynamical algebra for $H$, $[H \in \text{u}(2)]$.

## III. Coherent States

We define our normalized spin coherent state (SCS) by:

$$|\mu\rangle = (1 + |\mu|^2)^{-\frac{M}{2}} e^{\mu\eta} |0\rangle, \tag{9}$$

where the state $|0\rangle$ is the normalized filled pair state $|0\rangle = \frac{1}{M!}(\eta^+)^M |\text{vac}\rangle$. For comparison we note Yang's exact, normalized eigenstates of $H$:[4]

$$|\Psi_N\rangle = \beta(N, M)(\eta^+)^N |\text{vac}\rangle, \qquad N = 1, \ldots, M, \tag{10}$$

which are also eigenstates of the operator $N_2$ counting the number of doubly occupied sites, $N_2 = \sum_\mathbf{r} n_\mathbf{r}^{(a)} n_\mathbf{r}^{(b)}$, $([H, N_2] \neq 0)$,

$$H|\Psi_N\rangle = NE|\Psi_N\rangle, \tag{11}$$
$$N_2|\Psi_N\rangle = N|\Psi_N\rangle. \tag{12}$$

Here $\beta(N, M)$ is a normalizing factor

$$\beta(N, M) = \left(\frac{(M-N)!}{M! \, N!}\right)^{1/2}. \tag{13}$$

For this simple finite spin system, with "spin" $S = \frac{1}{2}M$ (Ref. 5) we may readily calculate quantities of physical interest; for example, the average energy

$$\langle H \rangle = \langle \mu | H | \mu \rangle = ME/(1 + |\mu|^2), \tag{14}$$

and the average number of pairs

$$\langle N_2 \rangle = \langle \mu | N_2 | \mu \rangle = \frac{1}{2}\left\langle \frac{\partial H}{\partial W} \right\rangle = \frac{M}{(1 + |\mu|^2)}, \tag{15}$$

which gives a physical interpretation of the parameter $\mu$. We may also calculate fluctuations, for example:

$$(\Delta H)^2 = \langle \mu | H^2 | \mu \rangle - (\langle \mu | H | \mu \rangle)^2 = ME^2 |\mu|^2/(1 + |\mu|^2)^2, \tag{16}$$

$$\frac{(\Delta H)}{\langle \mu | H | \mu \rangle} = \frac{|\mu|}{M^{1/2}}. \tag{17}$$

We note that the Yang functions, Eqs. (10), display off-diagonal long-range order (ODLRO) which in the thermodynamic limit is proportional to $n_2(1 - n_2)$ where $n_2 \equiv N/M$ is the density of pairs in the system. We can show that the coherent state $|\mu\rangle$ also has nonvanishing ODLRO for every $\mu \neq 0$ proportional to $\rho_2(1 - \rho_2)$ where $\rho_2 = \langle N_2 \rangle/M$, so $|\mu\rangle$ is superconducting.

We observe that the repeated use of Eq. (6) and its extension for any power-expandable $f(\eta^+)$

$$[H, f(\eta^+)] = E\eta^+ f'(\eta^+), \tag{18}$$

permits exact calculation of any matrix element of type $\langle\mu|(\eta^+)^r Q\eta^r|\mu\rangle$ provided $\langle\mu|Q|\mu\rangle$ is known, through ($Q$ arbitrary operator)

$$\langle\mu|(\eta^+)^r Q\eta^r|\mu\rangle = (1+|\mu|^2)^{-M} \frac{\partial^r}{\partial(\mu^*)^r} \frac{\partial^r}{\partial(\mu)^r} \left[(1+|\mu|^2)^M \langle\mu|Q|\mu\rangle\right]. \quad (19)$$

## VI. Ultracoherent States

We define the (first) ultracoherent state (UCSI) $|\gamma, M\rangle$ by:

$$|\gamma, M\rangle = \mathcal{N}_M^{-1/2}(\gamma) e^{\exp(\gamma\eta)-1}|0\rangle, \quad (20)$$

where $\gamma$ is a complex number and $\mathcal{N}_M(\gamma)$ is a normalization factor which can be calculated from the Bell identity[6-8]

$$e^{\exp z - 1} = \sum_{n=0}^{\infty} \frac{\Phi(n) z^n}{n!}, \quad (21)$$

where the Bell numbers $\Phi(n)$ are defined by:

$$\Phi(n) = \frac{1}{e} \sum_{k=0}^{\infty} \frac{k^n}{k!} \quad (22)$$

(first Dobinski identity) and $\Phi(n)$ is the number of all partitions of a set of $n$ elements [$\Phi(0) = 0$, $\Phi(1) = 1$, $\Phi(2) = 2$, $\Phi(3) = 5$, ...]. Then

$$\mathcal{N}_M(\rho) = \sum_{p=0}^{M} C_M^p \Phi^2(p) \rho^p, \quad \rho = |\gamma|^2, \quad (23)$$

where $G_M^p$ is binomial coeffient. The average energy in $|\gamma, M\rangle$ is:

$$\langle\gamma, M|H|\gamma, M\rangle = E \frac{M\mathcal{N}_M(\rho) - \rho[\partial\mathcal{N}_M(\rho)/\partial\rho]}{\mathcal{N}_M(\rho)}. \quad (24)$$

In addition to Eq. (20) we define the second ultracoherent state (UCSII) $|z, M\rangle$ by:

$$|z, M\rangle = \mathcal{N}_M^{-1/2}(z) e^{z(\exp(\eta)-1)}|0\rangle \quad (25)$$

From the second Bell identity[6-8] we get

$$\exp\left(x[\exp(z)-1]\right) = \sum_{n=0}^{\infty} \frac{L_n(x) z^n}{n!}, \quad (26)$$

where the exponential polynomial $L_n(x)$ is

$$L_n(x) = \sum_{k=0}^{n} \Phi(n,k) x^k \quad (27)$$

where $\Phi(n,k)$ are Stirling numbers of second kind.

For $L_n(x)$ we can use the second Dobinski identity

$$L_n(x) = e^{-x} \sum_{k=0}^{\infty} \frac{k^n}{k!} x^k \quad (28)$$

By analogy with Eqs. (23) and (24) we obtain

$$\mathcal{N}_M(z) = \sum_{p=0}^{M} C_M^p L_p^2(z), \tag{29}$$

and

$$\langle z, M|H|z, M\rangle = E \frac{\sum_{p=0}^{M}(M-p)C_M^p L_p^2(z)}{\mathcal{N}_M(z)}. \tag{30}$$

For both ultracoherent states UCSI and UCSII the calculation of the matrix elements of any operator reduce, as in the SCS case, to differentiation of the normalization factors $\mathcal{N}_M(|\gamma|^2)$ and $\mathcal{N}_M(z)$ respectively. This is a consequence of Eq. (18). As an example, we quote the results for $|\gamma\rangle$:

$$\langle\gamma|H^2|\gamma\rangle = M^2 E^2 + \rho E^2(1-2M)\mathcal{N}_M^{-1}(\rho)\frac{\partial \mathcal{N}_M(\rho)}{\partial \rho} + \rho^2 E^2 \mathcal{N}_M^{-1}(\rho)\frac{\partial^2 \mathcal{N}_M(\rho)}{\partial^2 \rho}, \tag{31}$$

from which one may obtain energy dispersion in the state $|\gamma\rangle$. Since for lattice size $M$ the expressions Eq. (23) and Eq. (29) contain $M$ terms, it is instructive to compare the expressions for the energy *per site*. To this end we compare the average energy $\overline{E} \equiv \langle |H|\rangle/4E$ for the same lattice size $M = 4$, parametrizing all states by the same variable $\mu$, with appropriate normalizations [Eqs. (9), (23), and 29].
Spin coherent state (SCS):

$$\overline{E} = \frac{1}{1+|\mu|^2}. \tag{32}$$

where

$$|\mu\rangle_4 = \mathcal{N}_4^{-1/2}\left[1 + \mu\eta + \frac{(\mu\eta)^2}{2!} + \frac{(\mu\eta)^3}{3!} + \frac{(\mu\eta)^4}{4!}\right]|0\rangle. \tag{33}$$

First ultracoherent state (UCSI):

$$\overline{E} = \frac{1+3\rho+12\rho^2+25\rho^3}{1+4\rho+24\rho^2+100\rho^3+225\rho^4}, \quad \rho \equiv |\mu|^2, \tag{34}$$

where

$$|\mu\rangle_4 = \mathcal{N}_4^{-1/2}\exp\left[\mu\eta + \frac{(\mu\eta)^2}{2!} + \frac{(\mu\eta)^3}{3!} + \frac{(\mu\eta)^4}{4!}\right]|0\rangle, \tag{35}$$

and finally second ultracoherent state (UCSII):

$$\overline{E} = \frac{1+\rho[3+3(1+\rho^{1/2})^2+(1+3\rho^{1/2}+\rho)^2]}{1+\rho[4+6(1+\rho^{1/2})^2+4(1+\rho+3\rho^2)^2+(1+7\rho^{1/2}+\rho^{3/2}+6\rho^2)^2]}, \tag{36}$$

where $\rho \equiv |\mu|^2$ and

$$|\mu\rangle_4 = \mathcal{N}_4^{-1/2}\exp\left[\mu\left(\eta + \frac{\eta^2}{2!} + \frac{\eta^3}{3!} + \frac{\eta^4}{4!}\right)\right]|0\rangle. \tag{37}$$

By comparing Eqs. (34) and (36) (for $E > 0$) we conclude that for all values of $\rho$ the energy of the UCSI and UCSII are lower than the energy of the SCS state. From Eqs. (34) and (36) we see that for $\rho = 1$ the energies of the UCSI and UCSII are equal. For $\rho < 1$ the energy of UCSII is lower than that of the UCSI. Similar results hold for larger $M$. In addition, it may be shown that both the UCSI and UCSII possess ODLRO and are thus superconducting.

## V. Discussion

The SCS of Eq. (9) is conceptually closely related to the BCS wave function $|\text{BCS}\rangle$. To see this, define the operator $\eta_C^+$ ('C' for Cooper pair) by

$$\eta_C^+ = \sum_{\mathbf{r}} a_{\mathbf{r}}^+ b_{\mathbf{r}}^+ = \sum_{\mathbf{k}} a_{\mathbf{k}}^+ b_{-\mathbf{k}}^+. \qquad (38)$$

Following Schrieffer[9] we may in fact define a more general operator

$$\eta_V^+ = \sum_{\mathbf{k}} g(\mathbf{k}) a_{\mathbf{k}}^+ b_{-\mathbf{k}}^+, \qquad (39)$$

where $g(\mathbf{k})$ are variational functions arising from the minimization of $\langle H_{\text{BCS}} \rangle$. The (normalized) BCS wave function is

$$|\text{BCS}\rangle = \frac{1}{\prod_{\mathbf{k}}(1 + |g(\mathbf{k})|^2)^{1/2}} \exp(\eta_V^+)|\text{vac}\rangle. \qquad (40)$$

The analogy between Eq. (9) and Eq. (40) is clear; there exist, however, important differences. Whereas in Eq. (5) the summation is over all the $\mathbf{k}$ values, in Eq. (39) and Eq. (40) the summation is over a shell $|\mathbf{k} - \mathbf{k}_F| < \delta$ where $\delta$ is related to the Debye frequency of phonons.

It should be noted that the algebraic relations, Eq. (6), are true for the exact Hamiltonian, unlike the BCS case where the analogous su(2) relations only hold for the mean-field situation Similarly, the BCS wave-function provides an excellent approximation to the ground state of the BCS-Hamiltonian, but is not an eigenfunction of an exact Hamiltonian, whereas our states are constructed from exact $(\eta)$ eigenstates, although presumably do not provide the true ground state.

Since the UCSI and UCSII states lower the energy of the Hubbard model, it would be of interest to see how they perform as variational wave functions.

We note that Eq. (6) holds for a related Hamiltonian[10]

$$H = T_0 + T_1 + V \sum_{\langle ij \rangle} \vec{\eta}_i \cdot \vec{\eta}_j - \tfrac{1}{2} V \sum_i n_i, \qquad (41)$$

which also describes real-space fermion pairing. Since $[H - T_0 - T_1, \eta^+] = V\eta^+$ the formalism may be extended to this case.

We dedicate this paper to Professor Karl-Heinz Bennemann on the occasion of his 65th birthday.

## Acknowledgments

A.I. Solomon wishes to thank Professor Richard Kerner and the members of his Laboratoire de Gravitation et Cosmologie Relativistes for their support and hospitality during his recent stay at the Université Pierre et Marie Curie where most of this work was carried out.

# References

* E-mail: penson@lptl.jussieu.fr
† Permanent address: Faculty of Mathematics and Computing, The Open University, Milton Keynes, MK7 6AA, United Kingdom. E-mail: a.i.solomon@open.ac.uk
1. J. Hubbard, *Proc. R. Soc. London Ser. A* **276**, 238 (1963); *ibid.* **277**, 237 (1964).
2. *The Hubbard Model—a reprint volume*, edited by A. Montorsi (World-Scientific, Singapore, 1992).
3. A. Danani, M. Rasetti, and A. I. Solomon, in *Theories of Matter*, edited by A. I. Solomon, M. Balkanski, and H. R. Trebin (World Scientific, Singapore, 1994).
4. C. N. Yang, *Phys. Rev. Lett.* **63**, 2144 (1989).
5. J. M. Radcliffe, *J. Phys. A* **4**, 313 (1971).
6. S. V. Yablonsky, *Introduction to Discrete Mathematics* (Mir Publishers, Moscow, 1989).
7. J. Riordan, *Combinatorial Identities* (John Wiley, New York, 1968).
8. L. Comtet, *Advanced Combinatorics* (D. Reidel, Dodrecht, Holland, 1974).
9. J. R. Schrieffer, *Theory of Superconductivity* (Benjamin, New York, 1964).
10. K. A. Penson and M. Kolb, *Phys. Rev. B* **33**, 1663 (1986).

# What is Noncollinear Magnetism?

Peter Weinberger

*Institut für Technische Elektrochemie and Center for Computational Materials Science*
*Technical University of Vienna*
*Getreidemarkt 9/158, A-1060 Vienna*
*AUSTRIA*

## Abstract

The terms "collinear" and "noncollinear" refer originally to a geometrical concept. Nowadays they are frequently used to classify magnetism in a rather vague sense. In the present paper an attempt is made to classify collinear and noncollinear by means of transformation properties of classical vectors (Heisenberg-like models of magnetism), and spinors and bispinors as used within a nonrelativistic and a relativistic description of magnetism based on the (local) density functional approach.

## I. Introduction

The words "collinear" and "noncollinear" magnetism are frequently used in a rather vague sense. In the following special attention is given to a classification of these two words in terms of transformation properties by viewing classical vectors, spinors, and bispinors by inspecting magnetism in terms of a semiclassical model (Heisenberg-like models of magnetism), and a nonrelativistic description within the (local) density functional approach (LDA). Quite clearly by using a different approach yet another type of definition of these two words can be given. However, it seems that presently most descriptions of magnetism are related either to semiclassical models, using in turn for example numerical tools such as Monte Carlo methods, or to applications of the LDA.

## II. Classical Vectors

Suppose the spin is viewed as a classical three-dimensional vector,

$$\mathbf{s}_i = (s_{i,x},\ s_{i,y},\ s_{i,z}), \qquad i = 1, 2, \ldots, N, \qquad (1)$$

where $i$ denotes site-indices, referring to location vectors $\mathbf{R}_i$ in real space. Spin models such as

$$\mathcal{H} = -\frac{J}{2}\sum_{ij}(\mathbf{s}_i\cdot\mathbf{s}_j) + \frac{\omega}{2}\sum_{ij}\left[\frac{(\mathbf{s}_i\cdot\mathbf{s}_j)}{R_{i,j}^3} - \frac{3(\mathbf{s}_i\cdot\mathbf{R}_{ij})(\mathbf{s}_j\cdot\mathbf{R}_{ij})}{R_{ij}^5}\right] - \lambda\sum_i s_{i,z}^2, \qquad (2)$$

where $\mathbf{R}_{i,j} = \mathbf{R}_i - \mathbf{R}_j$ and $J, \omega$, and $\lambda$ refer in turn to the exchange interaction parameter, the magnetic dipole-dipole parameter, and the spin-orbit interaction parameter, give then very interesting and useful information about collinearity or noncollinearity of magnetic systems.[1]

Quite clearly by the terms "collinear" or "noncollinear" implicitly transformation properties of classical vectors are implied. If $\mathcal{D}(E)$ and $\mathcal{D}(i)$ denote the following matrix representatives of $E$ (identity) and $i$ (inversion) $\in O(3)$.

$$\mathcal{D}^{(3)}(E) = \begin{pmatrix} 1 & 0 & 0 \\ 0 & 1 & 0 \\ 0 & 0 & 1 \end{pmatrix}, \quad \mathcal{D}^{(3)}(i) = \begin{pmatrix} -1 & 0 & 0 \\ 0 & -1 & 0 \\ 0 & 0 & -1 \end{pmatrix}, \qquad (3)$$

and $\mathcal{N} = \{\mathbf{n}_i\}$ denotes a set of unit vectors in one and the same (chosen) direction centered in the sites $i = 1, 2, \ldots, N$, then an arbitrary pair of spins, $\mathbf{s}_i$ and $\mathbf{s}_j$, is said to be parallel if

$$\hat{\mathbf{s}}_i = \mathcal{D}^{(3)}(E)\mathbf{n}_i, \qquad \hat{\mathbf{s}}_j = \mathcal{D}^{(3)}(E)\mathbf{n}_j, \qquad (4)$$

antiparallel if

$$\hat{\mathbf{s}}_i = \mathcal{D}^{(3)}(E)\mathbf{n}_i, \qquad \hat{\mathbf{s}}_j = \mathcal{D}^{(3)}(i)\mathbf{n}_j, \qquad (5)$$

and collinear if

$$\hat{\mathbf{s}}_i = \mathcal{D}^{(3)}(E)\mathbf{n}_i, \qquad \hat{\mathbf{s}}_j = \mathcal{D}^{(3)}(R)\mathbf{n}_j; \qquad R = E \text{ or } i, \qquad (6)$$

with $\hat{\mathbf{s}}_i$ given by

$$\hat{\mathbf{s}}_i = \frac{\mathbf{s}_i}{|\mathbf{s}_i|} \qquad i = 1, 2, \ldots, N.$$

If, however, in Eq. (6) $R$ refers to an arbitrary element $\in O(3)$ such that $R \neq E$ and $R \neq i$ then this pair of spins is called noncollinear.

It should be recalled that $\mathcal{H}$ in Eq. (2) is not an operator, since all terms on the right hand side of this equation are scalars $\mathcal{H}$ is a semiclassical Hamilton function, whose value is determined for a given set of spins $\{\mathbf{s}_i\}$ and location vectors $\{\mathbf{R}_i\}$ by the parameters $J$, $\omega$, and $\lambda$.

## III. Spinors

Suppose that one-particle (electron) wave functions are products of the following kind,

$$\Psi(\mathbf{r}, \sigma) = \psi(\mathbf{r})\,\phi(\sigma); \qquad \sigma \equiv m_s = \pm\tfrac{1}{2}, \qquad (7)$$

where

$$\phi(\tfrac{1}{2}) = \begin{pmatrix} 1 \\ 0 \end{pmatrix}, \qquad \phi(-\tfrac{1}{2}) = \begin{pmatrix} 0 \\ 1 \end{pmatrix}. \qquad (8)$$

Then the Pauli exclusion principle for a $N$ electron system is based on antisymmetric tensor products (Slater determinants), *i.e.*, tensor products that are symmetrized

# What is Noncollinear Magnetism?

with respect of the permutation group $S_N$ and the permutational irreducible representation $-1$, (see Ref. 3). Obviously the $\phi(\sigma)$, $\sigma = \pm\frac{1}{2}$, are not functions, but unit vectors in a two-dimensional vector space, usually termed spin space:

$$[\phi(\sigma) \cdot \phi(\sigma')] = \delta_{\sigma\sigma'}. \tag{9}$$

In principle the transformation properties[†] of $\Psi(\mathbf{r}, \sigma)$ are conceptually very easy, since

$$R\Psi(\mathbf{r}, \sigma) = \psi(R^{-1}\mathbf{r})\phi(\sigma) \equiv \phi(\sigma)\psi(R^{-1}\mathbf{r}); \qquad R \in O(3), \tag{10}$$
$$= \phi(\sigma)\mathcal{D}(R)\psi(\mathbf{r}),$$

where $\mathcal{D}(R)$ is a representation of $R \in O(3)$, and

$$U(\zeta, \varphi) \{\psi(\mathbf{r})\,\phi(\sigma)\} = \psi(\mathbf{r})\,\{U(\zeta, \varphi)\,\phi(\sigma)\}; \qquad U(\zeta, \varphi) \in \mathrm{SU}(2), \tag{11}$$

where $(\zeta, \varphi)$ is a quarternion, *i.e.*, where $\zeta$ denotes a unit vector specifying an axis of rotation and $\varphi$ specifies the angle of rotation around this axis.

Recalling now the definition of the vector of Pauli spin matrices,

$$\sigma = (\sigma_x, \sigma_y, \sigma_z), \tag{12}$$

where

$$\sigma_x = \begin{pmatrix} 0 & 1 \\ 1 & 0 \end{pmatrix}, \qquad \sigma_y = \begin{pmatrix} 0 & -i \\ i & 0 \end{pmatrix}, \qquad \sigma_z = \begin{pmatrix} 1 & 0 \\ 0 & -1 \end{pmatrix}, \tag{13}$$

within the local density functional the Hamiltonian is usually defined as

$$\mathcal{H}(\mathbf{r}) = I_2 \left[-\tfrac{1}{2}\nabla^2 + V(\mathbf{r})\right] + \sigma_z B(\mathbf{r}), \tag{14}$$

where $I_n$ is a $n \times n$ unit matrix and $V(\mathbf{r})$ is the (effective) potential. One obvious meaning of the second term on the right hand side of Eq. (14) is that the (effective) magnetization $\mathbf{B}(\mathbf{r})$ points along an arbitrary assumed $\hat{z}$-direction, say $\mathbf{n} \in \mathcal{R}_3$, *i.e.*, is of the form

$$\mathbf{B}(\mathbf{r}) = B(r)\mathbf{n}, \qquad \mathbf{n} = (0,\,0,\,1). \tag{15}$$

The general form of the scalar product between $\sigma$ and $\mathbf{B}(\mathbf{r})$ is of course given by

$$[\sigma \cdot \mathbf{B}(\mathbf{r})] = B(\mathbf{r})(\sigma \cdot \xi) = B(\mathbf{r})(\sigma_x \xi_x + \sigma_y \xi_y + \sigma_z \xi_z), \tag{16}$$

where $\xi \in \mathcal{R}_3$ is a vector of unit length in an arbitrary direction.

Quite clearly by keeping in mind Eqs. (10) and (11), a transformation of Eq. (14) of the following kind

$$U(\zeta, \varphi)\,\mathcal{H}(\mathbf{r})\,U^{-1}(\zeta, \varphi) = I_2\left[-\tfrac{1}{2}\nabla^2 + V(\mathbf{r})\right] + U(\zeta, \varphi)\,\sigma_z\,B_z(\mathbf{r})\,U^{-1}(\zeta, \varphi), \tag{17}$$

really means that only the second term on the right hand side of Eq. (17) is transformed as

$$U(\zeta, \varphi)\,\sigma_z\,B_z(\mathbf{r})\,U^{-1}(\zeta, \varphi) = B(\mathbf{r})\,U(\zeta, \varphi)\,(\sigma \cdot \mathbf{n})\,U^{-1}(\zeta, \varphi) = (\sigma' \cdot \mathbf{n}) \tag{18}$$

where

$$\sigma' = (\sigma'_x, \sigma'_y, \sigma'_z) = U(\zeta, \varphi)\,\sigma\,U^{-1}(\zeta, \varphi). \tag{19}$$

Reviewing Eq. (18), again it is obvious that the scalar product on the right hand side of this equation can be written also as

$$(\sigma' \cdot \mathbf{n}) = (\sigma \cdot \xi), \tag{20}$$

---

[†] If point group symmetry is present, O(3) has to be replaced in the following by the point group $P$.

where—as should be noted in particular

$$\xi = \mathcal{D}^{(3)}(R)\mathbf{n}, \qquad (21)$$

$\mathcal{D}^{(3)}(R)$ is a rotation in $\mathcal{R}$ such that the condition in Eq. (20) is met.

Since obviously a transformation in spin-space, spanned by the spinors $\phi(\sigma)$, $\sigma = \pm\frac{1}{2}$, corresponds to a similarity transformation for the Pauli spin matrices, such a transformation can be viewed also in terms of a transformation for the orientation of $\mathbf{B}(\mathbf{r})$.

If $\mathcal{N}_0 = \{\mathbf{n}_i | \mathbf{n}_i = (0, 0, 1), \forall i\}$ denotes a set of unit vectors in $\hat{\mathbf{z}}$-direction centered in sites $i = 1, 2, \ldots N$, and the set $\mathcal{N} = \{\xi_i\}$ specifies the actual orientations in these sites, and arbitrary pair of orientations, $\xi_i$ an $\xi_j$, is said to be parallel if,

$$\xi_i = \mathcal{D}^{(3)}(E)\,\mathbf{n}_i, \qquad \xi_j = \mathcal{D}^{(3)}(E)\,\mathbf{n}_j, \qquad (22)$$

antiparallel if

$$\xi_i = \mathcal{D}^{(3)}(E)\,\mathbf{n}_i, \qquad \xi_j = \mathcal{D}^{(3)}(i)\,\mathbf{n}_j, \qquad (23)$$

and collinear if

$$\xi_i = \mathcal{D}^{(3)}(E)\,\mathbf{n}_i, \qquad \xi_j = \mathcal{D}^{(3)}(R)\,\mathbf{n}_j, \qquad R = E \text{ or } i, \qquad (24)$$

where $\mathcal{D}^{(3)}(E)$ and $\mathcal{D}^{(3)}(i)$ are defined in Eq. (3).

It should be recalled that $\mathcal{D}^{(3)}(E)$ is induced by a transformation in spin space with

$$U(\zeta,\varphi) \equiv U(\hat{\mathbf{z}}, 0) = \begin{pmatrix} 1 & 0 \\ 0 & 1 \end{pmatrix} \to \mathcal{D}^{(3)}(E), \qquad (25)$$

and $\mathcal{D}^{(3)}(i)$ for example by

$$U(\zeta,\varphi) \equiv U(\hat{\mathbf{y}}, 0) = \sigma_y = \begin{pmatrix} 0 & -i \\ i & 0 \end{pmatrix} \to \mathcal{D}^{(3)}(i). \qquad (26)$$

The last equation can easily be checked using the properties of the Pauli spin matrices, namely

$$U(\zeta,\varphi)\,U^{-1}(\zeta,\varphi) = \begin{pmatrix} 1 & 0 \\ 0 & 1 \end{pmatrix} = \sigma_y^2, \qquad (27)$$

and

$$\sigma_y \sigma_z \sigma_y = -\sigma_z, \qquad (28)$$

from which immediately also follows that

$$U(\zeta,\varphi) \equiv U(\hat{\mathbf{x}}, 0) = \sigma_x \begin{pmatrix} 0 & 1 \\ 1 & 0 \end{pmatrix} \to \mathcal{D}^{(3)}(i). \qquad (29)$$

If, therefore, in Eq. (24) $R$ is induced by a rotation in spin space corresponding to an arbitrary quaternion $[\zeta, \varphi] \neq [\hat{\mathbf{x}}, 0], [\hat{\mathbf{y}}, 0]$, then this pair of orientations is called noncollinear.

It should be noted that the use of $\mathcal{D}^{(3)}(E)$ in Eq. (24) does not imply a loss of generality, since the same description applies also to a pair of orientations

$$\xi_i = \mathcal{D}^{(3)}(S)\,\mathcal{D}^{(3)}(E)\,\mathbf{n}_i, \qquad \xi_j = \mathcal{D}^{(3)}(S)\,\mathcal{D}^{(3)}(R)\,\mathbf{n}_j, \qquad (30)$$

with $\mathcal{D}^{(3)}(S)$ being induced by some rotation $U(\zeta,\varphi') \in SU(2)$.

Going back to Eqs. (14) and (17) it is essential to recall that

$$U(\zeta,\varphi)\,I_2[-\tfrac{1}{2}\nabla^2 + V(\mathbf{r})]U^{-1}(\zeta,\varphi) = I_2[-\tfrac{1}{2}\nabla^2 + V(\mathbf{r})], \qquad \forall [\zeta,\varphi], \qquad (31)$$

which, however, is nothing but another way of viewing Eq. (7): $U(\zeta, \varphi)$ is a rotation in spin space that does not change $V(\mathbf{r})$! $U(\zeta, \varphi)$ cannot be used to define a (rotational) invariance condition for $V(\mathbf{r})$, it is completely independent of any (possibly underlying) point group symmetry.

## IV. Bispinors

By using a relativistic description within the local density functional the Hamiltonian is given by

$$\mathcal{H}(\mathbf{r}) = c\alpha \cdot \mathbf{p} + \beta mc^2 + I_4 V(\mathbf{r}) + \beta \Sigma_z B(\mathbf{r}), \tag{32}$$

where $\alpha = (\alpha_1, \alpha_2, \alpha_3)$,

$$\alpha = \begin{pmatrix} 0 & \sigma_i \\ \sigma_i & 0 \end{pmatrix}, \quad \beta = \begin{pmatrix} I_2 & \\ & -I_2 \end{pmatrix}, \quad \Sigma_i = \begin{pmatrix} \sigma_i & 0 \\ 0 & \sigma_i \end{pmatrix}. \tag{33}$$

The transformation properties of $\mathcal{H}(\mathbf{r})$ are now slightly more complicated.[4] Consider a rotation (point-group operation) $R$, then invariance by $R$ implies that

$$S(R)\,\mathcal{H}(R^{-1}\mathbf{r})\,S^{-1}(R) = \mathcal{H}(\mathbf{r}), \tag{34}$$

where $S(R)$ is a $4 \times 4$ matrix transforming the Dirac matrices $\alpha_i$, $\beta$, and $\Sigma_i$

$$S(R) = \begin{pmatrix} U(R) & 0 \\ 0 & \det[\pm]U(R) \end{pmatrix}, \tag{35}$$

and $U(R)$ is a (unimodular) $2 \times 2$ matrix and $\det[\pm] = \det[\mathcal{D}^{(3)}(R)]$ with $\mathcal{D}^{(3)}(R)$ being the corresponding three-dimensional rotation matrix. Using now the invariance condition in (34) explicitly, one can see immediately that the condition

$$S(R)\,[I_4 V(R^{-1}\mathbf{r})]\,S^{-1}(R) = I_4\,V(R^{-1}\mathbf{r}) = I_4\,V(\mathbf{r}), \tag{36}$$

yields the usual rotational invariance condition for the potential, while the terms

$$S(R)\,[c\alpha \cdot \mathbf{p}]\,S^{-1}(R), \qquad S(R)\,[\beta\Sigma \cdot \mathbf{B}(R^{-1}\mathbf{r})]\,S^{-1}(R),$$

have to be examined with more care. Considering the scalar product in here explicitly term-wise, this reduces to the following common condition for both expressions,

$$U(R)\,\sigma\,U^{-1}(R) = \sigma. \tag{37}$$

As in the previous spinor-case the obvious meaning of Eq. (32) is that the magnetization $\mathbf{B}(\mathbf{r})$ points along an arbitrary assumed $\hat{z}$-direction, i.e., is of the form

$$\mathbf{B}(\mathbf{r}) = B(\mathbf{r})\mathbf{n}, \qquad \mathbf{n} = (0, 0, 1). \tag{38}$$

However, by comparing now the transformation properties in the spinor- and the bispinor-case,

$$\begin{array}{ll} \text{nonrelativistic:} & U(\zeta, \varphi)\,\mathcal{H}(\mathbf{r})\,U^{-1}(\zeta, \varphi), \\ \text{relativistic:} & S(R)\,\mathcal{H}(R^{-1}\mathbf{r})\,S^{-1}(R), \end{array} \tag{39}$$

one easily can see that in the bispinor-case a definition of noncollinearity has to include the conditions stated in Eqs. (36) and (37), i.e., an induced rotation for the orientation of $\mathbf{B}(r)$, such as defined in Eqs. (20) and (21) is restricted by the (rotational) invariance condition for $V(\mathbf{r})$,

## Table I
Concepts discussed in the text and their transformations properties.

| Concept | Transformation properties | Quantity | Meaning |
|---|---|---|---|
| Classical vector | O(3) | $s_i$ | Spin vector |
| Spinor | SU(2) | $\xi_i$ | Orientation |
| Bispinor | $\mathcal{G}$ | $\xi_i$ | Orientation |

the (effective) magnetization,

$$V(R^{-1}\mathbf{r}) = V(\mathbf{r}), \qquad (40)$$

$$B(R^{-1}\mathbf{r}) = B(\mathbf{r}), \qquad (41)$$

and the invariance condition for the kinetic energy operator $c\alpha \cdot \mathbf{p}$, whereby, because of the term $\beta \Sigma_z B(\mathbf{r})$, the sign of $\sigma$ has to be preserved. If therefore

$$\mathcal{G} = \left\{ R | S(R) \, \mathcal{H}(R^{-1}\mathbf{r}) \, S^{-1}(R) = \mathcal{H}(\mathbf{r}) \right\}, \qquad (42)$$

denotes the (rotational) invariance group of $\mathcal{H}(\mathbf{r})$ in Eq. (32), then in the bispinor-case noncollinearity has to be defined with respect to this group, which in principle will be different for different systems.

## V. Translational Properties

In the spinor-case translational invariance,

$$\mathcal{H}(\mathbf{r}+\mathbf{t}) = I_2 \left[ -\tfrac{1}{2}\nabla^2 + V(\mathbf{r}+\mathbf{t}) \right] + \sigma_z B_z(\mathbf{r}+\mathbf{t}) = \mathcal{H}(\mathbf{r}), \qquad t \in \mathcal{L}, \qquad (43)$$

where $\mathcal{L}$ is either a three-dimensional or a two-dimensional lattice, implies—as easily can be checked—that

$$\xi_i = \xi_0, \qquad \forall i \in I(\mathcal{L}), \qquad (44)$$

where $I(\mathcal{L})$ denotes the set of indices corresponding to $\mathcal{L}$ and $\xi_0$ is some arbitrary chosen orientation of $\mathbf{B}(\mathbf{r})$ such as for example $\hat{\mathbf{z}}$. Equation (44) also applies in the bispinor-case, since for a translation the matrix $S(R)$ in Eq. (34) has to be the unit matrix, (see Ref. 4).

## VI. Summary

Comparing now all three concepts at once (Table I), it is obvious that in each case vectors $\in \mathcal{R}_3$ are considered, whereby—as should be recalled—the term orientation arises from the following identity:

$$(\sigma' \cdot \mathbf{n}) = (\sigma \cdot \xi). \qquad (45)$$

If no restrictions with respect to translational symmetry, see Eq. (45), are present, it is obvious that the concept of classical vectors and spinors are isomorphic in the following sense,

$$\{\hat{\mathbf{s}}_i\} \longleftrightarrow \{\xi_i\} \qquad (46)$$

since both, O(3) and SU(2), are continuous groups. Quite clearly in the case of bispinors, in general this relation is not fulfilled, although in practical terms, *i.e.*, by comparing numbers, very often one is tempted to use this relation in a colloquial sense. The Table I also gives a justification for the almost automatic use of arrows, traditionally inherent to illustrations of spins and/or magnetism. Since translational symmetry can only be defined properly in terms of an invariance condition for the Hamilton operator, translational symmetry for classical vectors implies conceptually a constraint for a semiclassical Hamilton function such as in Eq. (2). In the presence of translational symmetry therefore the above relation can be misleading!

## Acknowledgments

This paper is dedicated to Prof. Karl Heinz Bennemann on occasion of his 65th birthday.

## References

1. L. Udvardi *et al.*, *J. Magn. Magn. Mater.*, submitted for publication; C. Uiberacker, Diplomarbeit, Technical University of Vienna, 1996.
2. For all group theoretical notations and aspects see S. L. Altmann and P. Herzig, *Point-Group Theory Tables* (Clarendon Press, Oxford, 1994).
3. See for example: L. Jansen and M. Boom, *Theory of Finite Groups: Applications to Physics* (North Holland, Amsterdam, 1967).
4. See also the detailed discussion in P. Weinberger, *Philos. Mag. B* **75**, 509 (1997).

# Spectral Properties of Transition Metal Compounds and Metal-Insulator Transition: A Systematic Approach within the Dynamical Mean Field Theory

P. Lombardo,[1] M. Avignon,[1] J. Schmalian,[2] and K. H. Bennemann[2]

[1] Laboratoire d'Etude des Propriétés Electroniques des Solides-CNRS
Associated with Université Joseph Fourier
BP 166, 38042 Grenoble Cedex 9
FRANCE

[2] Institut für Theoretische Physik
Freie Universität Berlin
Arnimallee 14, D-14195 Berlin
GERMANY

## Abstract

Starting from the two-band Hubbard Hamiltonian for transition metal-$3d$ and oxygen-$2p$ states with perovskite geometry and a generalization of the dynamical mean field theory, we investigate some of the important features of a large class of transition metal compounds. We succeed in reproducing the metal to insulator transition using an unified point of view, for the purely Mott-Hubbard transition as well as for the charge transfer transition. The physically interesting intermediate regimen is also investigated, leading to the complete corresponding phase diagram, in which we classified some of these materials.

## I. Introduction

Until now, the occurrence of a metal to insulator transition (MIT) in transition metal (TM) compounds and the evolution of electronic structure through the transition remains highly controversial. In many of these compounds, band-structure calculations fail to reproduce some important experimental results, in particular the existence of an insulating state and feature of photo-emission spectra. Photo-emission of various

## Table I
Some of the $\Delta$, $U$, and $(pd\sigma)$ values (in eV) from Saitoh et al.[6]

| Compound | $\Delta$ | $U$ | $(pd\sigma)$ |
|---|---|---|---|
| LaFeO$_3$ | 2.5 | 7.5 | −1.4 |
| LaTiO$_3$ | 6.0 | 4.0 | −2.4 |
| LaCrO$_3$ | 5.2 | 5.2 | −1.9 |

TM oxides shows that paramagnetic metals like LaTiO$_3$, SrVO$_3$, CaVO$_3$ (Refs. 1–3) present two important features: One close to the Fermi energy refered to as *coherent part* corresponding well to the density of states obtained by renormalized band structure calculations and an other one at higher binding energy, the *incoherent part*, which has no counterpart in band-structure calculations. The optical conductivity has been investigated by Arima et al.[4] The existence and the evolution under doping of the coherent quasi-particle and the transfer of spectral weight from high to low energy remains an important question greatly debated.

The simple one-band Hubbard model is not appropriate to describe most of real transition metal compounds. Many real materials have a band gap which is strongly determined by the nature of the anion. For instance NiS and CuS are metallic but NiO and CuO clearly show an insulating behavior. This difference can not be understood within a simple Mott-Hubbard (MH) framework since $U$ is mainly dependent of the MT. A model taking into account the electronegativity of the anion is then needed. Consequently, it is of importance for the understanding of these materials that one takes charge fluctuations, both on the transitions metal and the oxygen states, into account. One major difficulty is the competition between them. This has led to a classification of these materials in two main categories according to the origin of the band gap.[5] The first one is the Mott insulator, where the $d$-$d$ Coulomb interaction ($U$) drives the transition, inducing a splitting of the $d$-band and then opening the band gap. The second one is the *charge transfer* (CT) *insulator*; where the gap is essentially the charge transfer energy ($\Delta$), the energy difference between occupied oxygen 2$p$-states and unoccupied 3$d$-states of the upper Hubbard band. These two regimes are of course limiting regimes: For many real transition metal compounds we have to deal with intermediate regimes. Recently, by a configuration-interaction cluster model analysis of the metal 2$p$ core-level photo-emission spectral,[6] many early transition metal compounds, usually classified in the Mott-Hubbard regime, ($\Delta \gg U$) have been re-classified as intermediated between this regime and the charge transfer regime, or in the CT regime itself ($U \gg \Delta$) as shown in Table I.

Recently, the dynamical mean-field theory[7] applied to transition metal compounds with perovskite geometry has led to significant improvements in describing the effect of strong correlations on spectral function of these systems.[8] The starting point was the Hubbard Hamiltonian including 3$d$-transition metal electrons hybridized with 2$d$-oxygen states, but neglecting intrinsic oxygen bandwidth. On-site correlations $U$ are included on the metal sites only. We have defined a generalization of this model in the limit of large coordination numbers in a way to preserve the local lattice topology. Then, a selective treatment of different hopping processes is used in order to obtain a physically reasonable limit in accordance with the corresponding Hamiltonian for real systems in finite dimensions. This cannot be generated by a unique scaling of the hopping element, on one hand one has hopping processes leading to local hybrization between 3d-TM and the linear combination of 2d-states of the surrounding oxygen sites, while other hopping processes produce a finite bandwidth. The local TM-O hybridization is preserved by mapping the system onto a local TM-O cluster model embedded within an effective medium. Our local TM-O cluster corresponds to the cluster used in the analysis of photo-emission spectra[6] first proposed by Sawatzky and collaborators[9] for Cu dihalides and provides a qualitative picture of the electronic structure. The dynamical mean-field approach includes the dynamical character of the interactions.

## II. Dynamical Mean-Field Approach for Transition Metal Compounds

An appropriate description of the TM compounds consists in considering TM $d$-states, as correlated atomic-like states with given configurations $d^n$ hybridized with uncorrelated bandlike ligand states, while neglecting direct oxygen-oxygen hopping in a first step. Assuming that the ground-state ionic TM-configuration is the $d^n$, the closest excited configurations become $d^{n+1}$ and $d^{n-1}$. Excited states in the $n(n+1)$ configuration involves $d^{n+1}\underline{L}$, $d^{n+2}\underline{L}^2$, ..., $(d^{n+2}\underline{L}, \ldots)$, configurations, where $\underline{L}$ denotes a ligand hole.

It is convenient to define the energy $\varepsilon_d$ to create a hole in the $d^{n+1}$ to obtain the $d^n$ configuration,

$$\varepsilon_d = E(d^n) - E(d^{n+1})$$

and the energy $\varepsilon_p$ to create a ligand hole from a $d$ configuration

$$\varepsilon_p = E(d^n\underline{L}) - E(d^n).$$

Then the energy of the various configurations are well defined, and the charge transfer energy becomes

$$E(d^{n+1}\underline{L}) - E(d^n) = \varepsilon_p - \varepsilon_d = \Delta.$$

The intra-atomic correlation $U$ for the $n$, $n-1$, and $n+1$ configurations correspond to the charge transfer energy $d_i^n d_j^n \leftrightarrow d_i^{n+1} d_j^{n-1}$ then

$$U = E(d^{n+1}) + E(d^{n-1}) - 2E(d^n).$$

Local hybridization mixes states with a given number of electrons, shifting the above configurations. The energy-level diagram of the configurations is shown in Fig. 1.

Depending on the relative values of $U$ and $\Delta$, the ground state of the $(n-1)$ configuration can be either the $d^n\underline{L}$ or the $d^{n-1}$ configuration. This is the essence of the difference between charge transfer systems $(U > \Delta)$ and Mott-Hubbard systems $(\Delta < U)$. The multiplet structure of the different configurations could also be included without conceptual difficulty.

For the sake of simplicity and completeness, to describe the local configurations, we make use of the Hubbard model thus retaining only the spin degeneracy. The configurations involved are $d^0(d^1\underline{L}, d^2\underline{L}^2)$, $d^1(d^1\underline{L}^2)$ configurations (respectively 2, 1, and 0 holes) corresponding to $d^{n-1}$, $d^n$, and $d^{n+1}$. The configurations with 3 and 4 holes are restricted to $d^0\underline{L}$, $(d^1\underline{L}^2)$, and $d^0\underline{L}^2$, respectively. This situation is relevant for systems with ground state $d^1$ configuration and the evolution towards $d^0$ and $d^2$ under doping. The set of local energy levels resulting from hybridization is shown in Table II with the corresponding spin degeneracy and hole occupation number.

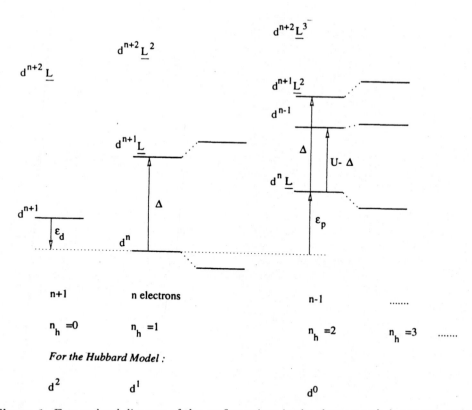

**Figure 1.** Energy-level diagram of the configurations in the electron or hole representation.

These $d$-states are hybridized with the appropiate linear combination of O $p$-states $p_{i,\sigma}$ and the resulting Hamiltonian in $D$-dimension reads:

$$H = \sum_{\mathbf{k}\sigma}(\varepsilon_d d^\dagger_{\mathbf{k}\sigma}d_{\mathbf{k}\sigma} + \varepsilon_d p^\dagger_{\mathbf{k}\sigma}p_{\mathbf{k}\sigma}) + \sqrt{2D}t \sum_{\mathbf{k}\sigma}(\gamma_\mathbf{k}\, d^\dagger_{\mathbf{k}\sigma}p_{\mathbf{k}\sigma} + \text{h.c.}) + U\sum_i n^d_{i\uparrow}n^d_{i\downarrow}. \quad (1)$$

Here $d^\dagger_{\mathbf{k}\sigma}$ is the Fourier transform of the $p$-state creation operator, and $p^\dagger_{\mathbf{k}\sigma}$ refers to the orthogonalized Fourier transform of the creation operator relative to the linear symmetric combination of neighboring $p$-states. $t$ is the amplitude of the nearest-neighbor TM-O hopping integral. $\varepsilon_d$, $\varepsilon_p$ and $U$ are as defined above. The orthogonalization factor is defined by:

$$\gamma^2_\mathbf{k} = 1 - \frac{1}{D}\sum_{\alpha=1}^{D}\cos k_\alpha. \quad (2)$$

As in the one band case the TM-O model is mapped onto an effective impurity model. Motivated by the above physical description, our impurity is just a cluster of hybridized TM and O-states embedded into an effective medium which results from the hybridization between neighbor clusters. The effective medium is coupled to the oxygen states only. We thus consider

$$H_\text{eff} = H_\text{loc} + H_\text{med}, \quad (3)$$

where the local part is given by

## Table II

The degeneracy, occupation and relative energy of all local eigenstates are given. Here, $\tan 2\phi = -4t/\Delta$, and $\tan 2\theta = 4t/(\Delta + U)$. The coefficients $u_i$, $v_i$, and $w_i$ can be obtained from the diagonalization of the tree dimensional space build up by $|S_0\rangle$, $|dd\rangle$, and $|pp\rangle$; see for example Ref. 12

| State label | State | Degeneracy | Occupation $n_0$ |
|---|---|---|---|
| 1 | $\|0\rangle$ | 1 | 0 |
| 2 | $\|d_\sigma\rangle = (\cos\phi\, d_\sigma^\dagger + \sin\phi\, p_\sigma^\dagger)\|0\rangle$ | 2 | 1 |
| 3 | $\|p_\sigma\rangle = (\cos\phi\, d_\sigma^\dagger - \sin\phi\, p_\sigma^\dagger)\|0\rangle$ | 2 | 1 |
| 4 | $\|T_0\rangle = \frac{1}{\sqrt{2}}(d_\uparrow^\dagger p_\downarrow^\dagger + d_\downarrow^\dagger p_\uparrow^\dagger)\|0\rangle$ | 1 | 2 |
| 5 | $\|T_\sigma\rangle = d_\sigma^\dagger p_\sigma^\dagger\|0\rangle$ | 2 | 2 |
| 6 | $\|S_0\rangle = [u_1 d_\uparrow^\dagger d_\downarrow^\dagger + v_1 p_\uparrow^\dagger p_\downarrow^\dagger - w_1(d_\uparrow^\dagger p_\downarrow^\dagger - d_\downarrow^\dagger p_\uparrow^\dagger)]\|0\rangle$ | 1 | 2 |
| 7 | $\|dd\rangle = [u_2 d_\uparrow^\dagger d_\downarrow^\dagger + v_2 p_\uparrow^\dagger p_\downarrow^\dagger - w_2(d_\uparrow^\dagger p_\downarrow^\dagger - d_\downarrow^\dagger p_\uparrow^\dagger)]\|0\rangle$ | 1 | 2 |
| 8 | $\|pp\rangle = [u_3 d_\uparrow^\dagger d_\downarrow^\dagger + v_3 p_\uparrow^\dagger p_\downarrow^\dagger - w_3(d_\uparrow^\dagger p_\downarrow^\dagger - d_\downarrow^\dagger p_\uparrow^\dagger)]\|0\rangle$ | 1 | 2 |
| 9 | $\|D_\sigma\rangle = (\cos\theta\, d_\sigma^\dagger p_\sigma^\dagger d_{-\sigma}^\dagger + \sin\theta\, d_\sigma^\dagger p_\sigma^\dagger p_{-\sigma}^\dagger)\|0\rangle$ | 2 | 3 |
| 10 | $\|P_\sigma\rangle = (\cos\theta\, d_\sigma^\dagger p_\sigma^\dagger p_{-\sigma}^\dagger + \sin\theta\, d_\sigma^\dagger p_\sigma^\dagger d_{-\sigma}^\dagger)\|0\rangle$ | 2 | 3 |
| 11 | $\|DP\rangle = d_\uparrow^\dagger p_\uparrow^\dagger d_\downarrow^\dagger p_\downarrow^\dagger\|0\rangle$ | 1 | 4 |

$$H_{\text{loc}} = \sum_\sigma \left[\varepsilon_d\, d_\sigma^\dagger d_\sigma + \varepsilon_p\, p_\sigma^\dagger p_\sigma + 2t(d_\sigma^\dagger p_\sigma + p_\sigma^\dagger d_\sigma)\right] + U n_\uparrow^d n_\downarrow^d,$$

and

$$H_{\text{med}} = \sum_{\mathbf{k}\sigma}(W_{\mathbf{k}}\, c_{\mathbf{k}\sigma}^\dagger p_\sigma + \text{h.c.}) + \sum_{\mathbf{k}\sigma}\varepsilon_{\mathbf{k}}\, c_{\mathbf{k}\sigma}^\dagger c_{\mathbf{k}\sigma}, \quad (4)$$

describes the coupling of the oxygen states with the effective medium, which can be described by the hybridization

$$\mathcal{J}(\omega) = \sum_{\mathbf{k}} \frac{|W_{\mathbf{k}}|^2}{\omega + i0^+ - \varepsilon_{\mathbf{k}}}. \quad (5)$$

This coupling of the effective mediums to the oxygen states, although not unique, is physically motivated by the local arrangement of the original lattice. $\mathcal{J}(\omega)$ is determinated by the condition that the Green's functions of the impurity model 3 becomes identical to that of the lattice system. Then, we have solved the model using a cluster generalization of the standard noncrossing aproximation (NCA)[10] in a basis in which the local Hamiltonian is diagonal:[8] The real $d$- and $p$- Green's functions are finally recovered from the local NCA one's.

The main electronic structure can be understood from transitions between the various many-body states of the TM-O cluster. Besides the general structure, the dynamical mean-field produces a coherent quasi-particle peak near the Fermi level which shows strong temperature and doping dependence.

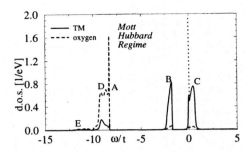

**Figure 2.** Transition metal (solid line) and oxygen (dashed line) density of states within our NCA method for a Mott-Hubbard transition metal compound. The temperature is $T = 300$ K

## III. Metal Insulator Phase Diagram

We have applied the above described calculation scheme to both Mott-Hubbard (MH) regime and charge transfer (CT) regime using a semi-elliptical density of states. At room temperature, which is the usual experimental situation, most of the systems are paramagnetic. We thus consider the paramagnetic phase only. This implies that we do not look for antiferromagnetism despite the fact that this phase is known to be favored at low temperature for high values of $U/t$ in the half-filled case.

### III.1 The Mott-Hubbard Regime

In the Mott-Hubbard regime, the band gap is purely driven by electronic correlations. The charge fluctuations $d_i^n d_j^n \leftrightarrow d_i^{n-1} d_j^{n+1}$ (where $i$ and $j$ are transition metal sites), are associated to the Coulomb energy $U$ which is a rather high energy scale for many transition compounds. Then, these fluctuations are supposed to have almost no incidence on the low-energy specific properties of these materials. This decoupling between low and high energy scales to the well-known description of these systems in terms of spin-only Hamiltonians. In our two-band model, this regime is reached when the CT parameter is much greater than $U$ and $t_{pd}$. Indeed, in the limit $\Delta \to \infty$, our model Hamiltonian becomes a one band Hamiltonian with a renormalized hopping going to zero like $t_{pd}^* = 4t_{pd}^2/\Delta$. The two band electronic properties in the Mott-Hubbard regime should therefore be close to those of the one band dynamical mean field theory given in details in the review article by Georges et al.[7] In figure 2, we have plotted the density of states (DOS) obtained for TM and oxygen in the Mott-Hubbard regime with $\Delta = 8$ eV, $U = 4$ eV, and $t_{pd} = 1$. (B) and (C) are dominated by TM states and correspond to the lower and the upper Hubbard bands separated by the energy $U$. Note that the numerical simulations have been done for a total electron occupation number slightly greater than the half-filled case. We have performed many calculations for various small electron doping. The value of the band gap, as well as the spectral weight and the energy position of each band (except very close to the Fermi level) are independent of this doping as soon as it remains lower than 5%. At half-filling ($n = 1$) the system would be insulator with a gap of $U$. Here, most of the oxygen states ($\approx 95\%$) constitute the high energy bands (E), (D), and (A). Then, these states are of minor importance concerning the metal insulator transition which is determined by the low-energy electronic properties.

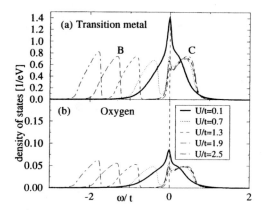

**Figure 3.** Low-energy behavior of TM (a) and oxygen (b) density of states for the Mott-Hubbard regime for various values of $U/t$ and a fixed $\Delta = 8$ eV.

In figure 3, the low-energy behavior of TM and oxygen states is shown for a Mott-Hubbard compound. For a large constant charge transfer parameter $\Delta = 8$ eV, we explore a line of the phase diagram corresponding to $U/t_{pd}$ going from 2.5 eV to 0. The uncorrelated case gives one unique band with a renormalized bandwidth $W^* = 4t_{pd}^2/\Delta = 0.5$ eV. For increasing $U$, we find a MIT which is purely a Mott transition. The critical values $U$ corresponding to this relation is $U_c = 1.6$ eV or $U_c^{\text{eff}} = U_c/W^* = 3.2$. This value is in good agreement with the one band case[7] where $U - c/W \approx 3$. In addition to the two Hubbard sub-bands, a coherent quasi-particle structure near the Fermi energy builds up at low temperature and is clearly seen. This low-energy part of the spectrum can be understood quit easily. Indeed, the high value $\Delta/t = 5$ leads to a very small oxygen states occupation and the correlated TM states are very close to half-filling. One thus retrieves a behavior similar to the one-band Hubbard model close to half-filling in which the quasi-particle peak lies at the bottom of one of the Hubbard sub-bands.

Another way to check the Mott character of this insulator is to inspect the doping dependent transfer of spectral weight from high to low-energy scale[11] in the electron doped case end in the hole doped case. Corresponding results are displayed in Fig. 4 where the transition metal DOS are showed. Concerning the transfer of spectral weight, this Mott insulator presents a hole-particle symmetry since the two side bands of the gap have the same mainly TM nature. Furthermore, the energy spectral weight is growing as $2x$ and the spectral weight of the upper Hubbard band is decreasing correspondingly, where $x$ is the percentage of hole from half-filling. There is no such a symmetry in the charge transfer limit since electrons will feel the strong repulsion of the TM sites and holes, which are mainly on oxygen sites, will behave more like free particles.

Here TM and O occupation numbers remains almost insensitive to the strength of the electronic correlation $U$. This establishes that the covalence between TM and O-states is completely determined by the local Hamiltonian. We could expect that an increase of $U$ would lead to an increase of the uncorrelated oxygen states occupation, but this does not occur in the MH regime because of the high charge-transfer parameter.

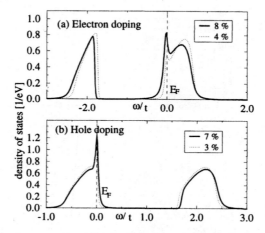

**Figure 4.** Transfer of spectral weight in hole doped (a) and electron doped MH compounds (b). In both cases the displayed DOS are transition metal DOS. Besides, solid line (dashed line) corresponds to higher (lower) doping value. Note that this figure represents the density of states per spin. The total DOS is obtained by multiplying this one by a factor 2.

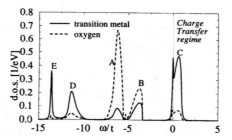

**Figure 5.** TM (solid line) and oxygen (dashed line) density of state for $U/t = 10$ and $\Delta/t = 5$. The electron doping is 3%.

### III.2 The Charge Transfer Regime

The charge transfer regime is the limiting case where $U$ becomes much larger than the charge transfer energy $\Delta$. Then, the occupied $p$ states take an active part into the electron dynamics. $\Delta$ is directly related to the electronegativity of the anion and can therefore explain the completely different behavior of some TM oxides and sulfides. In figure 5 we show the TM and O density of states for $U = 10$ eV, $t = 1$ eV, and $\Delta = 5$ eV. Like in the Mott-Hubbard regime the effect of a small electron doping is only sensible in the close neighborhood of $\varepsilon_F$. The bands (E) and (D) constitute the lower Hubbard band while (C) is the upper Hubbard band. Their energy positions are distant of $U$ and they present mainly TM character. The band gap is now of order $\Delta$ and the side bands are the empty higher Hubbard band (C) (TM character) and the electron occupied $p$-band (B) (O-character). This insulator has therefore all properties of a charge-transfer insulator.

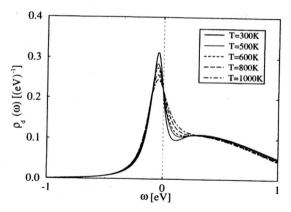

**Figure 6.** Density of states near the Fermi level for hole doping $x = 0.2$ and for various temperatures. Note the formation of a coherent quasi-particle peak resulting from Kondo like resonances between the singly occupied bonding TM-O state and the doubly occupied singlet state.

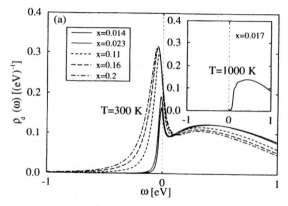

**Figure 7.** Doping dependence of the coherent quasi-particle peak for $T = 300$ K. For comparison we also show in the inset the result for low hole doping ($x = 0.017$), but high temperature $T = 1000$ K, which demonstrates that strong temperature dependencies occur in the low doping regime.

Again, a coherent quasi-particle peak occurs near the Fermi energy from the dynamical character of the interactions involving charge transfer from TM to O-states. Its strong temperature dependence is shown in Fig. 6 for a hole doping concentration $x = 0.2$. For electron doping, the corresponding low-energy part of the spectrum behaves very similarly. In figure 7 we present the doping dependence of the quasi-particle peak for hole doping at $T = 300$ K. The coherent peak is growing at the expense of the higher energy part of the spectrum when the hole concentration increases. In figure 8, we explore the transition line for fixed $U/t = 10$ and $\Delta/t$ going from 0.5 to 3. The band gap decreases for decreasing $\Delta$ and disappears for a critical value $\Delta_c \approx 1$ eV. Simultaneously we found an important transfer of spectral weight from low- to high-energy scale when $\Delta$ is decreasing. This leads to an increase of oxygen occupation number.

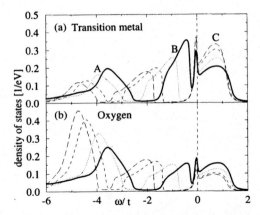

**Figure 8.** Low-energy behavior of transition metal (a) and oxygen (b) density of states in the charge transfer regime of our approach. $U/t = 10$ and $\Delta/t$ is going from 0.5 (solid line) to 3 (long dashed line). Intermediate values are 1.5 (dashed) and 2.5 (dotted).

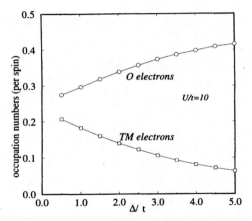

**Figure 9.** Occupation numbers of TM states (circle) and O-states (squares) for various $\Delta$ values and a fixed $U/t = 10$ for $T = 300$ K.

The variation of the occupation numbers with respect to the charge transfer energy are shown in Fig. 9. Therefore the covalence is independent of $U$ but strongly $\Delta$-dependent. The local Hamiltonian entirely decides the covalency between TM and O-states. The reason for this is that the upper Hubbard band is almost empty in our half-filled case.

### III.3 Phase Diagram

The phase-diagram obtained within our NCA approach is given in Fig. 10. These phases are found in the paramagnetic state of the two-band correlated system. The branches $\Delta/t \gg U/t$ and $U/t \gg \Delta/t$ correspond respectively to Mott and charge transfer insulator to metal transition. There are some differences between our phase diagram and the previous one obtained within a slave-boson approach[13] also in the

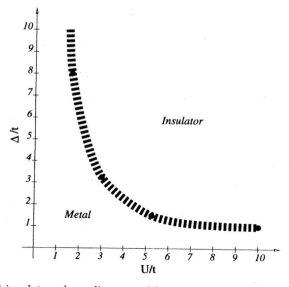

**Figure 10.** Metal insulator phase diagram with respect to the effective Coulomb $U/t$ and the charge transfer effective parameter $\Delta/t$.

paramagnetic state. The insulating state is reached for values of $U$ and $\Delta$ which are much smaller in the dynamical mean-field theory. We have a covalent insulator while it is purely ionic in the slave-boson approach. This results from the fact that in the slave-boson case the hybridization within a cluster and between different clusters is renormalized in the same way.

## IV. Real Materials

Saitoh et al.[6] have performed a systematic investigation of the electronic structure of 3d-transition-metal compounds. They have quantitatively estimated the model parameters $U$ and $\Delta$ using the 2d-core-level photo-emission spectra of these compounds. Using our NCA approach, we propose to determine for several materials, which kind of mechanism is involved in its dynamical properties. As a first step we use the values obtained from the cluster model by Saitoh et al.,[6] which are recalled in Table I.

Our approach is in good agreement with the experimentally observed insulating state of $LaFeO_3$. The spectrum of optical conductivity in $LaFeO_3$ ($3d^1$) shows two optical transitions.[4] The first one is weak and below 2 eV (Mott gap transition), and the second one is stronger and above 4 eV (CT gap transition). The corresponding density of states within our approach is shown in Fig. 11 where the parameters are taken from Table I. The insulating state is clearly of charge transfer nature, and the value of the band gap is in good agreement with the experimental one ($\approx 2.5$ eV).

$LaTiO_3$ compound has a Mott-Hubbard character. Experimental investigations[1,4] exhibit a Mott gap which is very small (around 0.5 eV) and a larger CT gap (above 4 eV). We show in Fig. 12 the obtained DOS as well as the resulting optical conductivity spectrum. Our spectral function reproduces rather nicely the experimental behavior with the peak at about 1.5 eV and the structure at the Fermi level. The insulating gap is a Mott gap. The two side bands are mainly TM bands with some oxygen character. The effective $d$-electron occupation is 1.18, a value consider-

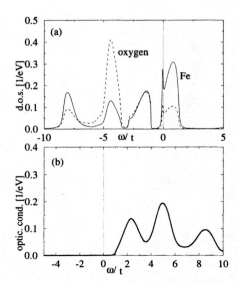

**Figure 11.** LaFeO$_3$ density of states (a) and optical conductivity spectrum (b) obtained within our approach. The parameters are $U_{\text{eff}} = 7.5$ eV and $\Delta_{\text{eff}} = 2.5$ eV. In part (a) the solid line is the Fe-DOS and the dashed line is the O-DOS

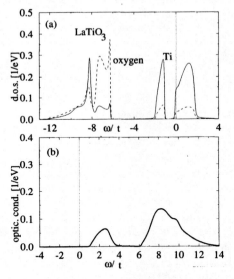

**Figure 12.** LaTiO$_3$-DOS (a) and optical conductivity spectrum (b). $U_{\text{eff}} = 3.5$ eV, $\Delta_{\text{eff}} = 5.5$ eV, and $t_{\text{eff}} = 1.5$ eV. In part (a) the solid line is the Ti-DOS and the dashed line is the O-DOS

ably smaller than the one given by Saitoh et al.[6] The CT transitions corresponding to higher energy scale like in the experimental data. Besides, the relative strength of these two transitions (the weights of the optical conductivity peaks) are comparable to the experimental ones. With the same set of parameters we obtain correctly the band

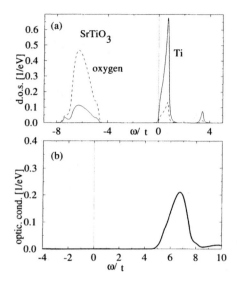

**Figure 13.** SrTiO$_3$-DOS (a) and optical conductivity spectrum (b). $U_{\text{eff}} = 3.5$ eV, $\Delta_{\text{eff}} = 5.5$ eV, and $t_{\text{eff}} = 1.5$ eV. In part (a) the solid line is the Ti DOS and dashed line is the O-DOS

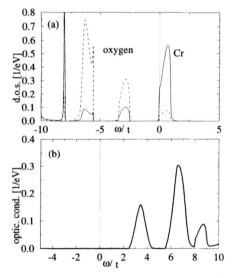

**Figure 14.** LaCrO$_3$ density of states (a) and optical spectrum (b) within our approach. In part (a) the solid line is the Cr-DOS and the dashed line is the O-DOS

insulator character of SrTiO$_3$, in which Ti has the nominal configuration $3d^0$. In this case the gap is between the O-band and the lower Hubbard band as shown in Fig. 13.

The nature of the insulating state in LaCrO$_3$ is more questionable because of the proximity of the two energy scales which are involved in charge transfer and Coulomb processes. Indeed, Saitoh et al.,[6] have proposed that $U = \Delta = 5.2$ eV. The optical conductivity spectrum[4] in LaCrO$_3$ exhibits an insulating phase with a strong charge transfer character. Our results concerning this compound are presented in Fig. 14. The density of states of part (a) proves the charge transfer character of the insulating state. The gap value is in good agreement with the experimental results. The corresponding optical conductivity spectrum (b) displays the two transitions and makes it possible to compare their respective strength. In the same intermediate regime, we have also been able to reproduce the qualitative evolution of the spectral function from $CaVO_3$ to $SrVO_3$ (Ref. 14).

## V. Conclusion

We showed that our two band approach within the framework of the dynamical mean field theory is able to deal with both charge transfer insulators and Mott insulators. Our results concerning the difficult intermediate regime of $LaCrO_3$ are very encouraging. This leads to interesting insights into the competition of these two mechanisms which is ambiguous for a lot of transition metal compounds, as well as an understanding of the covalence in these systems. We have proposed a classification of these compounds in the phase diagrams of the metal-insulator transition. The O-states contribution can be of primary importance in charge transfer materials or unimportant in Mott-Hubbard materials.

## References

1. A. Fujimori et al., *Phys. Rev. Lett.* **69**, 1796 (1992); Y. Tokura et al., *Phys. Rev. Lett.* **70**, 2126 (1993); A. Fujimori et al., *Phys. Rev. B* **46**, 9841 (1992).
2. I. H. Inoue et al., *Phys. Rev. Lett.* **74**, 1796 (1995).
3. K. Morikawa et al., *Phys. Rev. B* **52**, 13 711 (1995).
4. T. Arima, Y. Tokura, and J. B. Torrance, *Phys. Rev. B* **48**, 17 006 (1993).
5. J. Zaanen, G. A. Sawatzky, and J. A. Allen, *Phys. Rev. Lett.* **55**, 418 (1985).
6. T. Saitoh, A. E. Bocquet, T. Mizokawa, and A. Fijimori, *Phys. Rev. B* **52**, 7934 (1995); T. Saitoh et al., *Phys. Rev. B* **51**, 13 942 (1995); A. E. Bocquet et al., *Phys. Rev. B* **53**, 1161 (1996).
7. A. Georges, G. Kotliar, W. Krauth, and M. Rozenberg, *Rev. Mod. Phys.* **68**, 13 (1996).
8. P. Lombardo, J. Schmalian, M. Avignon, and K. H. Bennemann, *Phys. Rev. B* **54**, 5317 (1996).
9. G. van der Laan, C. Westra, C. Haas, and G. A. Sawatzky, *Phys. Rev. B* **23**, 4369 (1981).
10. N. E. Bickers, *Rev. Mod. Phys.* **59**, 845 (1987).
11. H. Eskes, M. B. Meinders, and G. A. Sawatzky, *Phys. Rev. Lett.* **67**, 1035 (1991).
12. C. Noce, *J. Phys. Condens. Matter* **3**, 7819 (1991).
13. C. Balseiro, M. Avignon, A. Rojo, and B. Alascio, *Phys. Rev. Lett.* **62**, 2624 (1989).
14. P. Lombardo, J. Schmalian, M. Avignon, and K. H. Bennemann, *Physica B* (to be published).

# Theoretical Study of the Metal-Nonmetal Transition in Transition Metal Clusters

F. Aguilera-Granja,[1] J. A. Alonso,[1] and J. M. Montejano-Carrizales[2]

[1] *Departamento de Física Teórica*
*Universidad de Valladolid*
*47011 Valladolid*
*SPAIN*

[2] *Instituto de Física*
*Universidad Autónoma de San Luis Potosí*
*Alvaro Obregón 64, 78000 San Luis Potosí, S.L.P.*
*MEXICO*

## Abstract

We have studied the metal-nonmetal transition in transition metal clusters assuming that the $d$ electrons determine the electronic properties of the system and using a rectangular band model. The average coordination number is calculated exactly as a function of the cluster size and the second-moment approximation is used to compute a size-dependent band-width. The metal-nonmetal transition occurs when the density of states near the Fermi level exceeds $1/k_\mathrm{B}T$ and the discrete energy levels begin to form a quasi-continuous band. We have found a sensitive dependence of the critical size with temperature. The critical sizes decrease fast with increasing $T$. At the liquid nitrogen temperature the critical sizes are of the order of 60 atoms or less.

## I. Introduction

Clusters of metallic elements are not simple fragments of the corresponding bulk crystal and their electronic, magnetic, and structural properties evolve in a peculiar and often unexpected way as the number of atoms in the cluster increases. The property we focus on here is the metallic behavior. The meaning of a metal is clear in the case of a bulk system, where the band theory concepts apply. However, for small clusters the characterization of the metallic behavior is more subtle.[1-7] Small clusters are insulators, having a discrete distribution of electronic states, and a critical cluster size $N_c$ is required before the distribution of electronic states turns into a quasi-continuous one in the region around the Fermi level. To make this criterion qualitative, Kubo[6] has proposed that a cluster presents metallic behavior when the average level spacing becomes smaller than $k_B T$, or in terms more convenient for our analysis in Sect. II, when the density of electronic states $\mathcal{D}(E)$ at the Fermi level exceeds $1/k_B T$.

The problem of the metal-nonmetal transition in clusters has been recently studied by many authors, both experimentally and theoretically.[1-8] In this paper we address this question from the theoretical side by presenting a simple rectangular band model based on a tight-binding approximation[9] and the Kubo criterion for metallic behavior.[6] This model has been used earlier by Zhao and coworkers,[8] but the average coordination number of the atoms in the cluster, which is the key ingredient of the model, was overestimated by a poor approximation that does not take into account the precise cluster geometry; furthermore some numerical inaccuracies are present in that work.[8] Here we improve the model by using the exact values of the coordination number. The model and the results are presented in Sect. II and III respectively, and the conclusions in Sect. IV.

## II. Model

It is well known that in materials formed by transition metal atoms the quasi-localized $d$-band dominates the electronic properties,[9,10] so we can consider only the $d$ electrons. The simple rectangular model of the $d$-band, introduced by Friedel within a tight-binding framework,[9] can be extended to clusters of $N$ atoms, and the total electronic density of states $\mathcal{D}(E)$ can be expressed as

$$\mathcal{D}(E) = \begin{cases} 10N/W(N), & \text{if } -\tfrac{1}{2}W(N) \leq E - E_d \leq \tfrac{1}{2}W(N); \\ 0, & \text{otherwise.} \end{cases} \quad (1)$$

Here the factor 10 is the total number of electrons in a full $d$ shell, and the $E_d$ is the energy of the $d$-atomic level, which can be considered as the reference energy. The dependence of the band-width $W(N)$ on the local atomic environment is introduced by using the second-moment approximation,[11] and for a cluster with an average coordination $Z(N)$, the band-width is given by

$$W(N) = W_b \sqrt{\frac{Z(N)}{Z_b}}, \quad (2)$$

where $W_b$ and $Z_b$ are the band width and the coordination number in the bulk, respectively. The average coordination number will be evaluated exactly by making use of a recent study of direct enumerations in clusters.[12] This represents an improvement compared to other calculations[8] that use the following approximation for the coordination number[13]

$$Z(N) = \frac{Z_b(N-1)}{Z_b + (N-1)}. \quad (3)$$

In this approximation $Z$ depends only on the number of atoms of the cluster and take into account the cluster geometry. For instance for a face-centered cubic ($fcc$) cluster with cubo-octahedral shape and $N = 55$ the average coordination number obtained from Eq. (3) is 25% too high with respect to the exact value 7.85. Apart from the overestimation of the $Z(N)$, another serious problem is the smooth (structureless)

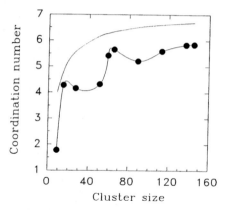

**Figure 1.** Comparison of the exact average coordination number for *bcc* spherical clusters (circles) and the approximated formula $Z(N) = Z_b(N-1)/[Z_b + (N-1)]$ with $Z_b = 8$ (dashed line).

behavior of $Z(N)$, which is illustrated in the Fig. 1 for the case of the spherical-like clusters with body-centered cubic (*bcc*) structure. Those clusters are constructed by starting with a central atom and adding successive coordination shells (with 8, 6, etc., atoms) around the central atom. In this case a non-monotonous increase of the $Z(N)$ can be expected arising from the formation of successive layers. The metal-nonmetal transition of transition metal clusters is reported experimentally to occur in the size range between 20 to 80 atoms and this is precisely the region where Eq. (3) gives a poor description of the average coordination number.

The insulator-metal transition takes place when the density of electronic states in the vecinity of the Fermi level is high enough so that the levels overlap by virtue of the thermal width.[6] Therefore in the case of atomic clusters the occurence of metallic behavior requiers the following inequality to be satisfied

$$\mathcal{D}(E_F) \geq \frac{1}{k_B T}. \tag{4}$$

Using Eq. (1) for the density of states, Eq. (4) can be written

$$\frac{10N}{W_b}\sqrt{\frac{Z_b}{Z(N)}} \geq \frac{1}{k_B T}, \tag{5}$$

which introduces the cluster size $N$ through the average coordination number $Z(N)$. The insulator-metal transition occurs for the critical size $N_c$ for which the equality is fulfilled. The form of Eq. (5) can be simplified by introducing the dimensionaless function of the temperature $G(T)$, defined as

$$\sqrt{G(T)} = \frac{1}{k_B T}\frac{W_b}{10\sqrt{Z_b}}. \tag{6}$$

Notice that the bulk band width $W_b$ and the coordination number $Z_b$ are fixed for a given metal. In this paper we take $W_b$ from Harrison.[14] Using Eq. (6) we can write the condition for metallic behavior

$$N^2 - Z(N)G(T) \geq 0. \tag{7}$$

We have considered clusters with two structural types: close-packing bcc and close-paking structures. For the *bcc* case we follow two different growth patterns: spherical clusters with 9, 15, 27, 51, 59, 65, 89, 113, 137, and 145 atoms, and the cube-type growth described by the series 9, 15, 27, 35, 59, 83, 91, 97, 121, and 145 atoms. The spherical clusters are formed by adding complete coordination shells around the central atom, that is, the 1st, 2nd, 3rd, 4th,..., neighbors of the atom. In the cube-type growth we also add atoms in order of increasing distance to the cluster's center (not complete coordination shells, however) in such a way that perfect cubes are obtained for 9, 35, 91, 187, atoms, etc. For the close-packing structures we consider fragments of a *fcc* lattice with cubo-octahedral (CO) shape and icosahedral (ICO) clusters that can be viewed as slightly deformed *fcc*-like structures.[12] The CO series is formed by clusters with 13, 19, 43, 55, 79, 87, 135, and 147 atoms and the ICO series by clusters with $N = 13, 43, 55, 75, 135$, and 147; in both cases we have perfect CO and ICO shapes for $N = 13, 55, 147, \ldots$. We stress that the close-packed cluster also grow shell-by-shell, in order of increasing distance to the center. We do not consider any other intermediate sizes except those given in the list above because the many possible sites for the atoms to be placed, a fact that complicates all evaluations of $Z(N)$. For a complete discussion on the geometrical structure of clusters, see Ref. 12.

Before presenting our results, let us comment on the results of Zhao and coworkers.[8] When $Z(N)$ from Eq. (3) is substituted into Eq. (7), one obtains

$$N^3 + (Z_b - 1)N^2 - Z_b(N-1)G(T) \geq 0, \tag{8}$$

and for $Z_b = 12$, the following approximate solution of this equation was proposed by Zhao et al.:[8]

$$N_c = [\,12G(T) + 171\,]^{1/2} - 6. \tag{9}$$

We call Zhao(1) to the results obtained from this approximation for $N_c$. On the other hand we can generalize Eq. (8), and write

$$N_c^3 + (\zeta - 1)N_c^2 - \zeta(N_c - 1)G(T) = 0, \tag{10}$$

where $\zeta$ does not necessarily coincide with the bulk coordination number $Z_b$ to allow the possibility that the cluster may have a different structure than the bulk [$Z_b$ is nevertheless, used in the evaluation of $G(T)$]. For instance $\zeta = 12$ is a CO Fe cluster, while $Z_b = 8$, since bulk Fe is *bcc*. The results obtained from exactly solving Eq. (10) will be called Zhao(2).

Equation (10) can be approximated, with an error of the order $1/N_c^2$, by the quadratic equation

$$N_c^2 + \zeta N_c - \zeta G(T) + \zeta - 1 = 0, \tag{11}$$

which has the solution

$$N_c = \sqrt{\zeta G(T) + \tfrac{1}{4}(\zeta - 2)^2} - \tfrac{1}{2}\zeta. \tag{12}$$

Even after this approximation there are differences with respect to Eq. (9). Those differences range between 5% and 17% depending on the temperature and structure.

## III. Results

To illustrate the predictions of Eq. (7) we present in Fig. 2 the behavior of $N^2 - Z(N)G(T)$ as a function of the cluster size for Fe clusters. Panel (a) corresponds to the *bcc*-spherical and panel (b) to the CO growth. The data points are for the clusters

## Table I

Critical cluster size for some transition metals at different temperatures, assuming close packing structures: cubo-octahedral (CO) and icosahedral (ICO). The results of Zhao and coworkers[8] are included for comparison.

| Structure | Temperature K | Fe | Co | Ni | Pd |
|---|---|---|---|---|---|
| CO | 44 | 135 | 95 | 83 | 120 |
|    | 77  | 75  | 55 | 43 | 66  |
|    | 110 | 50  | 31 | 27 | 43  |
| CO | 44 | 135 | 93 | 80 | 122 |
|    | 77  | 75  | 55 | 43 | 66  |
|    | 110 | 50  | 31 | 27 | 43  |
| fcc | Zhao(1) at 120 | 50 | 39 | 34 | 50 |
| fcc | Zhao(2) at 77  | 83 | 60 | 51 | 76 |

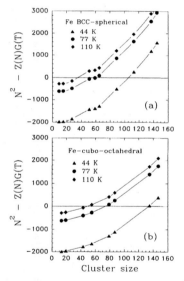

**Figure 2.** Size and temperature dependence of the function $N^2 - Z(N)G(T)$ for Fe clusters with spherical (*bcc*) and cubo-octahedral (*fcc*) geometries in (a) and (b), respectively.

studied. These have been connected by curves as an aid to the eye. We have not considered other intermediate sizes because of the many possibilities of placing atoms on the available sites. So, interpolation between the plotted points will be performed to calculate the critical size.

The results for the critical cluster size $N_c$ for three different temperatures (44 K, 77 K, and 110 K) are presented in Tables I and II. Table I contains the results for Fe, Co, Ni and Pd with close-packed CO and ICO structures, and the second one the results for the *bcc*-like Fe clusters. We observe that the critical cluster size decreases fast as the temperature increases. The difference in the critical sizes for CO and ICO

## Table II

Critical cluster size for Fe at different temperatures, assuming *bcc* structure. The results of Zhao's generalization are included for comparison.

| Structure | Temperature (K) | Fe |
|---|---|---|
| *bcc*-cube | 44 | 97 |
| | 77 | 59 |
| | 110 | 35 |
| *bcc*-spherical | 44 | 105 |
| | 77 | 59 |
| | 110 | 33 |
| *bcc*-Zhao(2) | 44 | 123 |
| | 77 | 69 |
| | 110 | 44 |

structures is negligible. Also, very small difference are found between the two *bcc* structures. However, the critical sizes for the Fe clusters with close-packed structures are larger than those for *bcc* structures. This can be understood from Eq. (5): $Z_b/Z(N)$ grows more slowly for the close-packed structures compared to *bcc* structures, so in the first case a larger cluster size is required in order to fulfill the inequality given by this equation. In Tables I and II, we have included the critical sizes calculated by Zhao et al.,[8] using Eq. (9) [called here Zhao(1)], and the corrected values called Zhao(2) (see Sect. II above). Zhao(2) overestimates the critical sizes with respect to our results by a factor 5% to 20% depending on the transition metal. On the other hand, the calculation performed by Zhao et al.[8] actually corresponds to 120 K, and not at 77 K that was the temperature reported by these authors. Consequently, Zhao(1) results should be compared with an extrapolation of our results to 120 K. Judging from our results at 110 K, we observe that Zhao(1) also overestimates the critical sizes.

Tunneling experiments probing the density of states near the Fermi level have been performed for Fe clusters supported on GaAs at room temperature.[1] The experiments indicate that the transition takes place for clusters with approximately 35 atoms. Although a *bcc* structure is reported by First[1] for large Fe clusters one cannot exclude the possibility that the small ones may present other structures.[15,16] Our results for the highest temperature studied (110 K) are consistent with those of First et al.[1] The model predicts the onset of metallic behavior to occur in the range 30–50 atoms. In the case of Pd clusters supported on an amorphous carbon sustrate, x-ray spectroscopy has been used to study the separation of the *d*-band from the Fermi level. The experiment indicates that the metal-nonmetal transition occurs for cluster radii in the range 7–10 Å (Ref. 5), while our model predicts a range 5–8 Å (40–120 atoms depending on the temperature).

It is important to mention that the critical size for the metal-nonmetal transition can also be estimated from ionization potential (IP) measurements. To that aim we replot the results of Parks and coworkers,[17] in Fig. 3(a). Metallic behavior in a cluster implies that the size variation of the ionization potential satisfies

$$\mathrm{IP}(R) = I_0 + \frac{C}{R}. \tag{13}$$

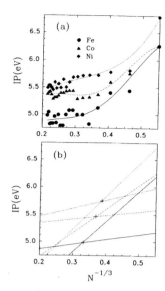

**Figure 3.** The ionization potential, as reported by Parks et al.,[17] as a function of the number of atoms is presented in (a). In (b) our estimation for the metal-nonmetal as the interception between the linear fits for large and small clusters and marked by a plus sign.

Here, $I_0$ is the bulk work function [(IP) of the macroscopic metal], $R$ is the effective radius of the cluster, and $C$ is a constant close to $\frac{3}{8}$ when atomic units are used. We can express Eq. (13) in terms of the number of atoms, since $R$ is proportional to $N^{1/3}$:

$$\text{IP}(R) = I_0 + \alpha N^{-1/3}, \qquad (14)$$

where the constant $\alpha$ depends now on the material. In practice the ionization potential of clusters of metallic elements follows this equation, except for small cluster sizes. We can identify the metal-nonmetal transition by the cluster size $N_c$ such that Eq. (14) provides a good fit to the experimental results for $N \geq N_c$. In fact, the experimetal data for large Fe, Co and Ni clusters plotted in Fig. 3(a) can be fitted by Eq. (14) (there is nevertheless, some scatter of the measured data about the fitted line). The fitted lines are plotted in Fig. 3(b), and extrapolated to $I_0 = 4.67$ eV, 5.17 eV, and 5.19 eV for Fe, Co and Ni respectively. These results are in good agreement with the experimental work functions:[21] 4.5 eV (Fe), 5.0 eV (Co), and 5.15 eV (Ni). The region of small cluster size deviates from Eq. (14), but can also be fitted by a linear function, with a different slope for each metal (clusters with $N \leq 6$ have been excluded from the fit, because of the large fluctuations of IP for these very small sizes). Those fits are also reported in Fig. 3(b), and we have identified the metal-nonmetal transition by the intercept between the fit to Eq. (13) and the fit for the small cluster region. The critical sizes estimated in this way are 17, 20 and 28 atoms, for Ni, Co, and Fe, respectively. The approximate cluster temperatures reported by Parks et al.,[17] are $225 \pm 50$ K. These results are to be compared with results of our model calculations at 225 K. These are 14 and 15 atoms for Ni and Co, respectively. In the case of Fe we obtain 21 atoms for fcc-like and 18 atoms for bcc-like clusters. The comparison is satisfactory, considering the uncertainties in our estimation of $N_c$ from measured IPs, as well as the approximations in the model.

## IV. Conclusions and Summary

We have studied the metal-nonmetal transition in clusters of transition metals, assuming that the $d$ electrons determine the electronic structure and using a simple rectangular band model. The band width has been calculated using the second-moment approximation, which makes the band width to depend on cluster size. The average coordination number required to determine the band width has been evaluated exactly from some model cluster structures. This contrasts with earlier work[8] where an analytical approximation was employed for the average coordination number. The calculated critical sizes depend stongly of the temperature, decreasing fast as the temperature increases. The dependence on the cluster structure is smaller, although one appreciates differences between close-packed and the $bcc$-type structures for the case of the Fe clusters (the precise structure of small Fe clusters is unknown). At liquid nitrogen temperature the predicted critical sizes are of the order of 60 atoms or lower, and this seems to be in rough agreement with experiment.

This work should be considered as a first step in a more rigorous study of the metal-nonmetal transition in transition metal clusters. It is our purpose to improve the description of the density of states by using a self-consistent tigh-binding method used recently in the calculation of ionization potentials of small Fe clusters.[16]

## Acknowledgments

We acknowledge to Dr. M. J. López for interesting discussions. This work has been supported by DGICYT Grants PB95-0720-C02-01 and SAB95-0390. Two of us (FAG and JMMC) also acknowledge to the Consejo Nacional de Ciencia y Tecnología Grants Ref. 961015 and 4920-E9406 (México).

## References

1. P. N. First et al., *Phys. Rev. Lett.* **63**, 1416 (1989).
2. K. Rademann, B. Kaiser, U. Even, and F. Hensel, *Phys. Rev. Lett.* **59**, 2319 (1987).
3. H. Haberland et al., *J. Chem. Soc. Faraday Trans.* **86**, 2473 (1990).
4. F. Yonezawa, S. Sakamoto, and F. Wooten, *J. Non-Cryst. Solids* **117-118**, 477 (1990).
5. G. K. Wertheim, *Z. Phys. D* **12**, 319 (1989).
6. R. Kubo, A. Kawabata, and S. Kobayashi, *Ann. Rev. Mater. Sci.* **14**, 49 (1984).
7. G. Pastor, P. Stampfli, and K. H. Bennemann, *Phys. Scripta* **38**, 623 (1988); *Europhys. Lett.* **7**, 419 (1988).
8. J. Zhao, X. Chen, and G. Wang, *Phys. Rev. B* **50**, 15 424 (1994).
9. J. Friedel, in: *The Physics of Metals*, edited by J. M. Ziman (Cambridge University Press, 1969) p. 340.
10. A. Vega, J. Dorantes-Dávila, L. C. Balbas, and G. M. Pastor, *Phys. Rev. B* **47**, 4742 (1993).
11. D. Tomanek, S. Mukherjee, and K. H. Bennemann, *Phys. Rev. B* **28**, 665 (1983); G. M. Pastor, J. Dorantes-Dávila, and K. H. Bennemann, *Chem. Phys. Lett.* **148**, 459 (1988).
12. J. M. Montejano-Carrizales, F. Aguilera-Granja, and J.L. Morán-López, *Proceedings of III Latin American Workshop on Magnetism, Magnetic Materials and*

*Their Applications*, edited by F. Leccabue and V. Sagredo, (World Scientific, Singapure, 1996) p. 320; *Nanostructured Mater.* **8**, 269 (1997).
13. B. N. Bhatt and T. M. Rice, *Phys. Rev. B* **20**, 466 (1979).
14. W. A. Harrison, *Electronic Structure and the Properties of Solids* (Freeman, San Francisco, 1980).
15. S. Yang and M. B. Knickelbein, *J. Chem. Phys.* **93**, 1533 (1990).
16. S. Bouarab, A. Vega, M. P. Iñiguez, and J. A. Alonso, *Phys. Rev. B* **54**, 3003 (1996).
17. E. K. Parks, T. D. Klots, and S. J. Riley, *J. Chem. Phys.* **92**, 3813 (1990).
18. M. M. Kappes and E. J. Schumacher, *Z. Phys. Chem. B* **156**, 23 (1988).
19. D. M. Wood, *Phys. Rev. Lett.* **46**, 749 (1981).
20. J. A. Alonso and N. H. March, *Electrons in Metals and Alloys* (Academic Press, London, 1989).
21. H. B. Michaelson, *J. Appl. Phys.* **48**, 4730 (1977).

# First-Principles Langevin Molecular Dynamics Studies of Metallic and Semiconductor Clusters

Luis Carlos Balbás

*Departamento de Física Teórica*
*Universidad de Valladolid*
*47011 Valladolid*
*SPAIN*

## Abstract

The combination of Langevin molecular dynamics for simulated annealing with realistic quantum-mechanical interactions obtained from first-principles supercell calculations (within the pseudopotential plane-wave method and the local density approximation) is applied in this paper to examine the structural and electronic properties of (i) bimetallic PbNa$_n$ ($n \leq 7$) clusters, and (ii) pure and mixed Ge$_n$Te$_m$ ($0 \leq n, m \leq 3$) clusters, as well as the diatomic molecules GeSe, PbSe and PbTe. In the case of bimetallic PbNa$_n$ clusters the aim is to explain the exceptional abundance of PbNa$_6$ observed in recent molecular beam experiments. It is found that adding another Na atom to PbNa$_6$ is energetically less favorable than adding it to a pure sodium cluster, in contrast to what it is obtained for smaller PbNa$_n$ clusters. In the case of semiconductor clusters the aim is to compute their permanent dipole moments in the equilibrium geometry and in selected geometries near to equilibrium corresponding to states higher in energy by a few tens of meV. These dipole moments are compared with those estimated from recent experiments at room temperature.

## I. Introduction

The structural determination of complex nonperiodical systems, such as clusters, is nowadays performed by means of first-principles molecular dynamics (FPMD),[1-6] which is based on realistic quantum-mechanical interactions obtained in the density functional theory (DFT) framework.[7,8] The dynamics of the system is then generated by integration of the motion equations through algorithms used in classical molecular dynamics simulations.[9] Car and Parrinello pioneered the ingenious approach of treating the parameters describing the electronic wave function as classical degrees of freedom and propagating them as such, instead of solving directly for the wave function at each step.[1] A limitation of this method is related to the use of a fictitious Lagrangian dynamics to describe the evolution of the electronic degrees of freedom which makes difficult the application of the method when energy transfer between electronic and ionic degrees of freedom occurs, as for example, in the simulation of metallic systems. By contrast, other workers[2-6] took the approach of solving the Kohn-Sham equations of DFT[8] for a new wave function each time step instead of performing fictitious dynamics on the wave functions. This approach is sometimes referred to as Born-Oppenheimer dynamics, since it ensures that the system remains on the Born-Oppenheimer potential energy surface throughout the propagation. Some review articles on the development of FPMD have appeared recently.[10-12]

A major difficulty of FPMD methods when applied to clusters is the existence of multiple local minima in the potential energy surface. For clusters sizes exceeding a few atoms, one generally relies on simulated annealing procedures[13] for global geometry optimization. Biswas and Hamman[14] implemented a Langevin molecular dynamics for simulated annealing, in which the system is coupled to an external bath. This means that, in addition to the internal forces acting on a particle, two terms were added: a friction force proportional to the velocity of the particle and a random force exhibiting a Gaussian distribution. The friction and the random force terms are not independent from each other but connected by the dissipation-fluctuation theorem.[15] The Langevin dynamics is a proper scheme to generate a *canonical ensemble* and can be exploited to perform *isothermal molecular dynamics simulations* as well as simulated annealing. It was combined for the first time with realistic DFT quantum-mechanical forces by Chelikowsky and coworkers[16] to examine small neutral and charged Si clusters. Applications to semiconductor[17,18,23] and metallic[19] clusters has been reported.

In this work I report on the application of first-principles Langevin molecular dynamics for simulated annealing to examine the structural and electronic properties of (i) bimetallic PbNa$_n$ ($n \leq 7$) clusters, and (ii) pure and mixed Ge$_n$Te$_m$ ($0 \leq n, m \leq 3$) semiconductor clusters. The structures presented were determined by the minimization of their total energy, in a process that combine simulated annealing followed by a conjugate gradient method.[20] For the smaller clusters I use directly the conjugate gradient minimization without any previous dynamical optimization. Results for the structure and dipole moments of the diatomic molecules GeSe, PbSe and PbTe are also presented.

In the case of bimetallic PbNa$_n$ clusters the aim is to explain the exceptional abundance of PbNa$_6$ observed in recent molecular beam experiments.[21] We have found previously[19] that adding another Na atom to PbNa$_6$ is energetically less favorable than adding it to a pure sodium cluster, in contrast to what it is obtained for smaller PbNa$_n$ clusters. I add here further details and also the dipole moments of these clusters are computed.

In the case of semiconductor clusters the aim is to compute their permanent dipole moments in the equilibrium geometry and in selected geometries near to equilibrium corresponding to states higher in energy by a few tens of meV. These dipole moments are compared with those estimated from recent experiments at room temperature.[22] Some of these results has been published previously.[23] For the dipole moment of diatomic molecules GeTe, GeSe, PbSe and PbTe it is obtained a good agreement with older experiments.[24] The trend shown by these dipole moments is rationalized in terms of the bond lengths and the interatomic charge transfer, which it is estimated using the atomic chemical hardness.[25]

In Section II it is given a brief account of the calculation method. In Section III the results are presented and discussed, including a previous motivation for each problem. Subsection III.1 is devoted to bimetallic PbNa$_n$ clusters. Subsection III.2 deals with the calculation of dipole moments of Ge$_n$Te$_m$ clusters. In Subsection III.3 the calculated dipole moment of some diatomic molecules is compared with experiments. In Section IV summary and conclusions are given.

## II. Method of Calculation

The method used in our investigation is the one introduced recently by Binggeli, Martins and Chelikowsky,[16] which combine Langevin molecular dynamics[14] with realistic quantum-mechanical interactions obtained in the local-density functional framework.[7,8] In this approach Langevin molecular dynamics is used for simulated annealing[13] and the atomic interactions are determined by means of efficient self-consistent pseudopotential plane-wave calculations. We use a fast iterative diagonalization procedure[26] and exploit the Broyden mixing scheme[27] to accelerate self-consistency. Metallic systems are conveniently handled by means of the Gaussian broadening scheme.[28] In this way it is possible to treat clusters which have occupied and empty orbitals that are degenerate or quasidegenerate, and properly include in the sampling metallic configurations which may occur in a cluster at $T \geq 0$.

In our simulations we use soft ionic pseudopotentials generated according to the procedure of Troullier and Martins[29] from the atomic ground state configurations [Ne]$3s$ for Na and [Xe]$6s^2(5f^{14}5d^{10})6p^2$ for Pb. The core radii for $3s$ and $3p$ orbitals of Na are 2.3 a.u. and 2.5 a.u. respectively and for both $6s$ and $6p$ orbitals of Pb the core radius is 3.18 a.u. A weighted average of the scalar relativistic potentials is used for Pb. For Ge and Te we assumed the atomic ground state configurations [Ar]$4s^2(3d^{10})4p^2$ and [Kr]$5s^2(4d^{10})5p^4$, respectively. The core radii for $4s$, $4p$, and (unoccupied) $4d$ orbitals of Ge are 2.5 a.u., 2.5 a.u., and 3.00 a.u., respectively. The $5s$, $5p$ and (unoccupied) $5d$ orbitals of Te have the same core radius 2.8 a.u. For Se we take the atomic ground state configuration [Ar]$4s^2(3d^{10})4p^4$ with the same core radius (1.99 a.u.) for $4s$, $4p$ and (unoccupied) $4d$ orbitals. The $p$-component is taken as local, and $s$-component nonlocality is treated via the Kleinman-Bylander construction.[30] We use the Ceperley-Alder exchange-correlation[31] as parametrized by Perdew and Zunger.[32] The partial-core correction for nonlinear exchange-correlation[33] has been included. The simulations for Na$_n$Pb clusters have been carried out in a body centered cubic supercell with an edge $a = 30$ a.u., and an energy cutoff of 9 Ry for the plane-wave set. For Ge$_n$Te$_m$ clusters we have taken a simple cubic supercell with an edge $a = 34$ a.u., and an energy cutoff of 14 Ry.

Only the $\Gamma$ point is considered for reciprocal space sampling, as appropriate for a cluster simulation, and we use an electronic level Gaussian broadening of 0.02 Ry. As a Langevin friction parameter we use $\gamma = 5 \times 10^{-4}$ a.u., which is similar to the value employed in the previous studies.[16] The equation of motion is integrated with a time step of 500 a.u. The Na$_n$Pb clusters were first heated to 500 K and then cooled to 10 K with four intermediate temperature values, each lasting 20 molecular dynamics steps. The Ge$_n$Te$_m$ clusters the initial temperature was 800 K. After this cooling process the cluster geometry is optimized using a conjugate gradient method.[20]

From the obtained ground state geometrical and electronic structures of all these clusters, we calculate the corresponding electric dipole moments. For a neutral cluster, the dipole moment $\mu$ is the sum of the ionic and electronic contributions,

$$\mu = \mu^{\text{ion}} + \mu^{\text{el}} = \sum_{i=1} Z_i \vec{R}_i - \int \vec{r} \rho(\vec{r}) \, d^3r, \tag{1}$$

$Z_i$ and $\vec{R}_i$ being the charge and the position of the $i$th core, and $\rho(\vec{r})$ is the electron density.

## III. Results and Discussion

### III.1 Bimetallic $Na_nPb$ Clusters

In a recent mass spectral study on the stability of bimetallic $Na_nPb$ clusters produced from the expansion of a mixed sodium/lead vapor, it was observed, under given experimental conditions, an intense molecular beam consisting almost exclusively of the cluster $Na_6Pb$ (Ref. 21). The physical origin of this striking stability has been studied by means of first-principles Langevin molecular dynamics calculations,[19] showing that formation of $Na_nPb$ with $n = 7$ is less favorable than formation of monoatomic sodium clusters of the same size, and that the exceptional stability of $Na_6Pb$ comes from a highly symmetrical configuration formed by a sixfold coordinated lead atom inside an octahedron with the six sodium atoms at the vertices, with an additional stabilization due to closed electronic shells. Similar conclusions have been obtained by means of Car-Parrinello type of calculations.[34]

With respect to bulk phases of intermetallic Na/Pb compounds the only first-principles calculation to our knowledge has been performed by Tegze and Hafner[35] for several structures of metallic and semiconducting alkali-metal/lead compounds, resulting in all cases that the electronic structure is dominated by the strong attractive Pb potential.

In figures 1 and 2 are shown the main results of our calculations for $PbNa_n$ ($1 \leq n \leq 7$) clusters. The ground-state structures are shown in Fig. 1. All the distances are given in angstroms with an estimated numerical accuracy of 0.01 Å. The average Pb-Na distance slowly decreases when the number of Na atoms increases and reaches the minimum value, 2.93 Å, for $Na_6Pb$. These distances are 7–15% smaller than the Na-Pb distance in the octet bulk compound $Na_{15}Pb_4$ (Ref. 35). Our calculated $Na_2Pb$ cluster has $C_{2v}$ symmetry with an apex angle of 75.5°, and a Na-Na distance of 3.70 Å, which is similar to the Na-Na distance in the equiatomic metallic NaPb compound.[35] The cluster $Na_2Pb$ can be considered as the building block of the remaining clusters. For example, $Na_3Pb$ can be viewed as formed by two $Na_2Pb$ structures (with apex angle 69.5°) sharing an edge, while the other edges forming an $Na_2Pb$ structure with apex angle 109° (which is nearly the exact $sp^3$ hybridization angle). The geometry of $Na_5Pb$ is a precursor of the $Na_6Pb$ structure, which has the very high symmetrical octahedral shape ($O_h$) with Na-Na distance of 4.14 Å, which is about 6% larger than those found in the smaller clusters. The form of $Na_7Pb$ resembles the $D_{4d}$ symmetry with one Na atom missing on the upper face. The average Na-Pb and Na-Na distances are 3.0 Å and 3.75 Å respectively.

In the lower part of Fig. 1 are shown contour and surface plots of the total pseudo-charge density of $Na_4Pb$, $Na_5Pb$ and $Na_6Pb$, as well as the binding charge density of $Na_6Pb$ defined as the difference between the total charge and the superposition of atomic charge densities. The almost spherical aspect of the total pseudo-charge density of $Na_6Pb$ reflects the delocalized nature of the bond and justifies an analysis

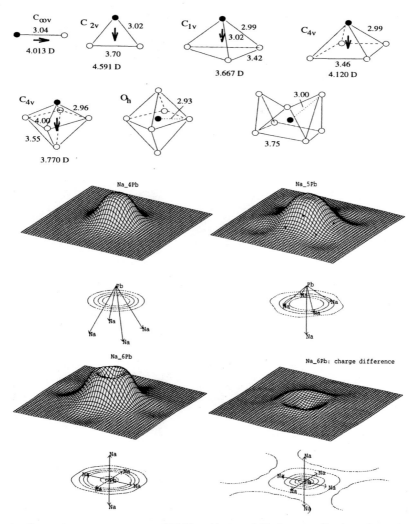

**Figure 1.** Ground-state structures of PbNa$_n$ ($1 \leq n \leq 7$) clusters. Dark and empty spheres represent Pb and Na atoms respectively. Interatomic distances are in angstroms and point group symmetries are indicated as well as the calculated dipole moments (in Debye). Below are shown surface and contour plots of the total electronic charge distribution of Na$_4$Pb, Na$_5$Pb, Na$_6$Pb and of the difference between the total charge distribution in Na$_6$Pb and the superposition of atomic charge densities. There are four contourlines, corresponding to 2000, 1500, 1000 and 500 a.u., from outside to inside.

of the orbitals in a "two steps" spherical jellium model.[36] The depletion of the pseudo-charge at the Pb site is a consequence of the absence of core states in pseudopotential calculations. The binding charge density shows the metallic character of the bond with delocalized charge accumulation in the region between the lead and the atomic shell of sodium atoms.

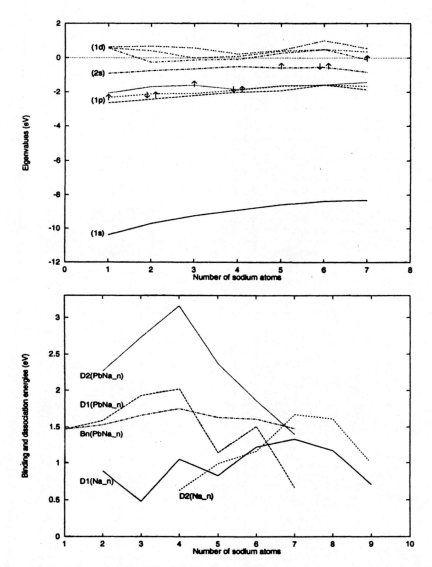

**Figure 2.** Upper: Energies (in eV) of Kohn-Sham orbitals for structurally optimized $Na_nPb$ clusters. Up and down arrows are placed at the uppermost occupied orbital. The lines are drawn only to guide the eye. Labels $(1s)$, $(1p)$, $(2s)$, and $(1d)$ make reference to typical shells in a spherical model. Lower: Binding and dissociation energies (in eV) for the structurally optimized $Na_nPb$ and $Na_n$ clusters (see text for details). $B_n$ is the atomization energy per number of Na atoms $D1_n = E(PbNa_{n-1}) + E(Na) - E(PbNa_n)$ and $D2_n = E(PbNa_{n-2}) + E(Na_2) - E(PbNa_n)$ are the energies for the fragmentation into a smaller Pb containing cluster and a sodium monomer or dimer respectively. Similar definitions hold for pure sodium clusters.

The calculated one-electron energy levels diagram are represented in the left panel of Fig. 2 using the potential in the interstitial sites as the vacuum level. The lowest eigenvalue (originating from the Pb 6s level) is well separated from the higher lying ones. The opening of a large gap between the s and p manifolds of Pb is due to the very attractive s part of the pseudopotential, which is a reflection of the relativistic corrections for Pb s core electrons. We can compare the levels diagram of Fig. 2 with the density of states for the octet bulk compounds $Na_{15}Pb_4$ obtained in Ref. 35 a very narrow band of pure Pb s character (containing just two electrons per lead atom) 8 eV below the Fermi level, and a band about 2.5–3.0 eV wide just below the Fermi level (and extending slightly above it). This band contains six electrons per lead atom. It has pure p character on the Pb sites and a completely mixed character on the alkali-metal sites. We see in Fig. 2 that the p levels are separated from the lowest s eigenvalue by about 8 eV for NaPb decreasing to about 6 eV for $Na_7Pb$; these p levels spread less than 1.0 eV and become degenerate for $Na_6Pb$ as a result of the $O_h$ symmetry. In the case of $Na_6Pb$ the occupied molecular orbitals, corresponding to $O_h$ symmetry, are $a_{1g}, t_{1u}, a_{1g}$ and the next unoccupied orbital, the doublet $e_g$, is not bound. After the p type levels we can distinguish an isolated level, which is filled in $Na_6Pb$, and after this well defined level appears a group of slightly bound or unbounds levels. The HOMO-LUMO gap is less than (or about) 1.0 eV, and the LUMO appears unbound for $Na_6Pb$ and $Na_7Pb$ clusters (which is probably an artifact of the LDA).

In the right panel of Fig. 2 are represented several quantities related to the calculated total energies $E(PbNa_n)$ of $PbNa_n$ clusters, namely, the binding energies per sodium atom, $B_n = [E(Pb) + nE(Na) - E(PbNa_n)]/n$ and the energies $D1_n = E(PbNa_{n-1}) + E(Na) - E(PbNa_n)$ and $D2_n = E(PbNa_{n-2}) + E(Na_2) - E(PbNa_n)$ for the fragmentation into a sodium monomer and dimer, respectively. For the sake of comparison, the values of $D1_n$ and $D2_n$ for pure $Na_n$ clusters, taken from the *ab initio* calculations of Ref. 37, are also shown. Notice the enhanced stability of $Na_6Pb$ relative to their neighbor clusters $Na_5Pb$ and $Na_7Pb$. The monomer dissociation channel is always the lowest for $Na_nPb$ clusters, whereas the dimer channel is preferred for $Na_5$ and $n \geq 7$ pure $Na_n$ clusters.

The stability of $Na_6Pb$ can be explained by the jellium model which is quite successful in predicting stability of alkali metal clusters by the closing of spherical electronic shells. For alkali clusters $A_n$ the ordering of the shells is[38] $1s, 1p, 1d, 2s,...$, resulting in increased stabilities[39] of $A_8, A_{20},...$ clusters with respect to the next in the series. Due to the strong attractive pseudopotential for the s electron of Pb one can not consider a constant jellium density, but assume a potential that is more attractive around the Pb atom. For the $Na_nPb$ clusters the result is that a $1s$ level is strongly split from the others and the ordering is $1s, 1p, 2s, 1d,...$ resulting in shell closings for $Na_4Pb$ and $Na_6Pb$. These clusters also display the most symmetrical forms, in particular $Na_6Pb$ has an octahedral shape ($O_h$ symmetry) with filled molecular orbitals $a_{1g}, t_{1u}, a_{2g}$, (the doublet state $e_g$ is the first unoccupied orbital) and high symmetry is often related to high stability (as in $C_{60}$).

In the early stages of growth process of Na-Pb clusters in a Na rich vapor, the most likely growth process is the addition of a Na monomer or dimer to an existing cluster. The energetic gains of adding an Na atom to a $Na_m$ cluster is around 1.0 to 1.2 eV for $m \approx 4,...,10$ according to LDA calculations.[37] The binding energy per atom in the Na solid is 1.30 eV according to LDA, which overestimates experiment by 15–20%. From the values of $D1$ in Fig. 2 we see that adding one Na atom to a $PbNa_n$ cluster results for $n \leq 6$ in a larger gain in binding energy than adding it to a $Na_m$ cluster. However the binding energy gain for the calculated reaction $Na_6Pb + Na \rightarrow Na_7Pb$ of 0.67 eV is approximatively half the value for a reaction $Na_m + Na \rightarrow Na_{m+1}$. On the other hand, although $Na_4Pb$ seems to be even more stable than $Na_6Pb$, the addition of Na atoms to the $Na_4Pb$ cluster is still energetically more favorable than addition to pure Na clusters. We propose that this is the reason why under certain growth conditions the $Na_nPb$ growth process seems to stop at $Na_6Pb$.

## III.2 Pure and Mixed $Ge_nTe_m$ Semiconductor Clusters

The interplay between the electronic and the geometrical structure of a cluster determines many of its physical and chemical properties. For example, it has been shown that the comparison of the calculated photoabsorption cross sections of small silicon clusters with the measured photoabsorption spectra would serve as a useful tool to discern the structure of small semiconductor clusters.[40] One molecular electronic property which depends critically on the underlying geometrical structure is the electric dipolar moment, as was illustrated in Ref. 37 for the case of the $Na_3$ molecule. With respect to germanium and tellurium semiconductor clusters, the recent measurements reported in Ref. 22 indicate the existence of a permanent dipole moment for the mixed $Ge_2Te$ and $Ge_2Te_2$ clusters, whereas no evidence of dipole moment is obtained for $GeTe_2$, pointing to a linear structure of this molecule.

There are a few *ab initio* calculations of the electronic and geometrical structure of small pure germanium[41] and pure tellurium[42,43] clusters, but we are not aware of similar calculations for mixed $Ge_nTe_m$ clusters. In this subsection we report on the structural and electronic properties of the small molecules $Ge_2Te$, $GeTe_2$, $Ge_2Te_2$, $Ge_3$, and $Te_3$ obtained by means of first-principles supercell calculations, using the local density approximation and the pseudopotential plane wave method. The cluster geometry is optimized using a conjugate gradient method. In order to compare with the dipole moments estimated in experiments for these molecules, we calculate their permanent dipole moments in the ground state and in the geometries near to equilibrium corresponding to states higher in energy by a few tens of meV. The results presented for some of the mixed clusters has been published in a previous work.[23]

The calculated equilibrium geometries and dipole moments of $Ge_2Te$, $GeTe_2$, $Ge_2Te_2$ are shown in the upper part (A) of Fig. 3, and in the lower part (B) is shown, for each of the $Ge_2Te$ and $Ge_2Te_2$ clusters, one selected geometry whose total energy is a few tens meV higher than the equilibrium one. In the upper left corner of Fig. 3 is indicated the coordinates frame used for the calculation of the dipole moments, whose values (in Debye) are given at the right of each geometry. The thick arrow points in the direction of the dipole moment. The contours of the electronic density of some of these clusters are shown at the right.

The equilibrium state of the $GeTe_2$ cluster has $D_{\infty h}$ symmetry and the calculated dipole moment is less than $10^{-6}$ Debye, which serves as a measure of the numerical accuracy of our computed dipole moments. The fact that $GeTe_2$ prefers a linear structure to bent configurations, confirms the experimental conjecture[22] based on the null value estimated for the effective dipole moment.

For the equilibrium geometry of $Ge_2Te$ we obtain an equilateral triangle with interatomic distance of 2.52 Å. The calculated dipole moment is 0.31 Debye, which is close to the measured effective dipole moment,[22] $0.31 \pm 0.03$ Debye, but it is smaller that the permanent dipole moment of 1.4 Debye estimated in Ref. 22 by means of a classical analysis. A value close to the permanent dipole moment is obtained for the $C_{2v}$ geometry represented in part (B) of Fig. 3. This geometry has the Ge-Te and Ge-Ge bond lengths slightly larger and shorter, respectively, than for the equilibrium one, and their total energy is 637 meV higher. The increase of the dipole moment due to a larger Ge-Te bond length is further enhanced by the shorter bond length between

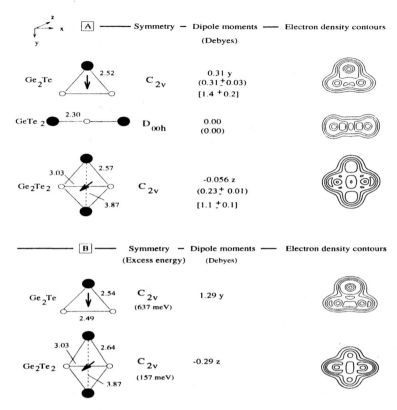

**Figure 3.** Equilibrium (A) and selected isomeric (B) structures of several Ge/Te mixed clusters. Interatomic distances are in angstroms and point group symmetries are indicated. The excess energy (in meV) of each isomeric structure is given in part (B) below the symmetry symbol. Thick arrows indicate the direction of the dipole moments and the calculated values (in Debye) are given at the right of each symmetry. Effective and permanent dipole moments estimated from experiments are given in parenthesis and square brackets, respectively, below the calculated moments. Calculated and experimental dipole moments (in Debye) are also shown. Contours of the electronic density are shown at the right. From outside to inside there are four contourlines corresponding to increasing density from 500 a.u. to 2000 a.u. in steps of 500 a.u. From the fourth line to inside the order is reversed. The plane for the contours of $Ge_2Te_2$ is parallel to Ge-Ge and Te-Te bonds and midway between them.

the Ge atoms. This effects can be better understood by comparing the contours of the electronic density represented in the right part of Fig. 3. It can be observed that the Ge-Ge bond is reinforced in the excited geometry and, simultaneously, the electronic charge around the Te atom becomes more localized. This facts help to the increase of the dipole moment with respect to the equilibrium one.

The equilibrium geometry of $Ge_2Te_2$ shows a butterfly shape (deformed tetrahedrom with $C_{2v}$ symmetry). The angle Ge-Te-Ge is 72.4°, the angle Te-Ge-Te is 98°, and the angle between the two triangular wings sharing the Ge-Ge base is 138.5°. The calculated dipole moment is 0.06 Debye, which is very small compared to the experimental value.[22] For the $C_{2v}$ geometry of $Ge_2Te_2$ shown in part (B) of Fig. 3,

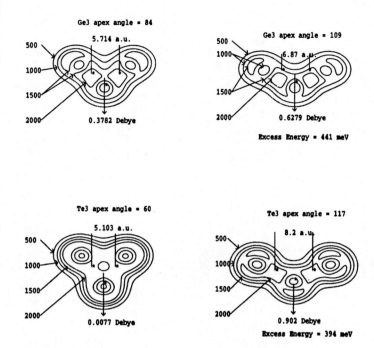

**Figure 4.** Contour density of the equilibrium (left) and isomeric structures (right) of Ge$_3$ and Te$_3$.

which have the same Ge-Ge and Te-Te distances than the equilibrium geometry, but larger Ge-Te bond length (the angle between the two triangular wings sharing the Ge-Ge base is 94.3°), the total energy is 157 meV higher than the equilibrium one, and the dipole moment is 0.29 Debye. This value is comparable to the measured effective moment, but still smaller than the experimental permanent dipole moment.[22] An explanation for this enlarged dipole moment can be given by comparing the contour charge densities, similarly to the case of Ge$_2$Te discussed above.

Our results for the trimers Ge$_3$ and Te$_3$ are shown in Fig. 4. The equilibrium geometry of Ge$_3$ is of the C$_{2v}$ type with apex angle of 84° and the basis Ge-Ge distance is 3.023 Å. Another local minimum is obtained at 441 meV higher energy with apex angle of 109.25° and larger Ge-Ge distance of 3.634 Å. These results can be compared with those obtained for several geometries of Ge$_3$ in Ref. 41 by means of *ab initio* molecular orbital computations using Gaussian-type basis sets and different levels of approximation. At the SCF level is obtained in Ref. 41 that the structure C$_{2v}$ with apex angle of 88° and basis Ge-Ge distance of 3.24 Å is 66 meV more bound than that with 109.5° and 3.76 Å, but, at the MP4 level, the reverse order is obtained, which illustrates the importance of correlation for Ge clusters. The dipole moments for the geometries of Ge$_3$ shown in Fig. 4 are 0.378 Debye and 0.628 Debye, respectively.

The equilibrium geometry of Te$_3$ is of the D$_{3h}$ type with bond length of 2.70 Å. Another local minimum with C$_{2v}$ symmetry is obtained at 394 meV higher energy with apex angle of 117° and basis Te-Te distance of 4.34 Å. This results can be compared with those of Ref. 42 obtained by means of *ab initio* molecular orbital computations using Gaussian-type basis sets and different levels of approximation. At the CASSCF

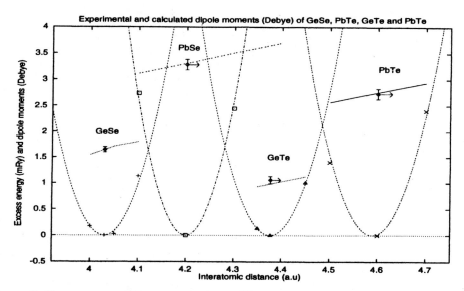

**Figure 5.** Excess energy with respect to the equilibrium state and calculated dipole moments for GeSe, PbSe, GeTe, and PbTe molecules at different bond-lengths. The calculated excess energy values are fitted to parabolic curves. The experimental dipole moments, taken from Ref. 24, are denoted by diamonds, and the attached horizontal arrows indicate that the experimentally unknown equilibrium bond length will be probably larger than the calculated ones.

level is obtained in Ref. 42 that the structure $D_{3h}$ with Te-Te distance of 2.83 Å is 100 meV more bound that that the structure $C_{2v}$ with apex angle of 113.2° and 3.20 Å for the basis. At the MRSDCI+Q level the reverse order is obtained, illustrating again the importance of correlation effects. The calculated dipole moments for the geometries of $Te_3$, shown in Fig. 4, are 0.008 Debye and 0.902 Debye, respectively, showing the sensitivity of the dipole moments to the molecular geometry.

### III.3 Diatomic Semiconductor Molecules

The energy with respect to the equilibrium energy (which is taken as the origin) and the dipole moments (in Debye) of GeSe, PbSe, GeTe and PbTe molecules are shown in Fig. 5 as a function of the interatomic distance.

To our knowledge there are not available measurements of the equilibrium bond lengths of these molecules. From the known reduced mass,[44] $\mu$, and the experimental frequencies,[24] $\nu_{\exp}$, for the transition between rotational levels of the fundamental band, we can use the simple formula $r = \sqrt{(J+1)/\nu_{\exp}\mu}$ to estimate the bond lengths. These estimations yields 4.038 a.u, 4.566 a.u, and 4.938 a.u. for GeSe, PbSe, and PbTe, respectively, to be compared with our calculated values 4.03 a.u., 4.20 a.u., and 4.60 a.u. With the exception of GeSe our calculated equilibrium bond lengths are smaller than these estimations. The underestimation of bond lengths is a well known drawback of the local density approximation.

We can observe in Fig. 5 very simple correlations between pairs of molecules having a common atom. The increase of bond distance between XY and XZ can be attached to the larger atomic size of Z with respect to Y (which is also reflected in the respective pseudopotential core radii). The difference of dipole moments between XY and XZ is the consequence of two effects, the different bond distances and different interatomic charge transfer. To estimate that charge transfer we can use the simple formula[25] $Q = 2(\eta_1 - \eta_2)/(\eta_1 + \eta_2)$ where $\eta$ is the experimental atomic hardness, $\eta = 0.5(I - A)/(I + A)$, with $I$ and $A$ the experimental ionization potential and electron affinity respectively.[45,46] In this way we obtain for $Q$ of GeTe, GeSe, PbTe and PbSe the values 0.23 a.u., 0.40 a.u., 0.35 a.u., and 0.36 a.u. respectively. These values, together with the bond lengths, qualitatively explain the trend of the dipole moments observed in Fig. 5.

## IV. Summary and Conclusions

The combination of Langevin molecular dynamics for simulated annealing with realistic quantum-mechanical interactions obtained from first-principles supercell calculations (within the pseudopotential plane-wave method and the local density approximation) is probed in this paper to be very useful in determining the properties of metallic and semiconductor clusters. We have optimized using that method the structures of $Na_n Pb$ clusters. Our numericals results explain the special stability observed for $Na_6 Pb$ as due to the combination of two related factors, one structural, that is, highly symmetric $O_h$ geometry, and other electronic, namely the closing of the $O_h$ molecular orbitals $a_{1g}$, $t_{1u}$, $a_{2g}$ (correspondingly the electronic shell closing of the spherical jellium model type orbitals $1s^2 1p^6 2s^2$). The absence of $PbNa_7$ signal in the experimental mass spectral distribution can be attached to the low energy gain by adding Na atoms to $Na_6 Pb$ clusters compared to the energy gain in pure $Na_n$ clusters.

With respect to semiconductor molecules we have optimized the structures of $Ge_2 Te$, $GeTe_2$, $Ge_2 Te_2$, $Ge_3$, $Te_3$, GeTe, GeSe, PbTe and PbSe. For $GeTe_2$ we confirm the linear equilibrium structure suggested by experiments.[22] The equilibrium structures of $Ge_2 Te$ and $Ge_2 Te_2$ results to be an equilateral triangle and a distorted rhombus (butterfly $C_{2v}$ symmetry) respectively. We have computed the dipole moment of these clusters in the equilibrium geometry and in selected geometries near to equilibrium corresponding to states higher in energy by several tens meV. For that selected excited states, whose geometries have larger Ge-Te bond lengths than the equilibrium ones, the computed dipole moments are in fair agreement with experimental measurements at room temperature.[22] For $Ge_3$ and $Te_3$ we obtain equilibrium and excited structures in agreement with previous determinations using quantum chemistry type *ab initio* methods.[41,42] There are not available experimental dipole moments for these trimers. With respect to the dipole moments of several semiconductor diatomic molecules our calculations are in good agreement with experiments.

The determination of molecular dipole moments is a very stringent test for all type of *ab initio* methods. In our case, the overall underestimation of the calculated permanent dipole moments with respect to experiments, could be attached to the well known failure of the local density approximation in reproducing the molecular bond lengths. For example, the bond length calculated for the $Ge_2$ dimer is 2.33 Å, to be compared with the experimental value of 2.43 Å. We have shown that by slightly increasing the bond-lengths of our cluster geometries, it is obtained an enhancement of the computed dipole moments, improving the agreement with experiments.

## Acknowledgments

I acknowledge financial support of DGICYT of Spain (Grants PB95-0720-C02-01 and PB95-0202) and of Junta de Castilla y León.

## References

* I wish to dedicate this work to Professor Karl-Heinz Bennemann for the many physics and other important matters that I have learned from him.
1. R. Car and M. Parrinello, *Phys. Rev. Lett.* **55**, 2471 (1985).
2. R. M. Wentzcovitch and J. L. Martins, *Solid State Commun.* **78**, 831 (1991).
3. R. N. Barnett, U. Landman, A. Nitzan, and G. Rajagopal, *J. Chem. Phys.* **94**, 608 (1991).
4. T. Arias *et al.*, *Phys. Rev. B* **45**, 1538 (1992).
5. D. M. Bylander and L. Kleinman, *Phys. Rev. B* **45**, 9663 (1992).
6. G. Kresse and J. Hafner, *J. Non-Cryst. Solids* **156-158**, 956 (1993).
7. P. Hohenberg and W. Kohn, *Phys. Rev. B* **136**, 864 (1964).
8. W. Kohn and L. J. Sham, *Phys. Rev. A* **140**, 1133 (1965); L. J. Sham and W. Kohn, *Phys. Rev. B* **145**, 561 (1966).
9. M. P. Allen and D. J. Tildesley, *Computer Simulation of Liquids* (Oxford University Press, Oxford, 1987).
10. M. C. Payne *et al.*, *Rev. Mod. Phys.* **64**, 1045 (1992).
11. D. K. Remler and P. A. Madden, *Mol. Phys.* **70**, 921 (1990).
12. G. Galli, *J. Phys. Condens. Matter* **5**, B107 (1993).
13. S. Kirkpatrick, C. D. Gelatt, and M. P. Vecchi, *Science* **220**, 671 (1983).
14. R. Biswas and D. R. Hamann, *Phys. Rev. B* **34**, 895 (1986).
15. R. Kubo, *Rep. Prog. Theor. Phys.* **29**, 255 (1996); S. A. Adelman and J. Doll, *J. Chem. Phys.* **64**, 2375 (1976).
16. N. Binggeli, J. L. Martins, and J. R. Chelikowsky, *Phys. Rev. Lett.* **68**, 2956 (1992).
17. N. Binggeli and J. R. Chelikowsky, *Phys. Rev. B* **50**, 11 764 (1994).
18. J. R. Chelikowsky, N. Troullier, and N. Binggeli, *Phys. Rev. B* **49**, 114 (1994).
19. L. C. Balbás and J. L. Martins, *Phys. Rev. B* **54**, 2937 (1996).
20. W. C. Davidon, *Math. Progr.* **9**, 1 (1975).
21. C. Yeretzian, U. Röthlisberger, and E. Schumacher, *Chem. Phys. Lett.* **237**, 334 (1995).
22. R. Schäfer, S. Schlecht, J. Woenckhaus, and J. A. Becker, *Phys. Rev. Lett.* **76**, 471 (1996).
23. L. C. Balbás, A. Rubio, and J. L. Martins, *Z. Phys. D* **40**, 182 (1997).
24. J. Hoeft, H. J. Lovas, E. Tiemann, and T. Törring, *Z. Naturforsch.* **25a**, 539 (1970).
25. J. A. Alonso and L. C. Balbás, in *Structure and Bonding*, edited by K. D. Sen, (Springer-Verlag, Berlin, 1987), Vol. 66, p. 41.
26. J. L. Martins and M. L. Cohen, *Phys. Rev. B* **37**, 6134 (1988).
27. C. G. Broyden, *Math. Comp.* **19**, 577 (1965).
28. C. L. Fu and K. M. Ho, *Phys. Rev.* **28**, 5480 (1983).
29. N. Troullier and J. L. Martins, *Phys. Rev. B* **43**, 1993 (1991).
30. L. Kleinman and D. M. Bylander, *Phys. Rev. Lett.* **48**, 1425 (1982).
31. D. M. Ceperley and B. J. Adler, *Phys. Rev. Lett.* **45**, 566 (1980).
32. J. Perdew and A. Zunger, *Phys. Rev. B* **23**, 5048 (1981).

33. S. G. Louie, S. Froyen, and M. L. Cohen, *Phys. Rev. B* **26**, 1738 (1982).
34. J. Chang, M. J. Stott, and J. A. Alonso, *J. Chem. Phys.* **104**, 8043 (1996).
35. M. Tegze and J. Hafner, *Phys. Rev. B* **39**, 8263 (1989).
36. C. Baladrón and J. A. Alonso, *Physica B* **154**, 73 (1978); C. Yannouleas, P. Jena, and S. N. Khana, *Phys. Rev. B* **46**, 9751 (1992).
37. I. Moullet, J. L. Martins, F. Reuse, and J. Buttet, *Phys. Rev. B* **42**, 11 598 (1990).
38. J. L. Martins, R. Car, and J. Buttet, *Surf. Sci.* **106**, 265 (1981).
39. W. D. Knight et al., *Phys. Rev. Lett.* **31** 1804 (1985).
40. A. Rubio et al., *Phys. Rev. Lett.* **77**, 247 (1996).
41. M. S. Islam and A. K. Ray, *Chem. Phys. Lett.* **153**, 496 (1988).
42. K. Balasubramanian and D. Dai, *J. Chem. Phys.* **99**, 5239 (1995).
43. G. Igel-Mann, R. Schlunk, and H. Stoll, *Mol. Phys.* **80**, 341 (1993).
44. G. Herzberg, *Molecular Spectra and Molecular Structure. Vol. I: Spectra of Diatomic Molecules*, Second Edition (D. van Nostrand Company, Inc. New York, 1965).
45. *Handbook of Chemistry and Physics*, 58th edition (CRC Press, Boca Raton, 1977).
46. For Te we take the recent experimental value of its electron affinity reported by G. Haeffler, A. E. Klinkmüller, J. Rangel, U. Berzinsh, and D. Hanstorp, *Z. Phys. D* **38**, 211 (1996).

# Theoretical Study of the Collective Electronic Excitations of the Endohedral Clusters Na$_N$@C$_{780}$

J. M. Cabrera-Trujillo,[1] R. Pis-Diez,[2] J. A. Alonso,[3] M. J. López,[3] M. P. Iñiguez,[3] and A. Rubio[3]

[1] Facultad de Ciencias
Universidad Autónoma de San Luis Potosí
78000 San Luis Potosí, S.L.P.
MEXICO

[2] QUINOR
Universidad Nacional de La Plata
1900 La Plata
ARGENTINA

[3] Departamento de Física Teórica
Universidad de Valladolid
47011 Valladolid
SPAIN

## Abstract

The low energy part of the photoabsorption spectrum of the two layer fullerene C$_{780}$ and the endohedrals Na$_{20}$@C$_{780}$ and Na$_{25}$@C$_{780}$ has been calculated using time-dependent density functional theory. The main feature in C$_{780}$ is a collective resonance at 6.60 eV that we interpret as a precursor of the $\pi$ plasmon of graphite. The presence of the endohedral Na clusters shifts this resonance by 0.2 eV to higher energies. A small feature also appears at 3 eV, associated to the well known collective resonance of free sodium clusters.

## I. Introduction

The collective electronic response of C$_{60}$ has been studied in great detail by applying different experimental[1] and theoretical techniques.[2] Similar studies have been performed for more complex fullerene materials like carbon multishells,[3-5] fullerenes coated by metal layers[6,7] and the endohedrals A@$C_{60}$, where A indicates an atom.[8,9] Recently, large fullerenes encapsulating clusters and nanocrystals have been produced[10] and it becomes interesting to study the changes in the dynamical response of a large hollow fullerene when a metallic particle of medium size is inside the cage. Here we report calculations for $Na_{20}$ and $Na_{25}$ inside the giant fullerene $C_{780}$. This is a cage formed by two concentric shells having 240 and 540 carbon atoms respectively. The calculations use the time dependent density functional theory in its local density version (TDLDA).[11,12] Since free sodium clusters have a collective resonance at about 3 eV, we only consider the $\pi$-electrons of the fullerene (the $\pi$ plasmon of $C_{60}$ occurs at 6 eV). Sodium and carbon atoms are described by pseudopotentials, and following early work on related systems[5,6,9,13] we use the spherically averaged pseudopotentials (SAPS) model.[14]

## II. Theoretical Background

If the endohedral onion is placed in an oscillatory external potential $V_{ex}(\mathbf{r},\omega)e^{-i\omega t}$, then the electron density changes, and this change can be described in linear response[11,12] as an induced density $\delta\rho(\mathbf{r},\omega)e^{-i\omega t}$ given by

$$\delta\rho(\mathbf{r},\omega) = \int \chi(\mathbf{r},\mathbf{r}',\omega) V_{ex}(\mathbf{r}',\omega)\, d\mathbf{r}'. \qquad (1)$$

Here, $\chi(\mathbf{r},\mathbf{r}',\omega)$ is the interacting electron susceptibility and $\rho(\mathbf{r})$ is understood to include only the valence electrons of the endohedral sodium cluster, and the $\pi$- electrons of the fullerene (one per atom). Since DFT is formulated as a single-particle-like theory in which all the many-body effects enter through an effective potential, common to all the electrons of the system, the induced density can also be written

$$\delta\rho(\mathbf{r},\omega) = \int \chi^0(\mathbf{r},\mathbf{r}',\omega) V_{\text{eff}}(\mathbf{r}',\omega)\, d\mathbf{r}', \qquad (2)$$

where $\chi^0$ is the independent-particle susceptibility and

$$V_{\text{eff}}(\mathbf{r}',\omega) = V_{ex}(\mathbf{r}',\omega) + \int K(\mathbf{r}',\mathbf{r}'')\delta\rho(\mathbf{r}'',\omega)\, d\mathbf{r}'', \qquad (3)$$

$$K(\mathbf{r}',\mathbf{r}'') = \frac{1}{|\mathbf{r}'-\mathbf{r}''|} + \frac{\delta V_{\text{xc}}(\mathbf{r}')}{\delta\rho(\mathbf{r}'')}\delta(|\mathbf{r}'-\mathbf{r}''|). \qquad (4)$$

In other words, Eq. (2) indicates that the electrons respond as independent particles to the potential $V_{\text{eff}}$, which is the self-consistent effective potential made up of the external potential and the induced Coulomb and exchange-correlation (xc) potentials. Eq. (4) shows that the kernel $K(\mathbf{r}',\mathbf{r}'')$ is the functional derivative of the effective potential. From the above equations one arrives at

$$\chi(\mathbf{r},\mathbf{r}',\omega) = \chi^0(\mathbf{r},\mathbf{r}',\omega) + \iint \chi^0(\mathbf{r},\mathbf{r}_1,\omega) K(\mathbf{r}_1,\mathbf{r}_2) \chi(\mathbf{r}_2,\mathbf{r}',\omega)\, d\mathbf{r}_1\, d\mathbf{r}_2, \qquad (5)$$

which is a self-consistent Dyson-type equation for $\chi$. The non-interacting susceptibility $\chi^0(\mathbf{r},\mathbf{r}',\omega)$ is calculated as follows:[12]

$$\chi^0(\mathbf{r},\mathbf{r}',\omega) = \sum_i^{\text{occ}} \phi_i^*(\mathbf{r})\phi_i(\mathbf{r}')\, G(\mathbf{r},\mathbf{r}',\epsilon_i+\hbar\omega) + \sum_i^{\text{occ}} \phi_i(\mathbf{r})\phi_i^*(\mathbf{r}')\, G(\mathbf{r},\mathbf{r}',\epsilon_i-\hbar\omega), \qquad (6)$$

where the sum is over all occupied single-particle orbitals in the ground state of the Kohn-Sham Hamiltonian and $G$ is the retarded one-electron Green's function corresponding to the ground state self-consistent Kohn-Sham potential, that is,

$$[\epsilon + \tfrac{1}{2}\nabla^2 - V_{\text{KS}}(\mathbf{r})]G(\mathbf{r},\mathbf{r}',\epsilon) = \delta(\mathbf{r}-\mathbf{r}'). \tag{7}$$

Once the dynamical susceptibility $\chi(\mathbf{r},\mathbf{r}',\omega)$ has been calculated, the photoabsorption cross section $\sigma(\omega)$ is obtained using Fermi's golden rule

$$\sigma(\omega) = \frac{4\pi\omega}{c}\,\text{Im}\left\{\iint V_{\text{ex}}^*(\mathbf{r},\omega)\,\chi(\mathbf{r},\mathbf{r}',\omega)\,V_{ex}(\mathbf{r}',\omega)\,d\mathbf{r}\,d\mathbf{r}'\right\}, \tag{8}$$

The symbol Im stands for the imaginary part of the quantity between parenthesis.

In order to follow step by step the influence of the electron-electron interactions in the absorption spectrum, a coupling constant $\lambda$ can be introduced in Eq. (5), that takes the form[15]

$$\chi(\mathbf{r},\mathbf{r}',\omega) = \chi^0(\mathbf{r},\mathbf{r}',\omega) + \lambda \iint \chi^0(\mathbf{r},\mathbf{r_1},\omega)\,K(\mathbf{r_1},\mathbf{r_2})\,\chi(\mathbf{r_2},\mathbf{r}',\omega)\,d\mathbf{r_1}\,d\mathbf{r_2}. \tag{9}$$

The coupling constant takes values between 0 and 1: $\lambda = 0$ means independent particle response, while $\lambda = 1$ means that the full electron-electron interaction is taken into account (full dynamical response). We will present our results for $\lambda = 0$ and $\lambda = 1$. In this work we are mainly interested in the dipolar response, in which case the perturbing potential takes the form

$$V_{\text{ex}}(\mathbf{r},\omega)e^{-i\omega t} = rY_1^0(\theta)e^{-i\omega t}. \tag{10}$$

In the calculations we have applied a Lorentzian broadening of the spectrum (full width at half maximum 0.1 eV) to model the effect of the temperature on the absorption spectrum.

## III. Structural Model

It is well known that simple models like an spherical jellium give a reasonable account of the electronic structure of free sodium clusters and of their dynamical electronic response.[16] On the other hand the full introduction of the detailed cluster geometry in a response calculation is a very difficult task. For this reason we use a simplified structural model. The $C_{780}$ cage is modeled as formed by two concentric shells with 240 and 540 atoms respectively. The two shells are assumed perfectly spherical, with radii $R_{240} = 13.46$ a.u. and $R_{540} = 19.85$ a.u., taken from *ab initio* DFT calculations[17] which suggest that large fullerenes prefer spherical rather than faceted structures. The separation between the two shells ($\simeq 6.39$ a.u.) is approximately equal to the interplanar separation in graphite. In each hollow cage the carbon atoms are distributed in a similar way as in the corresponding truncated-icosahedron fullerene. Response calculations are presented for $C_{780}$ in Sect. IV below. In those calculations we use the SAPS approximation[14] for obtaining the electronic structure. This means, first of all, that carbon is modeled as a monovalent atom (since we are only trying to evaluate the $\pi$-electron response), with an empty-core pseudopotential[18] of radius $r_c = 0.21$ a.u. (Ref. 5). Furthermore, in solving the Kohn-Sham equations,[19,20] the total potential of the pseudo-ions is spherically averaged about the cage center.

The ground state geometries of $Na_{20}$ and $Na_{25}$, have previously been obtained using a relaxation technique that also uses the SAPS approximation for the electron-ion interaction.[14] $Na^+$ ions are also described by an empty core pseudopotential with

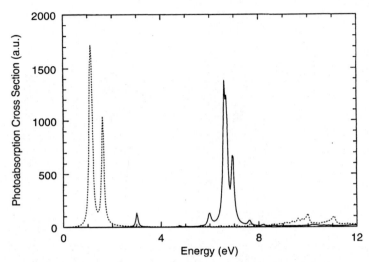

**Figure 1.** Calculated photoabsorption cross section of $C_{780}$ for $\lambda = 0$ (dotted line) and $\lambda = 1$ (solid line).

$r_c = 1.74$ a.u. In those two clusters the Na atoms are arranged into shells. In $Na_{20}$ the inner shell contains two atoms at a distance 2.68 from the cluster center, and the rest form a surface shell with a mean radius 7.94 a.u. and 1 a.u. width. In the case of $Na_{25}$ the respective radii are 3.46 a.u. and 8.84 a.u., the inner shell contains three atoms and the width of the surface shell is again 1 a.u.

Finally, the two endohedral systems $Na_{20}@C_{780}$ and $Na_{25}@C_{780}$ are constructed by simply placing the Na cluster at the center of the cage (the centers of mass coincide). The mean separations between the external shell of each Na cluster and the inner shell of the fullerene are about 5.5 a.u. and 4.6 a.u. for $Na_{20}$ and $Na_{25}$, respectively, and possible relaxation of the structures has been neglected. The calculations of the electronic structure again employ the SAPS approximation.

## IV. Results and Discussion

### IV.1 Photoabsorption Cross Section of $C_{780}$

The photoabsorption cross secction is given in Fig. 1 for two values of the parameter $\lambda$ of Eq. (9). In the $\lambda = 0$ curve (dotted line) the spectrum is contributed by one particle-hole excitations. There are two peaks at low energies. The first one is a double peak with excitation energies at 1.10 eV and 1.16 eV, respectively. These two features are indistinguishable in the figure due to the broadening we have used and have been assigned to dipole allowed transitions between bound frontier levels with large angular momentum in this spherical model: $(n=1, l=15) \to (n=1, l=16)$ and $(n = 1, l = 16) \to (n = 1, l = 17)$. The second peak, at 1.61 eV is assigned to the transition $(n = 2, l = 10) \to (n = 2, l = 11)$. As the electron-electron interaction is switched on, the spectrum evolves and for $\lambda = 1$ (solid line) it has a rather simple form: the main feature is a strong resonance centered at 6.60 eV. This is the collective resonance of the $\pi$ electron cloud. The size of this two-layer cluster is large and the

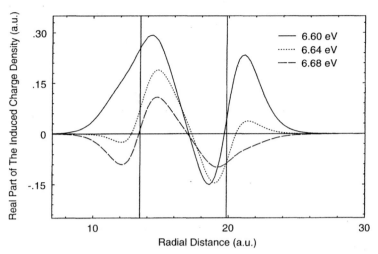

**Figure 2.** Real part of the induced charge density of $C_{780}$ for three energies (6.60 eV, 6.64 eV and 6.68 eV) in the region of the main resonance.

energy of the collective resonance is already very close to the energy of the $\pi$-plasmon in graphite (7 eV) measured by Saito et al.[21] The resonance is broadened as an effect of Landau damping: The calculated ionization threshold of $C_{780}$, measured by the binding energy of the highest occupied molecular orbital, is 3.95 eV so the resonance overlaps strongly with transitions to the continuum.

The real part of the induced charge density $\delta\rho(\mathbf{r},\omega)$, given by Eq. (1), is plotted in Fig. 2 as a function of $r$ for energies in the resonance region. The vertical lines in Fig. 2 indicate the radii of the two cages. The density oscillates in the two shells, which confirms the interpretation of the resonance as a collective one. We note that each of the two cages separately has a resonance energy close to the one of $C_{780}$, with the induced density located at its surface. When the two cages are put together to form the $C_{780}$, the two collective resonances interact and merge into a broad and fragmented collective resonance. This is clearly seen in Fig. 2 where for different energy values around the resonance frequency the induced density is located in both shells.

## IV.2 Photoabsorption Cross Section of $Na_n@C_{780}$

In figures 3 and 4 we compare the absorption cross section of $C_{780}$, $Na_{20}@C_{780}$ and $Na_{25}@C_{780}$. Since the main differences occur on the low energy region of the spectrum we have plotted the three spectra using a logarithmic scale for $\sigma(\omega)$, which enhances the low intensity features. It is convenient to compare the results for $C_{780}$ in Figs. 1 and 3, to appreciate the effect of the different scales. With this comparison in mind we notice that the main feature in the spectra of the two endohedral clusters (Fig. 4) is again a collective resonance centered near 7 eV. As for empty $C_{780}$, we interpret this resonance as the precursor of the $\pi$ plasmon of graphite. The resonance is affected a little by the presence of the Na cluster inside the cage; specifically the maximum is shifted by 0.2 eV to higher energies. The induced densities $\delta\rho(\mathbf{r},\omega)$ of $Na_{20}@C_{780}$ at 7.0 eV and 7.04 eV, plotted in Fig. 5, are also appreciable in the region $r < 10$ a.u. The

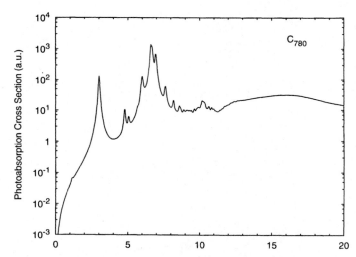

**Figure 3.** Calculated photoabsorption cross sections of $C_{780}$. The logarithmic scale enhances the low intensity features of the spectra.

origin of this difference between the filled and empty fullerenes is that the endohedral cluster is chemically bound to the cage. In a previous paper[13] we have studied the electronic structure of $Na_n$ clusters enclosed in a $C_{240}$ cage. The conclusion was that there is a substantial binding between the endohedral and the fullerene. This binding reaches values of 17 eV and 22 eV for $Na_{20}$ and $Na_{25}$, respectively, and is a mixture of ionic binding, arising from the transfer of electrons to the fullerene, and covalent binding between the valence electrons of the alkali cluster and the $\pi$-electrons of the fullerene, reflected in an accumulation of charge in a shell bound by inner and outer radii of $\simeq$ 7 a.u. and 10 a.u. respectively. The induced densities of Fig. 5 reveal that the $\pi$-electrons of $C_{780}$ are now tied up to the valence electrons of the endohedral. Qualitatively, the larger binding energy of the $\pi$-electrons in endohedral $C_{780}$ as compared to $C_{780}$ would increase a little the energy required to excite the collective mode. This is in agreement with the computed shift of 0.2 eV.

Another subtle difference between the empty and filled fullerenes can also be appreciated in the logarithmic representation. The *small* peak at 3 eV in $C_{780}$ becomes broadened and fragmented in the filled clusters. In the free $C_{780}$ the induced density at 3 eV is distributed over the region of the two carbon layers and excludes completely the inner region of the cluster ($\delta\rho$ is zero inside a sphere of radius 10 a.u.). On the other hand, the induced densities for the filled fullerenes at energies corresponding to the peaks near 3 eV are mainly distributed over the inner region ($r < 10$ a.u.). This is shown in Figure 6 for energies of 3.0 eV and 3.2 eV in $Na_{20}@C_{780}$. The region $r < 10$ a.u. is that associated to the electronic cloud of the endohedral cluster, and the surface plasmon of charged, free sodium clusters is only a few tenths of an eV below 3 eV (Ref. 22). We then interpret the fragmented feature near 3 eV in both endohedral clusters as a manifestation of the collective resonance of the alkali clusters. This resonance is shifted a little to higher energies because of the interaction between the valence electrons of the endohedral and the $\pi$- electrons of the cage (we observe that the induced densities of Fig. 6 also reach, although attenuated, the region of the fullerene shells). The fragmentation of the peak is due to the interaction with the feature at 3 eV in free $C_{780}$ (see Fig. 1).

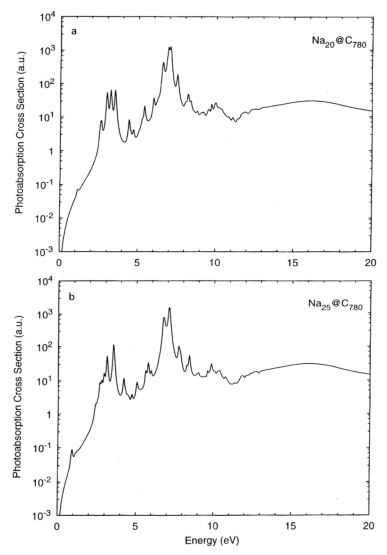

**Figure 4.** Calculated photoabsorption cross sections of $Na_{20}@C_{780}$ and $Na_{25}@C_{780}$. The logarithmic scale enhances the low intensity features of the spectra.

## V. Summary

Using the TDLDA and the SAPS model we have calculated and analyzed the low energy absorption spectrum of $C_{780}$ and the two endohedral systems $Na_{20}@C_{780}$ and $Na_{25}@C_{780}$. The main feature for $C_{780}$ is a collective resonance at 6.60 eV that we interpret as the precursor of the $\pi$ plasmon of graphite. The presence of the endohedral Na cluster affects a little this resonance (a shift of 0.2 eV to higher energies occurs due to the increase in binding energy of the $\pi$-electrons). A small new feature appears

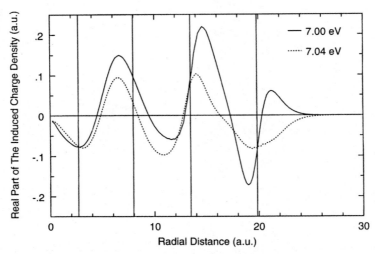

**Figure 5.** Induced charge density in $Na_{20}@C_{780}$ for the energies 7.0 eV and 7.04 eV near the maximum of the main resonance.

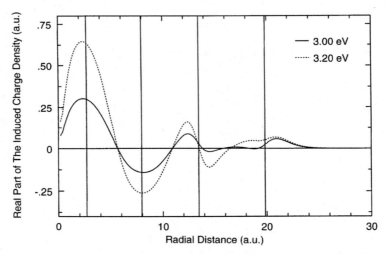

**Figure 6.** Induced charge densities in $Na_{20}@C_{780}$ for the energies 3.0 eV and 3.2 eV.

at 3 eV that can be associated with the collective resonance of free Na clusters, here shifted and fragmented by the interaction between the endohedrals and the cage.

## Acknowledgments

Work supported by DGICYT (Grant PB95-0720-C02-01) and Junta de Castilla y León (Grants VA25-95 and VA72-96). Two of us acknowledge postdoctoral grants from DGICYT (R.P.-D.) and CONACYT (J.M.C.-T.) during our stay at the University of Valladolid. J.M.C.-T. acknowledges also support from Fondo de Apoyo a la Investigación, UASLP, Mexico.

# References

1. See for instance, J. W. Keller and M. A. Coplan, *Chem. Phys. Lett.* **193**, 784 (1992); R. Abouaf, J. Pommier, and S. Cvejanovic, *Chem. Phys. Lett.* **213**, 503 (1993).
2. See for instance, F. Alasia et al., *J. Phys. B* **27**, L643 (1994).
3. A. A. Lucas, L. Henrad, and Ph. Lambin, *Phys. Rev. B* **49**, 2888 (1994).
4. P. A. Apell, D. Ostling, and G. Mukhopadhyai, *Solid State Commun.* **87**, 219 (1993).
5. R. Pis-Diez, M. P. Iñiguez, M. J. Stott, and J. A. Alonso, *Phys. Rev. B* **52**, 8446 (1995).
6. A. Rubio, J. A. Alonso, J. M. López, and M. J. Stott, *Phys. Rev. B* **49**, 17397 (1994).
7. S. Frank, et al., *Z. Phys. D* **40**, 250 (1997).
8. M. J. Puska and R. M. Nieminen, *Phys. Rev. A* **47**, 1181 (1993).
9. A. Rubio, J. A. Alonso, J. M. López, and M. J. Stott, *Physica B* **183**, 247 (1993).
10. D. Ugarte, *Chem. Phys. Lett.* **209**, 99 (1993).
11. M. J. Stott and E. Zaremba, *Phys. Rev. A* **21**, 12 (1980).
12. A. Zangwill and P. Soven, *Phys. Rev. A* **21**, 1561 (1980).
13. J. M. Cabrera-Trujillo et al., *Phys. Rev. B* **53**, 16059 (1996).
14. M. P. Iñiguez, M. J. López, J. A. Alonso, and J. M. Soler, *Z. Phys. D* **11**, 163 (1989).
15. K. Nuroh, M. J. Stott, and E. Zaremba, *Phys. Rev. Lett.* **49**, 862 (1982).
16. J. A. Alonso and L. C. Balbás, *Topics in Current Chemistry*, edited by R. F. Nalewajski, (Springer-Verlag, Berlin 1996), Vol. 182, p. 119.
17. D. York, J. P. Lu and W. Yang, *Phys. Rev. B* **49**, 8526 (1994); J. P. Lu and W. Yang, *ibid* **49**, 11421 (1994).
18. N. W. Ashcroft, *Phys. Lett.* **23**, 48 (1966).
19. R. G. Parr and W. Yang, *Density Functional Theory of Atoms and Molecules* (Oxford University Press, 1989).
20. W. Kohn and L. J. Sham, *Phys. Rev.* **140**, A1133 (1965).
21. Y. Saito, H. Shinohara, and A. Ohshita, *Jpn. J. Appl. Phys.* **30**, L1068 (1991).
22. P. G. Reinhard, O. Genzken, and M. Brack, *Ann. Physik* **5**, 576 (1996).

# Molecular Precursor of Soot and Quantification of the Associated Health Risk

K. Siegmann and H. C. Siegmann

*Swiss Federal Institute of Technology*
*CH-8093 Zürich*
*SWITZERLAND*

## Abstract

Particles generated in the combustion of organic materials are intrinsically toxic. Our work focuses on finding a way to quantify the health risk despite the complexity of the particles. First the current experimental research on the primary combustion aerosols is summarized. We take samples from various locations inside a laminar methane diffusion flame and freeze their physical and chemical state by rapid dilution with cold inert gas. Formation and growth of large molecules, mostly polycyclic aromatic hydrocarbons (PAH) is detected by mass spectroscopy and laser induced ionization. Fullerenes are also found. Particle size distribution are measured by standard aerosol techniques. To test the sampling, the flame is optionally seeded with palladium aerosol of known size distribution. We believe that the experiments on the model flame reveal some general principles of soot formation, in particular the fact that soot particles do not nucleate by agglomeration of large PAH. Photoemission is applied to study surface properties of soot particles from the flame. It is shown that the surface of the particles is covered with PAH. By heating the PAH can be removed and the properties of the carbon core are revealed. One can thereby distinguish a soot growth- from a soot burnout region in the flame. Time resolved desorption experiments of perylene (a PAH) from model aerosol particles are presented. It is shown that they follow a first order rate law. The photoelectric PAH sensor is introduced as a personal air quality monitor. The danger from inhaling combustion aerosol can be expressed in units of standard cigarettes.

## I. Introduction

The hypotesis that the particles suspended in ambient air are intrinsically toxic is supported by the famous Harvard six cities study[1] and subsequent similar work in 15

European cities. According to these epidemiological findings, human mortality from cardiopulmonary disease increases by 1% when the loading of the air with respiratory particles of a diameter below 2.5 microns increases by 10 $\mu g/m^3$. Even in allegedly clean cities such as the city of Zürich 3–4 times higher concentration of respiratory particles are generated by automotive traffic alone. Very recently, the epidemiological studies have been complemented by animal experiments in which bronchitic rats were exposed to particles of the ambient air. The animals died within a few hours of the exposure, although the particles concentration was at the legal limits.[2] So far, no previous study on animals had demonstrated pathological changes that might result in death. Even long-term exposure of animals to suspended particles has not produced lethal consequences unless the particle concentrations were excessive, resulting in overload conditions. The problem with the older animal experiments most likely is the fact that they had been carried out with simple materials such as sulphuric acid aerosols, silica, aluminum oxide, latex, iron oxide, and "carbon." The real particles of the ambient air are everything but simple. For instance, thousands of different chemicals have been identified in particles from combustion. Complete chemical characterization is impossible because of the very tiny amount of materials and the continuing changes of the chemicals in the atmosphere. The shape of the particles can also be quite bizarre and will change through agglomeration with other particles and through adsorption from the gas phase. However, surface and cluster science has developed the tools to characterize the early stage of nucleation of particles and to understand the principles on which the interaction with the surrounding gas and the sunlight takes place. Therefore, research on the formation of particles is at the forefront of solid state physics, and also is one of the most obvious applications of cluster science. K. H. Bennemann has contributed greatly to both surface and cluster science, and it is therefore appropriate to discuss this problem at a workshop dedicated to his 65th anniversary.

This paper focuses on the primary combustion aerosol, that is the aerosol that nucleates in the combustion zone. The carbon atom has the unique capability of forming clusters at temperatures of above 1000°C. Such temperatures occur in the combustion of organic materials and fuels used almost exclusively in our present civilization for power generation. The carbon clusters are very difficult to get rid of once they have formed. As the exhaust gas carrying the carbon clusters cools down, various other chemicals adsorb on the carbon clusters arising from wear and tear in the engine, or synthesized in the process of combustion, or simply from unburned fuel. In this way, the carbon clusters become the skeleton of the soot particles found suspended in great quantities in the air in which we live. The soot particles are of respiratory size, that is, they are deposited in the human respiratory tract where they are causing the above mentioned life threatening diseases. Apart from these short term health effects there are also long term health risks associated with the inhalation of respiratory particles; lung cancer is the most dreadful of them. There is a delay of as much as 20–30 years between the beginning of the exposure to the soot particles and the outbreak of the disease, and this delay very often leads younger people to believe that they are immune despite the clear-cut epidemiological evidence. However, very recently all doubts have been removed, as a direct etiological link between Benzo(a)pyrene (B(a)P) and human lung cancer has been established with the methods of molecular biology.[3] B(a)P belongs to the class of polycyclic aromatic hydrocarbons (PAH). These chemicals are at the center of our work. We will show how PAH are synthesized in combustion concomitantly with the carbon clusters. We will also show that the heavier PAH including the carcinogenic species such as B(a)P are adsorbed on particles at ambient temperature. Lastly, we will demonstrate that respiratory particles having adsorbed PAH on their surface are detected readily by the fact that they can be charged electrically with great efficiency through photoemission of electrons. Based on these findings, we determine the individual human exposure to particles from combustion and relate this exposure to the one obtained when smoking a cigarette. It turns out that for many people, the dose of B(a)P received from traffic fumes is equivalent to smoking a few cigarettes every day.

We are not the only people worrying about the health impact of particles from combustion. Mario Molina, winner of the Nobel prize 1995 in chemistry for his work on atmospheric ozone depletion, gave part of his prize money to a foundation supporting research on air pollution in Latin America.[4] At the first world congress on air pollution in San Jose in Costa Rica in November 1996 it was estimated that 460 000 people die world-wide every year from respiratory particles produced in automotive traffic. This figure shows that the danger of smoke from traffic now approaches the danger from cigarette smoke. The latter is estimated to kill more than 3 million people a year.

## II. Experiments on Carbon Formation in Combustion

The present work differs from the work of others in that we take samples of gas containing molecules and particles directly from the combustion zone. The small sampled volumes are immediately and highly diluted with cold inert gas such as Ar or $N_2$ and subsequently fed into a time of flight (TOF) mass spectrometer. Simultaneously, aerosol sizing equipment including an interaction tube with the light from pulsed lasers is used to determine the size distribution and surface properties of the particles. For the most important case of the PAH, the enthalpy of adsorption and the time needed to reach the desorption/adsorption equilibrium at the particle surface is also measured. As a model combustion, we have chosen a laminar methane diffusion flame. However, the technique may be applied to almost any combustion, be it wood fire, diesel engine, cigarette, or whatever else one likes to imagine. Yet the model system we present here has a few advantages for developing the basic understanding of carbon formation in combustion. First, methane ($CH_4$) is the simplest organic fuel. It produces comparatively few soot particles which are all burnt up before the combustion gases cool down. The small particle concentration means that the interaction between particles such as agglomeration is minimal. Second, a time scale is established by the laminar flow. The distance from the orifice of the burner can be translated into a time spent in the combustion zone. As the oxygen is supplied by diffusion from the outside, the symmetry axis in the middle of the flame has the lowest partial pressure of oxygen. The results discussed below are valid for the symmetry axis of the cylindrical flame. To avoid flickering of the flame, it is surrounded by a laminar flow of air.

In figure 1 we show the two sample extraction systems we have used and intercompared. In system (a), a capillary of ceramic material extends into the flame. Flame gases are extracted by maintaining a reduced pressure in the quartz tube to which the capillary is attached. The flame gases expand into the quartz tube where they are diluted with nitrogen or argon by factors ranging from 10–1000, depending on the experiment. The capillary does not visibly disturb the flame. The position of the orifice of the capillary in the flame is adjusted by moving the flame with a computer-driven mechanical motion. It is thus possible to take samples from any point in the flame. We have reasons to believe that the height resolution of the sampling is less than 0.1 mm

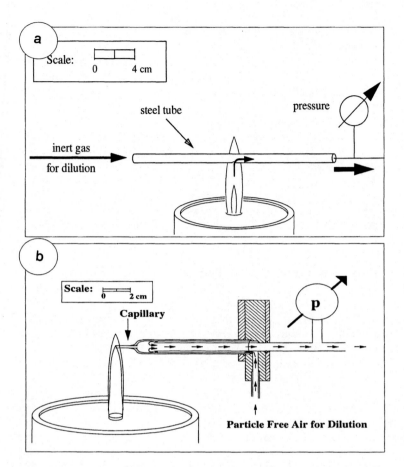

**Figure 1.** Sample extraction systems. (a) The ceramic capillary is glued to a double walled quartz tube. The dilution gas enters through the annular space between the tubes. (b) The flame gases enter through the hole in the bottom of the tube in which the dilution gas flows.

while we do not know the exact radial extension of the sampled volume. In system (b), a tube in which the inert dilution gas flows is put across the flame. Depending on the pressure difference between flame and the dilution gas stream, flame gases enter into the tube through a pinhole at the bottom. This device disturbs the flame, but essentially only after the sample has been taken. Its advantage compared to (a) is that it is simpler to construct and it samples with superior radial resolution. We have not observed any systematical differences between samples taken with system (a) or (b).[5,6]

## III. Mass Spectroscopy of Large Molecules and Clusters

The diluted flame gases pass by a pulsable valve which opens for a few microseconds, allowing the gases to expand into the vacuum chamber of a commercial Bruker-Franzen

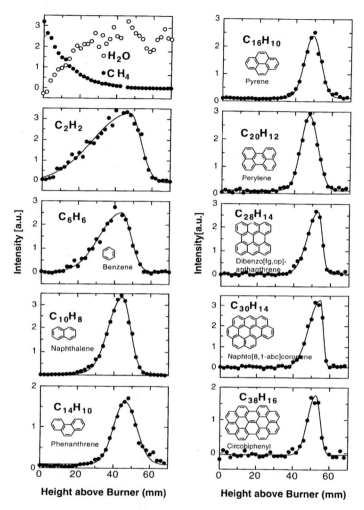

**Figure 2.** Height profiles of selected molecules found in the flame. $CH_4$, $H_2O$, and $C_2H_2$ were ionized by electron impact, the other molecules which are polycyclic aromatic hydrocarbons (PAH) by ionization with an excimer laser of a photon energy of 5 eV. The intensities depend on the ionization efficiency which is different for each molecule.

TOF mass spectrometer. The skimmer defines a beam in which the temperature is low due to the supersonic expansion. The light molecules such as $H_2O$, $CH_4$, or $C_2H_2$ are ionized by electron impact for TOF mass spectroscopy. With the PAHs, ionization by photoemission of electrons is used to avoid fragmentation. To this purpose, a light pulse of 20 ns duration from an excimer laser is fired across the supersonic jet.

Figure 2 shows examples of the height profiles of molecules found along the center of the flame. For the purpose of this work, it is not necessary to distinguish between various isomers. It should be noted that the visible height of the flame is 70 mm; one observes that the concentration of all the hydrocarbons is zero at that height. The concentration of two different species cannot be compared as the detection efficiency is different for each molecule. The concentration of the fuel $CH_4$ decreases sharply with height above the burner orifice; it is at the 10% level already at $h = 20$ mm. The concentration of the main combustion product $H_2O$ increases at the corresponding rate. Hence the main energy producing process is terminated already after about 20 mm. The remaining 50 mm of the flame are apparently needed to form and destroy a rich variety of PAH and carbon clusters. Acetylene (H–C≡C–H) with a triple carbon bond is present almost everywhere in the flame. Its concentration reaches a flat maximum at around 50 mm height. It is very plausible, as many authors have proposed, that acetylene is the main building block of the large and flat polycyclic aromatic hydrocarbons (PAH) found abundantly in the flame. The reason why such huge organic molecules can form in the combustion zone at more than 1000°C rests on the very principles of quantum mechanics: In a closed benzene ring where the electrons are delocalized, there exist no boundary conditions excluding the constant as a possible wave function. Hence two electrons can be accommodated with zero kinetic energy. This is not possible in linear molecules which are hence less robust at high temperatures. It is interesting to note that PAH are thought to be synthesized also when an oxygen deficient red giant expands in the cosmos.[7] An important conclusion from the height profiles shown in Fig. 2 is that the larger PAH are successively built from smaller units. While benzene occurs over a sizeable range of heights, the concentration profiles sharpen and their maxima shift to larger heights as one moves to the larger PAH. Yet we have observed that some PAH occur lower in the flame than expected from their size. This occurs for those species that are built from two small ones under dehydrogenation.[8] As an example we mention perylene which is obtained by adding two naphtalenes and releasing $H_2$. We note that we have detected the largest PAHs ever found in combustion, containing over 50 C-atoms in 20 coupled benzene rings. We also have found curved PAHs containing carbon rings with five corners which establish dishes.[8]

Extending the range of atomic mass units to 104 reveals new phenomena. Figure 3 shows that PAHs and their fragments come to an end at around 600 amu. New units appear after a clear cut minimum. The observed mass spectra depend on the photon energy $h\nu$. They are, apart from a superimposed beat structure, featureless for $h\nu = 6.4$ eV. But with $h\nu = 4$ eV, the spectra show a rich variety of fullerenes and their fragments. This arises with the identical average intensity of the excimer laser beam. The spectacular phenomenon is explained if one assumes that the masses detected in the 1000–10 000 amu range belong to carbon clusters. The photoelectric work function $\Phi$ of these clusters is larger than 4 eV but smaller than 6.4 eV. Hence for $h\nu = 6.4$ eV, the particles are charged by photoelectron emission which is impossible at $h\nu = 4$ eV. To understand charging at 4 eV, one must take into account that the intensity in a light beam from an excimer laser exhibits hot spots. We have shown that thermionic emission of electrons occurs from the particles in the hotspots.[9] With $h\nu = 4$ eV, the dominant charging mechanism is by thermionic emission. This mechanism selects the particles exposed to the hot spots for detection. Figure 3 proves that fullerenes and their fragments evaporate from those hot carbon clusters. Concluding the discussion of the phenomena in the range of the carbon clusters, we mention that a closer analysis of the beat structure obtained with $h\nu = 6.4$ eV might lead to deeper insight into the structure of the carbon clusters. The mass ratio of the C-atom to the $H_2O$-molecule is $\frac{2}{3}$. If carbon clusters were added in units of three carbon atoms, two adsorbed water molecules would be identical to adding one carbon unit, and a mass spectrum with peaks every 18 amu would appear, disturbed by the isotope effect as observed.

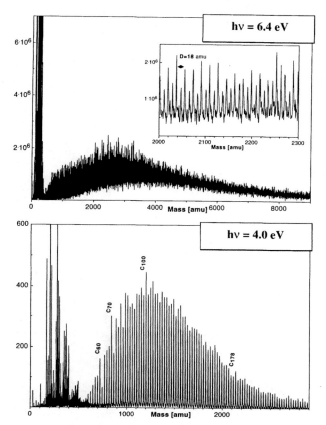

**Figure 3.** Mass spectra with samples taken at 55 mm above the burner orifice with a laser beam of 6.4 eV and 4.0 eV photons. The average intensity of the laser was kept constant.

The question now arises of whether the fullerenes can be generated from any kind of carbon particle. Figure 4 suggests that this is not the case. In this figure, the height profile of the PAH $C_{13}H_{10}$ and the height profile of the prototype fullerene $C_{60}$ are shown. In contrast to the PAH, the fullerenes all show the identical height profiles independently of their mass.[10] First, this key observation proves that the fullerenes are not built up successively from smaller units as is the case with PAHs. Second, as the particles are still fully developed even at 70 mm (see below), only the special particles found at or below 60 mm can emit fullerenes.

The conclusion from this height profile is that the fullerenes exist within the type of carbon cluster present at 60 mm. The fullerenes evaporate upon heating. It is possible that the heating with the laser beam supports or completes the formation of the fullerenes. We draw attention to the sharp upper boundary of the existence of the fullerenes. The sharpness of this feature proves the excellent height resolution of the sample extraction system.

To connect these findings with the large body of existing literature on soot formation, we propose to discuss the standard model of soot formation put forward by a number of authors.[11,12] According to this model, the first step is that the hydrocarbon

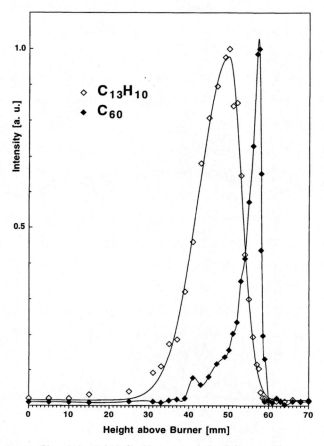

**Figure 4.** Height profile of the PAH $C_{13}H_{10}$ and of the fullerene $C_{60}$.

fuel is broken up into small radicals from which acetylene and benzene are synthesized. After that, larger PAHs are successively built from benzene and acetylene, or by polymerization under dehydrogenation. The PAHs can be flat or curved depending on the author. It is then supposed that the large PAHs agglomerate by van der Waals forces and/or occasional $sp^3$-bonds to form the first carbon clusters.

Although some observations made here such as the successive build-up of PAH and the depletion valley in the mass spectra before the onset of the carbon clusters seem to be in favor of the standard model of soot formation, we will show below that this model can really not be the right one as the first particles appear much before the heavy PAHs have reached their maximum concentration. To detect at which height the famous soot inception point is actually located, one needs to employ the far more sensitive aerosol techniques introduced below. The results from the aerosol techniques suggest that at least the large PAH are a byproduct rather than the cause of carbon cluster formation.

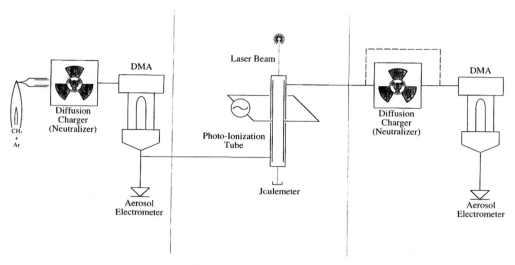

**Figure 5.** Experimental set-up for the measurement of size distributions and surface properties of gas suspended particles. DMA is the differential mobility analyzer.

## IV. Particles from Inside the Combustion Zone

In figure 5 is shown the setup needed to measure the size distribution of gas suspended particles and to determine their interaction with the light for instance emitted by a laser.

The gas stream carrying the particles enters first a diffusion charger. This device contains a radioactive source producing positive and negative charge carriers in the gas with the particles. If the particles are already charged upon entering, this charge will attract the opposite charge and be neutralized. After some time, the equilibrium charge will be established on the particles. With particles of a radius $r < 100$ nm, most will remain neutral since the electrostatic energy $e^2/(4\pi\varepsilon_0 r)$ exceeds the thermal energy $kT$. About 1% of the particles carry one elementary charge. The doubly charged particles are rare.

After the diffusion charger, the gas carrying the particles enters the differential mobility analyzer (DMA). It selects particles with one electrical mobility. To the extent that the assumption of singly charged particles holds, this also means that particles of one specific mobility diameter $d = 2r$ have been separated. If the particles are not spherical in shape, the mobility diameter is the diameter of the sphere having the same electrical mobility as the particle. After the DMA, the gas stream carries thus monodisperse singly charged particles. The intensity of this particle current is measured by letting the carrier gas pass through a filter catching all particles. The electric current flowing from the filter to ground potential is measured with a femto-amperemeter. By varying the voltage on the DMA, particles of different electrical mobility are selected. Plotting the current in the filter *vs.* the voltage applied to the DMA, the size distribution of the suspended particles is obtained. All this is by now standardized, commercially available equipment. We will call the filter with the amperemeter "the aerosol electrometer."

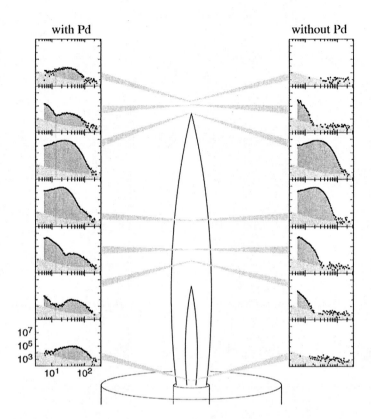

**Figure 6.** Particle size distributions with samples taken at various heights in the flame. The left side is with the flame seeded with Pd-particles, the right side with the unseeded flame.

To measure the photoelectric yield of the particles, the gas carrying the size selected particles flows through a tube before entering the aerosol electrometer. In the tube, the sample is exposed to ultraviolet light. The photoelectrons emitted from the particles are removed by an alternating electric field. This field precipitates the electrons at the electrodes whereas it leaves the particles having emitted the photoelectron within the gas stream. The differential current measured after the tube in the aerosol electrometer with the light switched on and off is proportional to the photoelectric yield. If one wants to observe any changes of particle size due to interaction with the light, a second DMA can be applied before admitting the particles to the aerosol-electrometer.

First of all one would like to test the extraction systems displayed in Fig. 1, in particular the question whether the particle size distribution measured in the extracted and diluted gas is the same as the one present in the flame. For this purpose, we have added a test aerosol of known size distribution to the fuel gas and checked whether we could retrieve this test aerosol after extraction from the flame. The test aerosol was produced by heating a Pd-wire in argon gas. The Pd-vapor condenses in the cold Ar to form an aerosol with a size distribution $10 \text{ nm} < d < 100 \text{ nm}$. We then mix the Ar with the methane flowing to the orfice of the burner. A particle filter can be inserted to remove the Pd-particles without affecting the $CH_4/Ar$ mixture.

One can do three different tests: (i) extract samples from the non-ignited Ar/CH$_4$ mixture, (ii) extract samples from the flame without Pd-particles, and (iii) extract samples from the flame with the Pd-particles added to the fuel gas. The summary of these tests is that the size distribution of the Pd-particles remains the same with the exception of shrinkage of the large Pd-particles in the flame. The large particles are agglomerates of a grapelike structure. The shrinkage occurs when these agglomerates are exposed to high temperatures. This has been observed before.[13]

In figure 6 we show size distributions of particles in the flame at various heights above the orifice of the burner. On the left side, Pd-particles were added while on the right side they were not. At $h = 0$, we see the original spectrum of the Pd-particles as generated by the hot Pd-wire, while no particles are of course present in the unseeded flame. However, already at $h = 30$ mm, one observes the first very small particles in the flame at $d < 20$ nm. As $h$ increases to 40 mm, the density and size of the particles formed in the flame increase dramatically, while the Pd-particle presence is still discernible. At still larger heights, the density of flame generated particles is so large that the Pd-particles disappear in the background. However, at $h > 70$ mm, first the large and later the small particles disappear and the Pd-particle size distribution emerges from the background with only the largest particles affected.[14]

The fact that the first particles appear at $h = 30$ mm contradicts the standard model of soot formation. Figure 2 shows that the larger PAH are not even present at $h = 30$ mm. Another mechanism different from condensation of PAH must be active to explain the formation of the particles at $h = 30$ mm.

Furthermore, the internal structure of the particles must change at $h = 60$ mm. Figure 4 shows that fullerenes are obtained only from particles sampled at $h \leq 60$ mm. It is indeed quite plausible that the particles change on switching from the build-up to the burn-out mode.

## V. Polycyclic Aromatic Hydrocarbons (PAH) as Tracers of Primary Combustion Particles

The aim is now to investigate the surface properties and the condensation of chemicals onto the particles nucleated in the flame. The best way to do this is to measure the photoemission of electrons from the particles in their natural gaseous environment. Our work is distinguished from numerous other experiments in that we do not precipitate the particles onto a substrate prior to photoemission spectroscopy. To introduce the spectroscopy of the photoelectric yield, we show in Fig. 7 results obtained with Pd-particles of a diameter of $d = 39$ nm. These particles were added to the fuel gas as described above. In the non-ignited gas flow, the photoelectric yield near photoelectric threshold does not change with the height above the burner orifice as expected. However, in the flame the Pd-particles start to reduce their yield already from $h = 10$ mm on. At approximately 30 mm, the curve reaches its minimum of $\frac{1}{5}$ of the initial yield, but recovers to a high value on increasing $h$. From 40 to 60 mm height, the Pd-particles are buried in the flame generated particles and therefore cannot be measured. When the Pd-particles reappear at 70 mm height, they are again in the low yield phase.[14]

Photoelectron yield spectroscopy can be interpreted on a phenomenological basis only. In any case, it is a very fine sensor of the surface properties responding to adsorbates at coverages at and below the monolayer level. The surface of Pd has been studied in detail since it is one of the key catalysts of heterogeneous reactions.

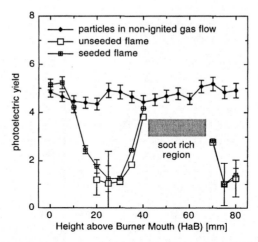

**Figure 7.** Photoelectric yield in relative units from Pd-particles of a mobility diameter of 39 nm. The light source is a mercury discharge. The particles were extracted at the heights indicated from the non-ignited gas flow and from the flame. For comparison, results with the unseeded flame are also shown.

Based on the extensive experience with this model surface, we propose the following explanation for the observations displayed in Fig. 7. The initial decrease of the yield indicates the availability of carbon in the combustion. The formation of graphitic carbon at the surface of Pd is known to be able to induce a substantial reduction of the photoelectric yield such as observed. The subsequent increase of the yield at the onset of PAH formation indicates that PAHs are adsorbed at the surface when the particles are in the photoemission tube. When the PAH have disappeared at the top of the flame, the yield is low, because the coverage of the Pd-surface with graphitic carbon persists. This is consistent with the well known fact that graphitic carbon, once deposited at the surface of Pd, is very hard to remove. In conclusion we see that the Pd-surface conveys a picture of the chemistry in the flame, in particular it clearly marks the soot inception point at $h = 30$ mm where carbon becomes available in large quantities.

We proceed now to the most interesting photoelectric yield of the particles generated in the flame. Figure 8 shows this yield for particles with $d = 35$ nm vs. the height above the burner orifice. The data labeled "Bypass" are valid for particles as extracted and diluted with inert gas. The photoelectric yield of these particles is quite high from the moment they appear in concentrations sufficient for the measurement. It hovers at a high level for extraction between 40 and 60 mm height, after which it plunges to reach $\frac{1}{5}$ of the maximum value at the end of the visible flame at $h = 70$ mm. The data labeled "Desorption" are valid for particles that have passed through a denuder prior to the measurement of the photoelectric yield. In the denuder, the gas carrying the particles is heated to 400°C, while the walls of the denuder contain active coal absorbing any species that have been desorbed from the particles. We see in Fig. 8 that the photoelectric yield of the "naked" particles having been stripped of the adsorbates that they have acquired in the flame or afterwards in the diluted gas stream is generally lower by an order of magnitude. Furthermore, it exhibits a different dependence on the height compared to the particles with adsorbates. The

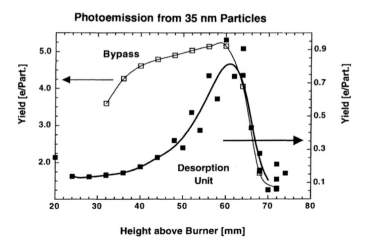

**Figure 8.** Photoelectric yield of particles generated in the flame, extracted with system (b) in Fig. 1, and of a mobility diameter of 35 nm. The dilution with $N_2$ was 1:10, the light source used for photoemission of electrons a mercury discharge. The curved labeled "Bypass" is for untreated particles, the curve labeled "Desorption" after passing the gas carrying the particles through a denuder kept at 400°C.

maximum yield occurs at $h = 60$ mm, and this maximum is sharp and reminiscent of the height profile maximum obtained with $C_{60}$ as shown in Fig. 4 (Ref. 15).

The interpretation is quite obvious from the many experimental observations we have made over the years in our laboratory: The yield of the adsorbate-covered particles is high due to the adsorption of PAHs. Indeed, the yield curve follows the integrated height profile of the PAHs.[8] However, in the case of the "naked" particles, the photoelectric yield curve is similar to the yield of fullerenes from the particles. Hence, the photoelectric yield of the "naked" particles demonstrates once more that the carbon clusters undergo a change in structure depending on the height above the burner orifice. The surface of highly oriented pyrolytic graphite (HOPG) has been extensively investigated by photoemission of electrons. It exhibits a very low photoelectric yield of electrons near photoelectric threshold of 4.8 eV. The yield increases as the number of defects and dangling bonds at the surface increases.[16] This might be related to the carbon clusters: In the locations of rigorous growth, there must be dangling bonds and defects at the surface, as a clean graphitic surface is inert and cannot bind any arriving species. But in the locations of the burnout, these active sites are destroyed first. Therefore, the photoelectric yield maximum separates the growth from the burnout mode.

Schlögl and collaborators have been able to obtain TEM-pictures from commercial soot and "fullerene" soot.[17] These pictures show the graphitic planes in the commercial soot but different, strandlike structures of the fullerene soot. The fullerene soot transforms into graphitic soot within hours. Hence we have to conclude that nascent soot is chemically different from old soot such as collected on a filter.

Lastly, we discuss a detailed study of the desorption of PAH from gas suspended particles. The adsorption and desorption of combustion generated PAHs on respiratory carrier particles is crucial for estimating the impact of combustion products on human health.

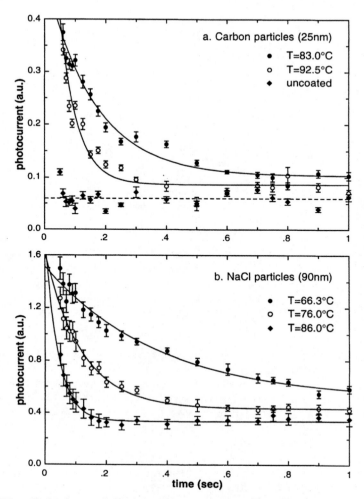

**Figure 9.** Time dependence of the desorption of the 5 ring PAH perylene at various temperatures as observed with a pulsed laser beam of 6 eV photons: (a) for graphite particles of a mobility diameter $d = 25$ nm; $E_d = (88 \pm 13)$ kJ/mole. (b) for NaCl particles of $d = 98$ nm, $E_d = (123 \pm 2.8)$ kJ/mole.

Aerosol photoemission makes it possible to observe directly the time dependence of the desorption of species from a surface. Hüglin has constructed a transport mechanism that can insert a small volume of gas containing particles into a particle free stream of gas at higher temperature.[18] While the particles being initially in a gas at low temperature are covered with the adsorbate at the instant of insertion into the hot gas, the adsorbates will start to evaporate as soon as they are in the hot gas environment. Photoemission of electrons induced by a sufficiently weak laser pulse, triggered a defined time after insertion of the particles into the hot gas will detect to what extent the adsorbates have desorbed because the photoelectric yield depends on the coverage of the surface.

Figure 9(a) shows results from a model experiment in which we have produced graphite particles in a spark generator and guided the gas stream carrying these particles into a tube in which the partial pressure of perylene was kept high by evaporation from an oven. In this way, the graphite particles adsorb some perylene in the submonolayer range. Perylene is not chronically toxic, but it is an isomer of the carcinogen B(a)P and therefore can closely simulate the adsorption characteristics of this dangerous chemical. Figure 9(a) shows that the desorption of perylene depends exponentially on time as expected. However, it is not possible to describe the observed desorption with one single adsorption site, rather one has to assume that at least two different adsorption sites exist. The heat of adsorption $E_d = (88 \pm 13)$ kJ/mole is in good agreement with other work if one takes into account that at least two adsorptions sites exist.[19] Important in the atmosphere are also salt particles. In figure 9(b) is shown the desorption of perylene from NaCl-particles.[18] These particles are produced by spraying an aqueous NaCl-solution and removing the water in a dryer. The resulting NaCl-particles are single crystals of a cubic shape. In this case, one adsorption site can describe the observations. The adsorption energy is $E_d = (123 \pm 2.8)$ kJ/mole which is in very good agreement with previous work.[20] Perylene might gain energy by desorbing from the combustion generated particles and re-adsorb on other particles. Our measurements suggest that the process of desorption and re-adsorption may take a time of the order of 10 min. Although the measurement were done with perylene, one has to assume that B(a)P will behave in the same manner. With this mechanism, a particle which is harmless for human health taken by itself may become the carrier of a carcinogen. Figures 9(a) and 9(b) also show clearly that with both, carbon- and NaCl-particles, PAH adsorbed on the surface are the dominant factor determining the photoelectric yield.

## VI. Quantification of the Health Risk

The key factor for evaluating the impact on human health is the size of the particles. Roughly speaking, particles with a diameter above 1 micrometer are deposited in the nose, whereas particles below 1 $\mu$m are deposited in the bronchial and pulmonary part of the respiratory system.[21] This can be understood by the fact that the large particles in the visible range above 1 $\mu$m are removed by impaction, whereas the smaller invisible ones reach the surface by diffusion. Particles generated in combustion initially are in the size range below 1 $\mu$m. Only at very high concentrations or long residence times in the atmosphere will these particles coagulate to the size range above 1 $\mu$m. Hence the main impact on human health is from the very fine particles deposited in the bronchies and the lung. Consequently, the following tasks need to be accomplished:

1. Measure the particles with diameters at or below 1 $\mu$m.
2. Distinguish particles from combustion or particles carrying combustion products from other harmless particles such as salt-particles, or other condensation nuclei.
3. Attribute particles separately to each source - as there may be a factor depending on the mix of chemicals characteristic for each source.
4. Compare to the particles in cigarette smoke as the danger from this source is best quantified so far.
5. Measure personal exposure rather than concentration of particles at a fixed location.

These requirements exclude many of the commonly used techniques such as: counting all particles; measuring the mass of all particles; measuring the blacken-

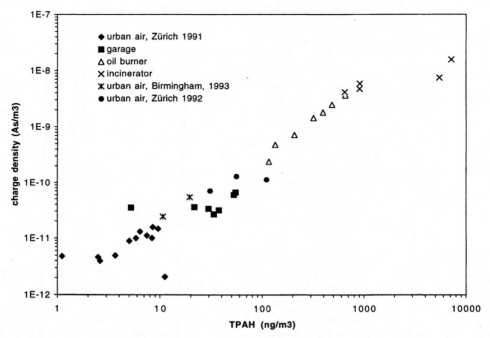

**Figure 10.** Photoelectric charge density in electrons per m$^3$ vs. total particle bound PAH-concentration in nanogram per m$^3$. From Ref. 22.

ing of a filter; sampling methods with long sampling times as they cannot distinguish the sources; any method that is unable to detect the main killer which is cigarette smoke.

We have previously proposed that the measurement of the photoelectric charge density generated by irradiating a volume of air for instance with the light from a Hg-discharge or an excimer lamp is eminently suited to determine human exposure to combustion products.[22] The two main reasons are that photoelectric charging occurs with particles below 1 $\mu$m only, hence the relevant size range is automatically selected, and that photoelectric charging is dominated by the PAHs adsorbed on the particle surface. The fact that photoemission selects particles below 1 $\mu$m is explained by the fact that recombination of the photoelectron with the positively charged particle is improbable when the mean free path of the electron exceeds the particle size. We have identified three different mechanisms which can explain the yield enhancement induced by adsorption of PAH.[16] The PAHs are the most characteristic byproduct of the combustion of organic materials and contain the best investigated carcinogen B(a)P.[3] Therefore, adsorbed PAHs are ideal tracers of combustion particles. There are also some practical reasons why one would like to use a sensor based on photoelectric charging: it can detect small concentrations of particle bound PAH, it is a continuous method, and it can also be constructed in a portable, battery operated form as we have demonstrated. The latter feature allows to determine the individual exposure and allows one to relate this exposure to the actual state of health. In figure 10 we show the comparison of the gas chromatographic chemical analysis of filter extracts with the photoelectric charge density. The remarkable result is that over a large range

## Table I

Personal exposure of a physicist in various locations

| | | |
|---|---|---|
| Live in countryside in Suitzerland, travel by train to office near motorway | 5.12.95 | 0.6 DACEE |
| Live in the city of Zürich, travel by public transportation to office in countryside(ETH-Höngg) | 6.3.96 | 1.2 DACEE |
| Vacations at lake Zürich near "Seestrasse" | Spring 96 | 2.0 DACEE |
| Live in San Francisco, travel by car on 280 to work at Stanford University | 2.9.94 | 3.6 DACEE |
| Live in Tokyo, walk to office at the University of Tokio[1] | 16.1.96 | 3.8 DACEE |
| Live in Campus at Tsinghua University, Bejing [2] | 20.4.96 | 2.7 DACEE |
| Average daily exposure to combustion particles [3] of total mass 10 $\mu g/m^3$ | | 5.5 DACEE |
| 1 h in street tunnel (Gubrist) near Zürich | | 6.0 DACEE |

[1] About 30% of this exposure is from cigarette smoke.
[2] This is Saturday 6 a.m. to Sunday 6 a.m. Most of the exposure is at night when delivery trucks are allowed to run.
[3] Harvard six cities study shows that death rate to cardiopulmonary disease rises by 1% a day when there is an increase in the fine particle loading of 10 $\mu g/m^3$.

of concentrations and with very different aerosols a linear relationship is found between the total mass of PAH adsorbed on the particles and the photoelectric charge density. This demonstrates that photoemission of electrons is able to monitor the presence of particle bound PAH in ambient air. The detection limit is about 1 $nm/m^3$ of total mass of particle bound PAH, at a time resolution of the order of 1 s (Ref. 23).

As the impact on human health from smoking cigarettes is well known by many people, we propose to use, instead of nanograms/$m^3$, a unit that can instantly be interpreted: the daily cigarette exposure equivalent (DACEE). To do this, we define a standard cigarette smoker. Naturally, any real smoker may be different from this standard smoker, yet we believe our definition comes close to reality for someone smoking contemporary, low tar filter cigarettes. The standard smoker inhales a total mass of $M = 200$ ng of particle bound PAH with each cigarette. The minimum volume of air needed by one human being in a day is $A = 11$ $m^3$. Hence the average concentration of TPAH in the air for 1 DACEE is:

$$a = M/A = 18.2 \text{ ng/m}^3.$$

From the current $i$ read in the photoelectric sensor, calibrated in ng/$m^3$ of TPAH according to Fig. 10, one obtains the exposure $x$ in DACEE: $x = i/a$.

The table shows some examples of personal exposures. With this table one can obtain a feeling of the individual health risk imposed by suspended particles from combustion in various locations and activities.

## Acknowledgments

We would like to acknowledge the help of Bingzhang Xue and Ernest Weingartner with the measurement of personal exposure. We also are grateful to H. Hepp, Ch. Hüglin, and M. Kasper for supplying results prior to publication

## References

1. D. W. Dockery et al., New Engl. J. Med. **329**, 1753 (1993).
2. J. J. Godleski et al., Proc. 2nd Coll. on Particulate Air Pollution and Human Health, Dec. (1996); 4, 136-143.
3. M. F. Denissenko, A. Pao, M. S. Tang, and G. P. Pfeifer, Science **274**, 430 (1996).
4. Physics Today, May, 63 (1996).
5. H. Hepp, Diss. Nr. 11666, ETHZ 1966.
6. M. Kasper, K. Siegmann, and K. Sattler, subm. J. Aerosol Sci., (1996).
7. T. Allain, S. Leach, and E. Sedlmayr, Astron. Astrophys. **305**, 602 and 616 (1996); Michel Brayer et al., Chem. Phys. Lett. **198**, 128 (1992).
8. K. Siegmann, H. Hepp, and K. Sattler, Combust. Sci. Tech. **109**, 165 (1995).
9. M. Kasper, H. Hepp, and K. Siegmann, subm. to J. Aerosol Sci., (1996).
10. K. Siegmann, H. Hepp, and K. Sattler, Surface Rev. Lett. **3**, 741 (1996).
11. J. T. McKinnon and J. B. Howard, Combust. Sci. Tech. **74**, 175 (1990).
12. K. H. Homann, Bull. Soc. Chim. Belg. **99**, 441 (1990).
13. A. Schmidt-Ott, J. Aerosol Sci. **19**, 553 (1988)
14. M. Kasper, K. Siegmann, and K. Sattler, subm. to J. of Aerosol Sci., (1996).
15. H. Hepp and K. Siegmann, subm. to Combustion and Flame, (1996).
16. T. Greber et al., Surf. Sci. Lett. **343**, L1187 (1995).
17. Thilo Belz et al., Angew. Chem. Int. Ed. Engl. **33**, 1866 (1994).
18. Ch. Hüglin, Diss. Nr., 11975, ETHZ (1996.
19. D. Steiner and H. K. Burtscher, Environ. Sci. Technol. **28**, 1254 (1994).
20. M. E. Loepfe, Diss. Nr. 9873, ETHZ (1992).
21. W. C. Hinds, Sem. Resp. Med. **1**, 197 (1980).
22. H. C. Siegmann, Vierteljahresschrift der Naturforsch. Ges. in Zürich **135**, 197 (1990).
23. H. Burtscher and H. C. Siegmann, Combust. Sci. Tech. **101**, 327 (1994).

# Magnetism of Transition Metal Clusters: Overview and Perspectives

G. M. Pastor

*Laboratoire de Physique Quantique*
*Université Paul Sabatier*
*118 route de Narbonne F-31062, Toulouse Cedex*
*FRANCE*

## Abstract

Recent theoretical work on itinerant magnetism and electron correlations in small clusters is reviewed. First, self-consistent tight-binding results of free and embedded clusters are discussed. The environment dependence of the magnetic behavior resulting from surface and interface proximity-effects, bond-length relaxations and changes in structure is analyzed concisely within this framework. A rigorous study of electron correlations, magnetism and cluster structure within the topological Hubbard model is then surveyed. Current perspectives are discussed giving emphasis to finite temperature phenomena.

## I. Introduction

Low dimensional systems constitute one of the most active research fields in solid-state physics and materials sciences. The interest of these systems is motivated by both fundamental and practical reasons. From the later point of view, the possibility of manipulating the chemical composition and the structure at an atomic scale has opened new prospects in the development of novel materials for specific technological purposes. From a fundamental point of view, it is of considerable interest to understand how the electronic properties depend on the composition, structure and dimensions of the system. The study of magnetic transition metals (TM) is particularly challenging, since the behavior of itinerant magnetism is known to be very sensitive to the local environment of the atoms.

Clusters play a very special role among low dimensional systems. First, since they are finite systems, they may show specific phenomena which do not have an equivalent in the thermodynamic limit. Let us mention, for instance, the variations of the magnetic moments resulting from temperature-induced structural changes, the coupling between magnetization, rotations and vibrations in an external magnetic field, or the consequences of the discreteness of the excitation spectra at finite temperatures. Moreover, the study of the evolution of the cluster properties as a function of size, from the atom all over to the solid, provides a new perspective to condensed matter behavior. From this point of view, it is particularly interesting to investigate many-body effects which are know to be very sensitive to the number of particles. For example, in the context of transition-metal magnetism, one would like to understand how the characteristic properties of itinerant magnetism develop starting from the localized magnetic moments of the atoms. Due to their extremely low dimensionality, small clusters are also ideal systems to study the effects of reducing the system dimensions on the electronic correlations. In the following Sections we shall discuss several electronic and magnetic properties which illustrate the previous general considerations.

Theorists were the first to study the magnetism of clusters. Salahub et al.,[1] determined the magnetic moments of small 3d-TM clusters using the X-$\alpha$ approximation. They predicted that the magnetic moment per atom $\bar{\mu}_N$ in small bcc-like Fe$_N$ clusters should be $\bar{\mu}_9 = 2.89\mu_B$ and $\bar{\mu}_{15} = 2.67\mu_B$, both values that are considerably larger that the bulk moment $\mu_b = 2.2\mu_B$. These results were confirmed qualitatively by the local-spin-density calculations of Lee, Callaway and coworkers[2] who obtained $\bar{\mu}_9 = 2.89\mu_B$ and $\bar{\mu}_{15} = 2.93\mu_B$. An alternative to the ab initio approach was provided by the self-consistent tight-binding (SCTB) studies on Cr, Fe and Ni clusters ($N \leq 51$).[3] This method yields results in good agreement with ab initio calculations and, at the same time, it takes advantage of the flexibility of the parametrized, minimal-basis approximation to explore a variety of structures and sizes which remain, even nowadays, unaccessible to first-principles techniques. Besides its practical interest, the main physical appeal of the SCTB method is its local character. It combines a self-consistent treatment of intra-atomic Coulomb interactions[4] with a recursive real-space expansion of the local Green's functions[5] and thus allows to relate the size and structural dependence of the magnetic properties with the changes in the local environment of the atoms. The SCTB approximation found many applications in the field of low-dimensional magnetism, some of which shall be surveyed in Sects. II and III.[6–11] A couple of years later, the theoretical research on cluster magnetism was considerably boosted by remarkable experimental findings.[12,13] An important number of ab initio calculations on magnetic TM clusters were then performed, in most cases within the framework of the local spin density approximation.[14]

The first experiments on the magnetic properties of free TM clusters were performed by Cox et al.[15] As in all the other experimental studies to come, the key principle of the experiment is the Stern-Gerlach (SG) deflection. Three main steps can be distinguished: (i) A beam of neutral clusters containing a more or less broad distribution of sizes is produced, typically using a laser vaporization source. (ii) The neutral clusters pass through a SG magnet where they are deflected along the field-gradient direction, depending on the value of the projection of the magnetic moment per atom onto the field $H$. Clusters with moment parallel (antiparallel) to $H$ are deflected in the direction of increasing (decreasing) field. (iii) Finally, the clusters are ionized in order to be detected after mass selection. Cox et al.,[15] observed that the beam intensity at the (undeflected) beam axis is depleted upon switching on the SG field. This was the first experimental indication that small Fe$_N$ clusters are magnetic ($2 \leq N \leq 17$). The measurement were interpreted assuming that magnetic clusters in an external magnetic field $H$ should be deflected essentially like atoms carrying a large magnetic moment $N\bar{\mu}_N$, i.e., with equal probability in the direction of increasing and decreasing field, depending on whether $\bar{\mu}_N$ is parallel or antiparallel to $H$.

However, later experiments,[12,13] revealed that this assumption was incorrect and that performing only on-axis detection (ionization) is a strong limitation. In particular, measuring only the on-axis depletion factor is insufficient for determining $\bar{\mu}_N$.

The next important experimental progress was done by de Heer, Milani, and Châtelain.[12] They performed SG deflection experiments on Fe$_N$ ($15 \leq N \leq 650$) and measured the cluster intensity all along the direction of the field gradient, which is perpendicular to the beam axis. In this way they discovered that the clusters deflect uniquely in the direction of *increasing* field showing a somewhat broad spatial distribution. This implies that the magnetic moments of *isolated* Fe clusters relax in external magnetic field tending to align parallel to the field, a remarkable effect which was not expected.[12] In this work, the average deflection $D$ was related directly to the average magnetic moment per atom $\mu$ by using the classical Stern-Gerlach relation $\mu \propto Dmv^2/H$. The values of $\mu$ derived in this way increase with increasing $H$ and are much *smaller* than the bulk magnetization per atom $\mu_b$. De Heer and coworkers already recognized that the "measured" magnetic moments $\mu$ are a lower bound for the magnitude of the intrinsic magnetic moments of the cluster, the $\bar{\mu}_N$ that theory predicted to be *larger* than $\mu_b$.[1-3] However, at that time there was no clear way of relating the experimental observations to the intrinsic magnetic behavior of the cluster. The comparison between theory and experiment seemed quite controversial, in particular concerning the temperature dependence of $D$. It was not until the work of Bucher, Douglass, and Bloomfield[13] that the relation between the measured deflections $D$ and $\bar{\mu}_N$ became evident.

Bucher et al.,[13] performed experiments on Co$_N$ clusters ($20 \leq N \leq 200$) and showed that the observed small average deflections were consistent with a superparamagnetic relaxation of the intrinsic magnetic moment of the cluster,[13] as it had been already observed in supported magnetic nanoparticles and as proposed for free clusters by Khanna and Linderodth.[16] Similar conclusions were also obtained by independent theoretical studies.[17] Assuming that clusters undergo superparamagnetic relaxation, Bucher et al., inferred the value of the intrinsic magnetic moment per atom $\bar{\mu}(Co_N) = (2.1 \pm 0.2)\mu_B$ which, in qualitative agreement with existing calculations,[1-3] is larger than the bulk magnetization. Working in the superparamagnetic regime opened the possibility of a series of systematic experimental studies.[18-23] Measurements on clusters of nonmagnetic TMs yield so far negative results for V$_N$, Cr$_N$ and Pd$_N$, *i.e.*, no magnetic deflection.[19] In contrast, Rh$_N$ clusters were found to bear rather large magnetizations.[20] This was the first experimental observation of a transition from nonmagnetic to magnetic behavior upon reduction of the system size. Theoretical results already indicated the possibility of this effect.[6,24] Billas et al., determined the magnetization of Fe$_N$ clusters as a function of size and temperature for $25 \leq N \leq 700$ and $100 \text{ K} \leq T \leq 900 \text{ K}$ (Ref. 21). They obtained that the low-temperature $\bar{\mu}_N$ increase with decreasing cluster size, presenting some oscillations as a function of $N$ and reaching a value of about $3\mu_B$ for the smallest sizes ($T \simeq 100$ K). A similar behavior is also observed in Co and Ni clusters.[22,23] These new experiments on Fe$_N$ (Ref. 21) also revealed the breakdown of the superparamagnetic regime in the case of supersonically cooled clusters. The later effect was explained as the result of a resonant coupling between cluster rotations and Zeeman splittings (see also Ref. 25).

In the following Sections a brief overview is presented on recent theoretical work on itinerant magnetism and electron correlations in small clusters. First, self-consistent tight-binding studies on free and embedded clusters are discussed (Sects. II and III). The environment dependence of the magnetic behavior resulting from surface and interface proximity-effects, bond-length relaxations and changes in structure is analyzed concisely within this framework. Section IV provides a first insight on the finite temperature behavior of small TM clusters from the point of view of spin-fluctuation theory. Finally, in Sect. V, we examine the role of electron correlations and cluster structure on the ground-state and excited-state magnetic properties. Rigorous results obtained in the framework of the topological Hubbard model are surveyed.

## II. Free Clusters: Surface Effects

A systematic way of analyzing the onset of magnetism in clusters and its relation to the electronic structure is provided by the determination of the magnetic properties as a function of the ratio between the intra-atomic $d$-electron exchange integral $J$ and the band width $W$ of the corresponding solid. The variations of $J/W$ can be regarded as resulting from changes in the interatomic distances $R_{ij}$ (*e.g.*, for TMs $W \propto t_{ij} \propto R_{ij}^{-5}$) or from changes in the spatial extension of the atomic-like $d$-electron wave functions, as it is the case for elements within the same group. Results for small Fe clusters have been reported in Ref. 6. Magnetism sets in for values of $J$ larger than a critical $J_c$ which depends on the size and structure of the cluster. $J_c$ is smaller in $Fe_9$ and $Fe_{15}$ than in the solid. Consequently, the magnetic moments of clusters are larger than the bulk magnetization. These effects result from the reduction of the local coordination numbers as the size of the system decreases[3] and are in good agreement with cluster experiments and well known surface properties.[13,21] A particularly interesting result is that the clusters can be magnetic for values of $J/W$ for which the macroscopic solid is non-magnetic. A transition form paramagnetic to ferromagnetic behavior is therefore possible by reducing the system size. Indeed, this phenomenon has been experimentally observed in the case of Rh clusters.[20] For more recent specific calculations on $Rh_N$ see Ref. 26. In any case, it is important to recall at this point that the Hartree-Fock approximation tends to overestimate systematically the stability of ferromagnetism and the formation of magnetic moments due to the strong overestimation of the energy of nonmagnetic states. Taking into account electron correlation effects is therefore crucial to a profound and systematic understanding of the magnetic behavior of $4d$-TM clusters. This problem will be also addressed in Sect. V.

As $J$ is increased beyond $J_c$, the average magnetic moment per atom $\bar{\mu}_N$ increases by discrete steps $\Delta \bar{\mu}_N = 2l/N$, where $l$ is the number of electrons which spin is flipped. The resulting redistributions of the spin-polarized density as a function of $J/W$ are accompanied with strong changes in the local magnetic moments $\mu(i)$. Close to a spin flip, the local densities of electronic states present large peaks at the Fermi energy $\varepsilon_F$, which indicates that the states changing occupation are highly degenerate.[6] The electronic structure is thus very sensitive to the interatomic distances. Concerning the magnetic order within the cluster one finds that *bcc*-like clusters are in general ferromagnetic [$\mu(i) > 0, \forall i$]. However, one may sometimes find, in a limited range of small values of $J/W$, that the local magnetic moments at the center of the cluster point opposite to the average magnetization. In these situations of unsaturated magnetic moments, $\mu(i)$ and $\bar{\mu}_N$ depend strongly on the nearest-neighbor (NN) distances. In particular for $Rh_N$ a remarkable size dependence of the average magnetic moments results.[24] The local magnetic moments tend to increase as we go from the center to the surface of the cluster. As expected, the enhancement of the average magnetic moment of the cluster is dominated by the surface contributions. Qualitatively this can be understood as a consequence of the reduction of the local coordination number and the associated reduction of the effective $d$ band width.[3] The structural dependence of the magnetic properties can be clearly illustrated by comparing the behaviors of clusters having fcc- and bcc-like structures. In contrast to bcc clusters, fcc $Fe_{13}$ and $Fe_{19}$ always present antiferromagnetic-like order as well as much smaller local magnetic moments $\mu(i)$.[3,6] Moreover, the presence of antiferromagnetic order in compact structures raises quite naturally the question of magnetic frustrations and possible noncollinear spin arrangements in finite systems. Such more complicated magnetic structures are expected to be important in $Cr_N$ and in alloy clusters like $(Cr_xFe_{1-x})_N$.[8,27] The remarkable role played in free clusters by the surface atoms let us expect that the physics at the interface of embedded clusters should be at least as fascinating.

## III. Embedded Clusters: Interface Effects

The magnetic properties of free clusters are a subject of considerable interest to materials science in view of magnetic recording technology. However, for many practical applications involving clusters, these are not isolated but embedded in a matrix or deposited on a surface. It is therefore important to extend our knowledge on the magnetic properties of free clusters to situations where the clusters are in contact with a macroscopic environment. Moreover, the studies on free and embedded clusters yield complementary informations which contribute to the characterization of the various specific behaviors of these novel materials.

In this Section we shall discuss the magnetic properties of Fe clusters embedded in Cr. These materials are relevant from a technological point of view since they show magnetic behaviors which are very sensitive to the structural and chemical environment of the atoms. Moreover, the competition between the antiferromagnetic order of the matrix and the tendency of $Fe_N$ clusters to order ferromagnetically[1-3] leads to a physical situation where the magnetic interactions between the cluster and the matrix are particularly interesting. Therefore, one expects to encounter a variety of novel magnetic effects. The problem is to some extent similar to that of Fe/Cr multilayers, which deserved considerable attention in past years.[28-30] It is one of the purposes of this Section to compare the properties of embedded clusters and multilayers from the point of view of interface magnetism, in order to stress the local aspects of the magnetic interactions between Fe and Cr.

In the following, we summarize the main conclusions obtained in a recent study on $Fe_N$ clusters ($N \leq 51$) embedded as substitutional impurities in the Cr bcc matrix. For further details the reader should refer to the original papers:[9]

(i) The magnetic moment of the atomic impurity, $\mu(1) = 0.26\mu_B$ ($N = 1$), is drastically reduced with respect to the bulk magnetization of pure Fe $\mu_b = 2.2\mu_B$. This can be interpreted as a consequence of the smaller exchange splittings of the Cr atoms with which the Fe-impurity orbitals hybridize strongly. Moreover, the local magnetic moment at the Fe impurity is antiparallel to its NNs, i.e., the magnetic coupling is consistent with the spin-density wave of the matrix.

(ii) As long as the size of the embedded cluster is very small ($N \leq 4$), the magnetic order *within* the $Fe_N$ cluster is antiferromagnetic. For $N \leq 4$, the size of the local magnetic moments $\mu(i)$ are strongly reduced, as in the case of the atomic impurity. In other words, the Fe clusters adopt the magnetic order prescribed by the spin-density wave of the matrix. This contrasts with the behavior observed in free clusters, which show ferromagnetic order and $\mu(i) > \mu_b$ (see Sect. II).

(iii) For $N \geq 6$ one observes a transition from antiferromagnetic to ferromagnetic order within $Fe_N$. This is followed by a considerable increase of the local magnetic moments at the Fe atoms, although close to the interface with the matrix the $\mu(i)$ are still smaller than $\mu_b$.
(iv) There is a strong tendency to antiferromagnetic coupling between Fe and Cr moments at the interface. The magnetic order among Cr atoms, given by the sign of $\mu(i)$, is not affected by the presence of the Fe cluster, even in cases where the shape of the cluster prevents a perfect antiferromagnetic coupling at the interface. However, one observes important quantitative changes in $|\mu(i)|$ at Cr atoms close to the cluster.
(v) The magnetic moments $\mu(i)$ of the Fe atoms are very sensitive to the local atomic environment. For example, $\mu(i)$ is found to be roughly proportional to the number of Fe atoms $Z_{Fe}(i)$ found in the first NN shell of atom $i$ [$\mu(i) = 0.61\mu_B$, $0.99\mu_B$, $1.29\mu_B$ and $1.42$–$1.59\mu_B$ for $Z_{Fe}(i) = 1$–$4$, respectively]. For larger clusters ($N \geq 9$), the magnetic moment at the center of the cluster is quite close to the Fe-bulk value. However, the average magnetization per atom is always smaller than $\mu_b(Fe)$ due to the contributions of interface Fe atoms. The trend is thus opposite to that of free clusters or infinite surfaces.
(vi) The electronic densities of states reflect very clearly the cluster-matrix hybridizations and the transition from antiferromagnetic to ferromagnetic order within $Fe_N$. Small embedded Fe clusters ($N \leq 4$) present an electronic structure which is very similar to that of pure Cr. Larger clusters ($N \geq 9$) show a ferromagnetic-like exchange splitting and the first signs of convergence towards Fe-bulk behavior. Nonetheless, the changes induced by the cluster-matrix interactions are very significant, as can be observed, for example, by comparing the results for free and embedded $Fe_{15}$.[7,9]

Fe clusters embedded in a Cr matrix have many properties in common with Fe/Cr multilayers. The antiferromagnetic order in Cr even close the interfaces, the antiferromagnetic coupling between Fe and Cr moments at the interface, the strong reduction of the local magnetic moments at Fe atoms close to the interface, the possibility of a slight enhancement of the Fe moment beyond the bulk value at the center of a cluster or film, and the strong changes of $\mu(i)$ associated to frustrations, are all characteristics shared by Fe-Cr systems both in the form of embedded clusters or multilayer structures.[9,30] One concludes that, at least in this case, the immediate local environment of the atoms gives the dominant contribution to the magnetic behavior, even though, for quantitative predictions, the details of the electronic structure and therefore the geometrical structure at a larger scale are also important.

The results discussed in this Section have shown that magnetic properties of Fe particles embedded in a matrix are qualitatively different form those of free (unsupported) clusters. Now, in the case of clusters on surfaces, one has to take into account both the reduction of the local coordination number, as in free clusters, and the cluster substrate interactions, as in embedded clusters. A variety of interesting magnetic behaviors is expected to result from these competing effects. Research in this direction is certainly worthwhile, particularly in view of potential applications.

## IV. Spin Fluctuations at Finite Temperatures

The temperature dependence of the magnetic properties of free clusters is a subject of fundamental importance, specially since the critical behavior of finite systems is intrinsically different from that of the bulk.[31] Strong deviations from bulk-like behavior are expected to occur when the correlation length $\lambda(T) \sim (T - T_c)^{-\gamma}$ becomes of the order of the cluster radius $R$. This is the case at temperatures $T \leq T^*$, where $(T^* - T_c) \sim R^{-1/\gamma}$. The divergence at $T_c$ in the specific heat $C_V(T)$ and magnetic susceptibility $\chi(T)$ disappear, since the long wave-length magnetic fluctuations are suppressed by the finite size of the cluster (i.e., $k \geq k_{\min} \sim 1/R$). Instead, these properties present a peak at $T = T_c$ with a size dependent width. Notice that it might be difficult to define a unique critical temperature $T_c(N)$, since the position of the peak in $C_V(T)$ and $\chi(T)$ can be different.[31] In addition to the temperatures $T^*(N)$ and $T_c(N)$ one is interested in the size dependence of the temperature $T_{SR}(N)$ at which the short-range correlations between local magnetic moments [e.g., between NN $\mu(i)$] are destroyed. Let us recall that in solids (in the bulk or near surfaces of Fe, Co and Ni) there is a significant degree of short-range magnetic order (SRMO) even for $T > T_c$, i.e., even after the average magnetization $M(T)$ vanishes [$T_{SR}(b) > T_c(b)$]. For small clusters having a radius $R$ smaller than the range of SRMO, it is no longer possible to increase the entropy without destroying the energetically favorable local magnetic correlations. Therefore, one expects that for sufficiently small $N$, $T_c(N) \to T_{SR}(N)$. However, it seems difficult to infer *a priori* the trends in the size dependence of $T_{SR}(N)$ and $T_c(N)$. On the one side, taking into account the enhancement of the local magnetic moments $\mu(i)$ and of the $d$-level exchange splittings $\Delta\varepsilon_x(i) = \varepsilon_{i\downarrow} - \varepsilon_{i\uparrow}$, one could expect that $T_{SR}(N)$ and $T_c(N)$ should be larger in small clusters than in the bulk. However, on the other side, it should be energetically easier to disorder the local magnetic moments in a cluster [e.g., by flipping or canting $\mu(i)$] since the local coordination numbers are smaller. If the later effect dominates, $T_c(N)$ should decrease with decreasing $N$. In addition, recent model calculations[32] indicate that structural changes or fluctuations at $T > 0$ are likely to affect the temperature dependence of the magnetization, in particular for systems like Fe$_N$ and Rh$_N$ which show, already at $T = 0$, a remarkable structural dependence. In order to derive reliable conclusions concerning the size dependence of $T_c(N)$ and $T_{SR}(N)$, the electronic theory must take into account both the fluctuations of the magnetic moments and the itinerant character of the $d$ electron states. Simple spin models, for example based on the Heisenberg or Ising model, are not expected to be very predictive, at least until they incorporate the electronic effects responsible for the size dependence of the interactions between the magnetic moments. In fact, surface studies of itinerant magnetism have already shown that the effective exchange interactions $J_{ij}$ between NN moments $\mu(i)$ and $\mu(j)$ depend strongly on the local coordination numbers of sites $i$ and $j$.[33] A similar behavior is also found in clusters, as it will be discussed below.[34]

Recently, a simple relation has been derived between the low and high-temperature values of the magnetization per atom of a $N$-atom cluster [$\bar{\mu}_N(T = 0)$ and $\bar{\mu}_N(T > T_c)$] which allows to infer the degree of SRMO within the cluster from the experimental results for $\bar{\mu}_N(T = 0)$ and $\bar{\mu}_N(T > T_c)$.[35] Let us measure the degree of SRMO by the number of atoms $\nu$ involved in a SRMO domain. The average magnetization per atom at $T > T_c(N)$ is then approximately given by[35]

$$\bar{\mu}_N(T > T_c) \simeq \bar{\mu}_N^0 \sqrt{\nu/N}, \qquad (1)$$

which represents the average $\sqrt{\langle \mu^2 \rangle}$ of $N/\nu$ randomly oriented SRMO domains, each carrying a magnetic moment $\nu \mu_N^0$ [$\mu_N^0 = \bar{\mu}_N(T = 0)$]. Notice that in order that Eq. (1) and the underlying physical picture of the magnetic cluster at $T > T_c(N)$ remain valid, it is implicitly necessary that $\nu \ll N$ and $T_{SR}(N) > T_c(N)$. By increasing $\nu$ we may go from the disordered-local-moment regime, where $\nu = 1$ and SRMO is negligible, to the limit where $\nu$ and $N$ are comparable and SRMO dominates ($T_c \to T_{\text{SR}}$). A first estimate of the actual value of $\nu$ is provided by bulk and surface results.[33,36,37] For bulk Fe, Haines et al.,[36] have retrieved a range of SRMO near $T_c$ of about 5.4 Å (at least 4 Å), which corresponds to $\nu \simeq 15$ (up to next-nearest neighbors). Similar values are obtained in calculations of SRMO in Fe bulk and Fe surfaces.[33] Moreover, for Fe$_N$ clusters one finds a particularly high stability of ferromagnetism for $N \simeq 15$. This concerns both the energy gained upon the formation of magnetic moments, $\Delta E(N) = E(\mu) - E(\mu = 0)$, and the energy $\Delta F_i(\xi)$ involved in flipping a local magnetic moment within Fe$_{15}$ (Refs. 34 and 35). Therefore, $\nu = 15$ seems a reasonable a priori estimate of the degree of SRMO in Fe clusters. For Ni, the SRMO is generally expected to be stronger than for Fe.[37]

As shown in Ref. 35, the comparison between experiment and the results derived from Eq. (1) neglecting SRMO (i.e., with $\nu = 1$) is very poor. This rules out the disorder-local-moment picture for Fe and Ni clusters, as it is already known to be the case for the bulk and near surfaces.[33,36,37] On the contrary, the results including SRMO are in very good agreement with experiment for both Fe$_N$, with $\nu = 15$, and Ni$_N$, with $\nu = 19$–43 (Ref. 35) This provides a clear evidence for the existence of SRMO in these clusters above $T_c(N)$ and is also consistent with known surface and bulk properties of itinerant magnetism.[33,36,37] Equation (1) has also been used to infer the degree of SRMO in clusters from the experimental results for $\bar{\mu}_N(T=0)$ and $\bar{\mu}_N(T > T_c)$. For example, assuming that $\nu$ is independent of $N$, one obtains $\nu = 13$–15 for Fe$_N$ ($25 \leq N \leq 700$). A more detailed analysis of Eq. (1) shows that these values of $\nu$ are not overestimated.[35] In fact, the degree of SRMO in the clusters could be somewhat larger than in the corresponding solids, possibly as a consequence of the enhancement of the local magnetic moments and exchange splittings.[1–3] One concludes that SRMO plays a significant role in the finite-temperature behavior of magnetic TM clusters and that it should be taken into account in the interpretation of experiments as well as in future theoretical developments.

The self-consistent tight-binding theory discussed in Sect. II has been extended to finite temperatures[34] by using a functional integral formalism.[38–40] In order to calculate the partition function of the $d$-electron Hamiltonian, the many-body interaction terms are linearized by means a two-field Hubbard-Stratonovich transformation within the static approximation. A charge field $\eta_i$ and an exchange field $\xi_i$ are thus introduced at each cluster site $i$, which represent the local fluctuations of the $d$-electron energy levels and exchange splittings at $T > 0$. The saddle-point approximation for the charge fields yields a set of self-consistent equations for $\bar{\eta}_i = \bar{\eta}_i(\xi_1, \ldots, \xi_N)$. The finite temperature properties are then obtained as an average over all possible exchange-field configurations. For example, the local magnetization at atom $i$ is given by

$$\mu_i(T) = \langle\langle S_i^z \rangle\rangle = \frac{1}{\mathcal{Z}} \int \xi\, e^{-\beta F_i(\xi)}\, d\xi, \tag{2}$$

where

$$F_i(\xi) = -k_{\text{B}} T \ln \int \prod_{j \neq i} d\xi_j\, e^{-\beta F(\xi_1 \ldots \xi \ldots \xi_N)}$$

is the local free energy associated to an exchange field $\xi$ at atom $i$. $\mathcal{Z} = \int e^{-\beta F_i(\xi)}\, d\xi$ is proportional to the partition function. $P_i(\xi) = e^{-\beta F_i(\xi)}/\mathcal{Z}$ represents the probability for the value $\xi$ of the exchange field at atom $i$.

A first insight on the finite-temperature magnetic behavior of Fe$_N$ and Ni$_N$ clusters can be obtained by considering the low-temperature limit of $F_i(\xi)$.[34] For these clusters one always obtains $\Delta F_i(\xi) = F_i(\xi) - F_i(\xi = \mu_i^0) > 0$, which indicates, as expected, that the ferromagnetic order is stable at low temperatures. For Fe$_9$, $F_i(\xi)$ shows two minima located at $\xi^+ = \mu_i^0$ and $\xi^- \simeq -\mu_i^0$, where $\mu_i^0$ refers to the local magnetic moment at $T = 0$. This double minimum structure indicates that the flips of the local magnetic moments keeping the modulus approximately constant are the dominant magnetic excitations. Only small fluctuations of the modulus of $\xi$ are possible within an excitation energy $\Delta F_i(\xi)$ smaller than the energy $\Delta F_i(\xi^-) = F_i(\xi^-) - F_i(\xi^+)$ required to flip a local magnetic moment. At finite temperatures the probability $P_i(\xi)$ has two sharp maxima at $\xi \simeq \xi^+$ and $\xi \simeq \xi^-$ with $P_i(\xi^-) \gg P_i(\xi = 0)$. The importance of moment flips is also characteristic of bulk Fe.[38–40] The fact that very small clusters show such a Heisenberg- or Ising-like behavior is not surprising since the kinetic energy loss caused by flipping a local magnetic moment $[\xi \simeq \mu_i \to \xi \simeq -\mu_i]$ is smaller when the local coordination number $Z_i$ is smaller ($E_K \propto \sqrt{Z_i}$), whereas the exchange energy $\Delta E_x = (J/4)\sum_i \mu_i^2$, being a local property, is much less affected by the change of sign of $\xi$. Even Ni, which in the solid state presents a single minimum in $F_i(\xi)$ and which is therefore dominated by fluctuations of the modulus of the local magnetic moments,[40] shows a double minimum structure in $F_i(\xi)$ for sufficiently small $N$ (Ref. 34). The existence of a transition from Heisenberg-like to itinerant-like behavior with increasing cluster size has been recently proposed. Physically, it has been interpreted as a consequence of the competition between the Coulomb interaction energy, which is relatively more important in small clusters, and the kinetic $d$-band energy, which is most important in the bulk.[34]

Upon relaxation of the NN distances, *i.e.*, bond-length contraction,[3] one finds no qualitative but strong quantitative changes in $F_i(\xi)$ as a function of $\xi$. Besides the shift of the position of the minimum at $\xi^+$ reflecting the reduction of the local magnetic moments $\mu_i$ at $T = 0$, one observes a remarkable reduction (about a factor 10) of the energy $\Delta F_i(\xi^-) = F_i(\xi^-) - F_i(\xi^+)$ required to flip a local magnetic moment or exchange field. A similar large reduction of the Curie temperature $T_c(N)$ is expected to occur upon relaxation, since in first approximation $T_c(N) \propto \Delta F_i$ (Ref. 41). These results clearly show, once more, the strong sensitivity of the magnetic properties of 3$d$-TM clusters to changes in the local environment. Recent calculations including correlations effects exactly within the single-band Hubbard model[32] have also revealed the importance of structural changes and structural fluctuations to the temperature dependence of the magnetic properties of clusters (see Sect. V.2). Finally, let us also point out that $F_i(\xi)$ depends strongly on the position of the atom $i$ within a given cluster, indicating that the temperature dependence of the local magnetization is different at different cluster sites. A similar behavior is found near the surface of macroscopic systems. Further details will be published elsewhere.[34]

## V. Electron Correlations and Geometry Optimization

The results discussed in the previous Sections have revealed a variety of cluster specific magnetic properties. Nevertheless, two central aspects still remain to be considered in some detail, namely, the role of electronic correlations, which are fundamental for magnetism, and the effect of the optimization (or sampling at $T > 0$) of the cluster geometry, on which the magnetic properties of itinerant electrons depend strongly. Since these two problems are very difficult, and their combination even more so, it is understandable that most theoretical studies performed so far have attempted to deal with one of them at a time.[1–3,14,42–45] It is the purpose of this Section to recall the main results of a recent study of electron correlations and cluster structure[32] where these two important features are treated rigorously within the framework of a many-body model approach. The point of view is therefore complementary to that of the previous Sections, in which the description of the electronic structure is more realistic but the effects of Coulomb interactions are treated in a mean-field approximation.

The electronic dynamics can be simplified by considering a model which allows an exact, or at least very accurate solution, and which at the same time contains a sufficient degree of complexity to be able shed light on the physics of real systems. A model with such characteristics is given by the well known Hubbard Hamiltonian[46] in which framework numerous significant results on the magnetic properties of itinerant electrons have been achieved. At low energies, the electronic behavior results from a delicate balance between the tendency to delocalize the valence electrons in order to reduce their kinetic energy, and the effect of the Coulomb repulsions associated to local charge fluctuations. The relative importance of these contributions depends strongly on the value of the total spin $S$ and on the ratio between the Coulomb interaction strength $U$ and the nearest-neighbor hopping $-t$, $(t > 0)$. Therefore, the determination of the structure and magnetic behavior requires an accurate treatment of electron correlations, particularly for $U \gg t$. Moreover, the ground-state corresponding to different cluster structures may be of very different nature.[42-44] For this reason, it is also necessary to optimize the geometry if rigorous conclusions about the magnetic properties of these clusters are to be achieved.

## V.1 Cluster Structure and Total Spin at $T = 0$

In the following, the cluster geometry has been determined by taking into account only NN hoppings with fixed NN bond length. This implies in fact a discretization of the configurational space, since only the topological aspect of the structure is relevant for the electronic properties. Therefore, the geometry optimization can be performed within the graph space,[32,45] although for the study of clusters we must restrict ourselves to graphs which can be represented as a true structure in space. It should be noted that the number of geometry configurations increases extremely rapidly with the number of sites $N$. For this reason, full geometry optimizations of this kind have only been done for $N \leq 8$ [$N \leq 9$ for $U = 0$ (Ref. 45)]. The optimal cluster structure is determined as a function of $U/t$ and the number of valence electrons $\nu$ by comparing the ground-state energies of *all* possible nonequivalent structures. The energies are calculated exactly within a full many-body scheme by means of Lanczos' iterative method. Once the optimal structure is identified, the spin correlation functions and total spin $S$ are calculated, from which the ground-state magnetic order is determined.

For low concentrations of electrons or holes (*i.e.*, $\nu/N \leq 0.4$–$0.6$ and $2 - \nu/N \leq 0.3$–$0.6$) the optimal cluster structure is independent of $U/t$, *i.e.*, the structure which yields the minimal kinetic energy (uncorrelated limit) remains the most stable one irrespectively of the strength of the Coulomb interactions. Furthermore, no magnetic transitions are observed: the ground-state is always a singlet or a doublet. This indicates that for low carrier concentrations the Coulomb interactions are very efficiently suppressed by correlations, so that the magnetic and geometric structure of the clusters are dominated by the kinetic term. These trends are physically plausible since the effects of electron-electron interactions should be roughly proportional to $(\nu/N)^2$. Furthermore, they can be justified by general analytical results.[32] Nonetheless, the fact that this holds for finite values of $\nu/N$ and for $U/t \to \infty$, seems not obvious *à priori*.

For small $\nu$, compact structures are obtained which have maximal average coordination number $\bar{Z}$ ($t > 0$). For example, for $\nu = 2$, the ground-state structures are the triangle, tetrahedron, triangular bipyramid, caped bipyramid, and pentagonal bipyramid for $N = 3$–$7$, respectively. These are all substructures of the icosahedron and have the largest possible number of *triangular* loops. In contrast, for large $\nu$ (small $\nu_h = 2N - \nu$) open ground-state structures are found. In particular for $\nu_h = 2$ we obtain bipartite structures which have the largest possible number of *square* loops ($N \geq 4$). As indicated above, this can be qualitatively understood in terms of the single-particle spectrum. In the first case (small $\nu$) the largest stability corresponds to the largest band-width for bonding states having a negative energy $\varepsilon_b \leq -\bar{Z}t$. This is achieved by the most compact structure. In the second case (small $\nu_h$) the largest stability is obtained for the largest band-width for antibonding (positive-energy) states, i.e, for the most compact *bipartite* structure.

A much more interesting interplay between electronic correlations, magnetism and cluster structure is observed around half-band filling (i.e., $|\nu/N - 1| \leq 0.2$–$0.4$). Here, several structural transitions are found as a function of $U/t$. Starting from the ground-state structures for $U = 0$,[45] one typically observes that as $U$ is increased, first one or more of the weakest bonds are broken. This change to more open structures occurs for $U/t \simeq 1$–$4$ and is most often seen for $\nu < N$. In this case the $U = 0$ structures are more compact, while for $\nu > N$ the structures are rather open already for $U \to 0$ (see Ref. 32). The trend to break bonds as $U/t$ increases seems physically plausible since the Coulomb interactions usually tend to weaken the kinetic energy contributions. However, if $U$ is further increased ($U/t > 5$–$6$) it is energetically more advantageous to create additional new bonds. This can be interpreted as a consequence of the electronic correlations. Higher coordination gives the strongly correlated electrons more possibilities for performing a mutually avoiding motion which lowers the kinetic energy without increasing excessively the Coulomb energy due to local charge fluctuations. Moreover, these compact structures are more symmetric. Therefore, the electron-density distribution is more uniform, which also contributes to lower the static Coulomb-repulsion energy.

The structural changes at larger $U/t$ are often accompanied by strong changes in the magnetic behavior. One may say that such structural changes are driven by magnetism. For half-band filling ($\nu = N$) the optimal structures have the minimal total spin $S$ and for large $U/t$ strong antiferromagnetic correlations. The optimal antiferromagnetic structures are *nonbipartite* (e.g., the rhombus is more stable than the square for $N = \nu = 4$). The bonds which are "frustrated" in a static picture of antiferromagnetism yield an appreciable energy lowering when quantum fluctuations are taken into account. Therefore, Hubbard clusters with one electron per site and large $U/t$ can be regarded as frustrated quantum antiferromagnets.

For all studied cluster sizes ($N \leq 8$) the most stable structures for $\nu = N + 1$ show ferromagnetism for large enough $U/t$ (typically $U/t > 4$–$14$). Increasing $U/t$ further leads to a fully-polarized ferromagnetic state [$S = \frac{1}{2}(N-1)$] which is in agreement with Nagaoka's theorem.[47] The value of the Coulomb interaction strength $U_c/t$, above which saturated ferromagnetism dominates, increases with increasing $N$. This is a consequence of the fact that the antiferromagnetic correlations tend to dominate as we approach half band-filling ($\nu/N = 1 + 1/N$). $U_c/t$ is approximately proportional to $N$ and diverges in the thermodynamic limit. For the smaller clusters (viz., $N = 3$, 4, and 6) the $\nu = N + 1$ band-filling is the only case where the optimal structures are ferromagnetic. For $N = 5$ unsaturated ferromagnetism ($S = 1$) is also found for $\nu = 4$, though at rather large values of $U$ ($U/t > 30$). However, for larger clusters, ferromagnetism extends more and more throughout the $U/t$-$\nu/N$ phase diagram. Indeed, for $N = 7$ ferromagnetism is found for $\nu = 4$, 6, 8, and 10; clusters with $N = 8$ can be ferromagnetic for $\nu = 9$–12 (see Ref. 32). The tendency towards ferromagnetism is much stronger for more than half-band filling than for $\nu \leq N$. This consequence of the asymmetry of the single-particle spectra of compact structures is qualitatively in agreement with experiments on 3$d$-TM clusters. In fact one observes, as already discussed in previous Sections, that the magnetic moments per atom $\bar{\mu}$ in V and Cr clusters are very small if not zero ($\mu < 0.6$–$0.8\mu_B$),[19] while Fe, Co and Ni clusters show large magnetizations.[13,21,23] Even small Rh clusters were found to be ferromagnetic.[20] Let us finally remark that the appearance of ferromagnetism is much less frequent than what one would expect from mean-field Hartree-Fock arguments (Stoner criterion). This reflects, once more, the importance of correlations in low-dimensional systems.[32,42–44] However, the Hubbard model, with the restriction of one orbital per site, tends to exaggerate the effects of quantum fluctuations, being one of the most extremely low-dimensional systems one can consider. Extensions of the model by including either several bands or nonlocal interactions would tend to weaken strong fluctuation effects.

## V.2 Electronic Excitations and Structural Changes at $T > 0$

The low-lying electronic excited states of different $S$ and the optimal structures corresponding to different $S$ can be calculated by using appropriate projectors.[32] This allows to quantify, for values of $U/t$ and $\nu$ for which the ground-state is ferromagnetic, the stability of cluster ferromagnetism and to examine the origin of the relevant magnetic excitations at finite temperatures. The excitation energy $\Delta E$ from a ferromagnetic ground state to the lowest lying nonferromagnetic state, $\Delta E = E(S = 0, \frac{1}{2}) - E(S \geq 1)$, gives a measure of the temperature $T_c$ above which the ferromagnetic correlations are strongly reduced by thermal excitations. The excitations $\Delta E$ can be mainly of two different kinds. One may consider a purely *electronic excitation* $\Delta E_{\text{el}}$, where the cluster structure remains fixed to the optimal structure at $T = 0$, or a purely *structural change* $\Delta E_{\text{st}}$, where the electrons remain in the ground state and the structure is changed until the ground state corresponding to this new structure shows no ferromagnetism. As in the case of the exchange-field fluctuation energy $\Delta F_i(\xi)$ (Sect. IV), $\Delta E_{\text{el}}(N)$ and $\Delta E_{\text{st}}(N)$ are very sensitive to changes in the interatomic distances in particular when the reduction of $U/t$ implies a transition from saturated to unsaturated ferromagnetism in the ground state. Concerning the size dependence, one finds that $\Delta E$ shows strong even-odd oscillations as a function of $N$, which are due to the size dependence of the final-state singlet or doublet energy $E(S = 0, \frac{1}{2})$. The doublets, which correspond to an even number of sites ($\nu = N+1$), are relatively more stable than the singlets (odd number of sites). This trend is opposite to the one observed at half-band filling ($\nu = N$) where the singlets ($\nu$ even) are relatively more stable than the doublets ($\nu$ odd). Moreover, the amplitude of the even-odd oscillations *increases* with decreasing $J = 4t^2/U$ (i.e., increasing $U/t$). Therefore, they are not related to the coupling constant $J$ but to an interference effect present even in the limit of $U/t \to \infty$. Besides these oscillations we observe that $\Delta E$ decreases with increasing $N$. This is due to the fact that as $N \to \infty$, the band filling approaches the half-filled case ($\nu/N = 1 + 1/N$) where antiferromagnetic correlations dominate. Therefore, for any finite $U/t$ there is always a finite size $N_c \propto U/t$ such that the ground-state is no longer ferromagnetic for $N > N_c$ (i.e., $\Delta E = 0$ for $N = N_c$).

A particularly interesting result is that $\Delta E_{el}$ and $\Delta E_{st}$ have similar values and that $\Delta E_{st}$ can be even smaller than $\Delta E_{el}$.[32] This indicates that *structural changes* are at least as important as the *electronic excitations* for determining the temperature dependence of the magnetization of small clusters. A complete description of the finite-temperature magnetic properties of small clusters within the Hubbard model requires that the electronic and structural effects are treated simultaneously and at the same level. Experimentally, there seem to be indications that structural changes could be related to the drop of the magnetization observed in Fe$_N$ for increasing $T$.[21] However, it still remains to be proven whether the conclusions obtained here from the calculation of $\Delta E_{el}$ and $\Delta E_{st}$ within the single-band Hubbard model and for $\nu = N + 1$ are also valid for more realistic $d$-band models and for other band fillings. If this would be so, the mechanisms responsible for the temperature dependence of the magnetization of small $3d$-TM clusters could be intrinsically different from those known in the solid state. The results obtained so far within the $d$-electron spin-fluctuations theory (Sect. IV and Ref. 34) do not allow to derive definitive conclusions on this matter. Nevertheless, the already reported strong dependence of the spin-flip energy on the cluster structure and NN distances lets us expect that the trends derived within the Hubbard model should remain valid in more realistic calculations.

## Acknowledgments

It is a pleasure to thank Prof. K. H. Bennemann and Prof. J. Dorantes-Dávila for numerous enthusiastic and helpful discussions. The Laboratoire de Physique Quantique (Toulouse) is Unité Mixte de Recherche of the CNRS.

## References

1. D. R. Salahub and R. P. Messmer, *Surf. Sci.* **106**, 415 (1981); C. Y. Yang, *Phys. Rev. B* **24**, 5673 (1981).
2. K. Lee, J. Callaway, and S. Dhar, *Phys. Rev. B* **30**, 1724 (1985); K. Lee et al., *ibid.* **31**, 1796 (1985); K. Lee and J. Callaway, *ibid.* **48**, 15 358 (1993).
3. G. M. Pastor, J. Dorantes-Dávila, and K. H. Bennemann, *Physica B* **149**, 22 (1988); *Phys. Rev. B* **40**, 7642 (1989).
4. R. H. Victora, L. M. Falicov, and S. Ishida, *Phys. Rev. B* **30**, 3896 (1989).
5. H. Haydock, *Solid State Physics*, (Academic, London, 1980), Vol. **35**, p. 215 ff.
6. J. Dorantes-Dávila, H. Dreyssé, and G. M. Pastor, *Phys. Rev. B* **46**, 10 432 (1992).
7. A. Vega, J. Dorantes-Dávila, L. C. Balbás, and G. M. Pastor, *Phys. Rev. B* **47**, 4742 (1993).
8. A. Vega, J. Dorantes-Dávila, G. M. Pastor, and L. C. Balbás, *Z. Phys. D* **19**, 263 (1991).
9. A. Vega, L. C. Balbás, J. Dorantes-Dávila, and G. M. Pastor, *Phys. Rev. B* **50**, 3899 (1994); *Computational Materials Science* **2**, 463 (1994).
10. G. M. Pastor, J. Dorantes-Dávila, S. Pick, and H. Dreyssé, *Phys. Rev. Lett.* **75**, 326 (1995).
11. See also the contributions by J. Dorantes-Dávila and H. Dreyssé to this volume.
12. W. A. de Heer, P. Milani, and A. Châtelain, *Phys. Rev. Lett.* **65**, 488 (1990).
13. J. P. Bucher, D. C. Douglass, and L. A. Bloomfield, *Phys. Rev. Lett.* **66**, 3052 (1991).

14. See, for instance, B. I. Dunlap, *Z. Phys. D* **19**, 255 (1991); J. L. Chen, C. S. Wang, K. A. Jackson, and M. R. Perderson, *Phys. Rev. B* **44**, 6558 (1991); M. Castro and D. R. Salahub, *Phys. Rev. B* **49**, 11 842 (1994); P. Ballone and R. O. Jones, *Chem. Phys. Lett.* **233**, 632 (1995).
15. D. M. Cox et al., *Phys. Rev. B* **32**, 7290 (1985).
16. S. N. Khanna and S. Linderoth, *Phys. Rev. Lett.* **67**, 742 (1991).
17. P. J. Jensen, S. Mukherjee, and K. H. Bennemann, *Z. Phys. D* **21**, 349 (1991).
18. D. C. Douglass, A. J. Cox, J. P. Bucher, and L. A. Bloomfield, *Phys. Rev. B* **47**, 12 874 (1993).
19. D. C. Douglass, J. P. Bucher, and L. A. Bloomfield, *Phys. Rev. B* **45**, 6341 (1992).
20. A. J. Cox, J. G. Louderback, and L. A. Bloomfield, *Phys. Rev. Lett.* **71**, 923 (1993); A. J. Cox, J. G. Louderback, S. E. Apsel, and L. A. Bloomfield, *Phys. Rev. B* **49**, 12 295 (1994).
21. I. M. L. Billas, J. A. Becker, A. Châtelain, and W. A. de Heer, *Phys. Rev. Lett.* **71**, 4067 (1993).
22. I. M. L. Billas, A. Châtelain, and W. A. de Heer, *Science* **265**, 1682 (1994).
23. S. E. Apsel, J. W. Emert, J. Deng, and L. A. Bloomfield, *Phys. Rev. Lett.* **76**, 1441 (1996).
24. R. Galicia, *Rev. Mex. Fís.* **32**, 51 (1985).
25. P. J. Jensen and K. H. Bennemann, *Z. Phys. D* **26**, 246 (1993); A. Maiti and L. M. Falicov, *Phys. Rev. B* **48**, 13 596 (1993).
26. B. V. Reddy, S. N. Khanna, and B. I. Dunlap, *Phys. Rev. Lett.* **70**, 3323 (1993); Y. Jinlong, F. Toigo, and W. Kelin, *Phys. Rev. B* **50**, 7915 (1994); Z. Q. Li, J. Z. Yu, K. Ohno, and Y. Kawazoe, *J. Phys. Cond. Matter* **7**, 47 (1995); B. Piveteau, M. C. Desjonquéres, A. M. Olés, and D. Spanjaard, *Phys. Rev. B* **53**, 9251 (1996); J. Dorantes-Dávila, P. Villaseñor-González, H. Dreyssé, and G. M. Pastor, *Phys. Rev. B* **55** 15 084 (1997).
27. M. A. Ojeda-López, J. Dorantes-Dávila, and G. M. Pastor, *J. Appl. Phys.* **81**, 4170 (1997), and in this volume.
28. J. Unguris, R. J. Celotta, and D. T. Pierce, *Phys. Rev. Lett.* **69**, 1125 (1992); T.G̃. Walker, A. W. Pang, H. Hopster, and S. F. Alvarado, *Phys. Rev. Lett.* **69**, 1121 (1992); F. U. Hillebrecht et al., *Europhys. Lett.* **19**, 711 (1992); R. W. Wang et al. *Phys. Rev. Lett.* **72**, 920 (1994).
29. P. Grünberg et al., *Phys. Rev. Lett.* **57**, 2442 (1986); C. Carbone and S. F. Alvarado, *Phys. Rev. B* **36**, 2433 (1987); M. N. Baibich et al., *Phys. Rev. Lett.* **61**, 2472 (1988); J. J. Krebs, P. Lubitz, A. Chaiken, and G. Prinz, *Phys. Rev. Lett.* **63**, 4828 (1989); S. S. P. Parkin, N. More, and K. P. Roche, *Phys. Rev. Lett.* **64**, 2304 (1990); S. Demokritov, J. A. Wolf, and P. Grünberg, *Europhys. Lett.* **15**, 881 (1991).
30. P. M. Levy et al., *J. Appl. Phys.* **67**, 5914 (1990); F. Herman, J. Sticht, and M. van Schilfgaarde, *J. Appl. Phys.* **69**, 4783 (1991); K. Ounadjela et al., *Europhys. Lett.* **15**, 875 (1991); D. Stoeffler, and F. Gautier, *Phys. Rev. B* **44**, 10 389 (1991); Z. P. Shi, P. M. Levy, and J. L. Fry, *Phys. Rev. Lett.* **69**, 3678 (1992); J-h. Xu and A. J. Freeman, *Phys. Rev. B* **47**, 165 (1993).
31. P. G. Watson, *Phase Transitions and Critical Phenomena*, edited by C. Domb and M. S. Green, (Academic, London, 1972), p. 101; M. N. Barber, *ibid.* Vol. 8, edited by C. Domb and J. L. Lebowitz, (Academic, London, 1983), Vol. 2 p. 145.
32. G. M. Pastor, R. Hirsch, and B. Mühlschlegel, *Phys. Rev. Lett.* **72**, 3879 (1994); *Phys. Rev. B* **53**, 10 382 (1996).

33. J. Dorantes-Dávila, G. M. Pastor, and K. H. Bennemann, *Solid State Commun.* **59**, 159 (1986); *ibid.* **60**, 465 (1986).
34. J. Dorantes-Dávila and G. M. Pastor, to be published.
35. G. M. Pastor and J. Dorantes-Dávila, *Phys. Rev. B* **52**, 13 799 (1995).
36. E. M. Haines, R. Clauberg, and R. Feder, *Phys. Rev. Lett.* **54**, 932 (1985).
37. V. Korenman and R. E. Prange, *Phys. Rev. Lett.* **53**, 186 (1984).
38. J. Hubbard, *Phys. Rev. B* **19**, 2626 (1979); *ibid.* **20**, 4584 (1979); H. Hasegawa, *J. Phys. Soc. Jpn.* **49**, 178 (1980); *ibid.* **49**, 963 (1980).
39. *Electron Correlation and Magnetism in Narrow-Band Systems*, edited by T. Moriya, Springer Series in Solid State Sciences **29**, (Springer, Heidelberg, 1981).
40. Y. Kakehashi, *J. Phys. Soc. Jpn.* **50**, 2251 (1981).
41. From bulk calculations one estimates $T_c \simeq 2\Delta F_i$.[38]
42. L. M. Falicov and R. H. Victora, *Phys. Rev. B* **30**, 1695 (1984).
43. Y. Ishii and S. Sugano, *J. Phys. Soc. Jpn.* **53**, 3895 (1984).
44. J. Callaway, D. P. Chen, and R. Tang, *Z. Phys D* **3**, 91 (1986); *Phys. Rev. B* **35**, 3705 (1987).
45. Y. Wang, T. F. George, D. M. Lindsay, and A. C. Beri, *J. Chem. Phys.* **86**, 3493 (1987).
46. J. Hubbard, *Proc. R. Soc. London Ser. A* **276**, 238 (1963); **281**, 401 (1964); J. Kanamori, *Prog. Theor. Phys.* **30**, 275 (1963); M. C. Gutzwiller, *Phys. Rev. Lett.* **10**, 159 (1963).
47. Y. Nagaoka, *Solid State Commun.* **3**, 409 (1965); D. J. Thouless, *Proc. Phys. Soc. London* **86**, 893 (1965); Y. Nagaoka, *Phys. Rev.* **147**, 392 (1966); H. Tasaki, *Phys. Rev. B* **40**, 9192 (1989).

# Magnetic Moments of Iron Clusters: A Simple Theoretical Model

F. Aguilera-Granja,[1] J. M. Montejano-Carrizales,[2] and
J. L. Morán-López[2]

[1] Departamento de Física Teórica
Universidad de Valladolid
47011 Valladolid
SPAIN

[2] Instituto de Física
Universidad Autónoma de San Luis Potosí
Alvaro obregón 64, 78000 San Luis Potosí, S.L.P.
MEXICO

## Abstract

We present a simple model for the study of ferromagnetic transition metal clusters. Based on Fridel's model and on the geometrical characteristics, we obtain analytical expressions for the cluster average magnetic moment ($\bar{\mu}_N$) as a function of the number of atoms in $bcc$-structures. The oscillatory behavior of $\bar{\mu}_N$ as a function of cluster size reported experimentally in Fe clusters is reproduced by considering *spherical bcc*-clusters.

## I. Introduction

The golden key for the design of the next-generation high-tech components is the understanding of the physico-chemical properties of nanostructures and its evolution to the bulk behavior. In particular, in the case of magnetic materials a large amount of experimental and theoretical work has being devoted to address this important problem.[1–15] However, there are many open questions on how the magnetic properties depend on the size and dimensionality of the system.

In the case of thin films it is experimentally and theoretically well established that the magnetic moment of the atoms in the surface region is larger than the corresponding in the bulk.[1,10] It is expected that stronger effects will show in the case of clusters since a larger fraction of the total number of atoms in the system are located

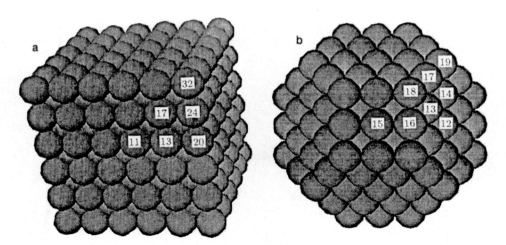

**Figure 1.** Two *bcc*-clusters with approximately the same number of atoms: (a) a 341-atom cubic *bcc*-cluster and (b) a 339-atom rounded *bcc*-cluster. The numbers indicate the *n*th shell of neighbors to the central atom forming the surface.

at the surface. Furthermore, additional unexpected features may produce the fact that clusters are finite in all directions. It was not until very recently that *clean* and *reliable* experimental techniques were developed and reports on the richness of the cluster magnetic behavior started to appear.[2-5] Some of these observations have been answered but many other still wait for a satisfactory explanation. For instance it has been reported that the average magnetic moment in Fe clusters shows an oscillatory behavior as a function of size.[5] Here we address that question and give a possible explanation.

In this communication, we study how the local magnetic moment depends on the site location in the cluster and how it changes as a function of the cluster size. To that aim we use a simple electronic model based on the Hubbard Hamiltonian and solve it within the square band approximation. We extend a study published recently[8] and provide analytical expressions for the average magnetic moment as a function of the number of atoms in terms of the cluster geometrical characteristics.

The clusters studied here are grown around a central site and follow the body-centered cubic structure. We consider two cluster shapes: one is obtained by constructing the cluster in the form of a cube. The smallest cube is obtained by locating eight neighbors around the central site. The next size can be obtained by covering totally the 9-atom cluster with one layer of atoms or a crust. This leads to a 35-atom cube. One can generalize this sequence and obtain the formula:[16]

$$N = (2\nu + 1)(\nu^2 + \nu + 1), \qquad (1)$$

where $\nu = 1, 2, \ldots$, is the number of outer layers or crusts, called also the order of the cluster. This cubic sequence of clusters is well characterized. For example, the number of surface sites $N_\sigma$ is given by:

$$N_\sigma = 6\nu^2 + 2. \qquad (2)$$

Other form to grow the cluster is in an onion-like structure, formed by concentric *shells*. A shell consists of the set of all atoms that, following the *bcc*-arrangement, are equidistant from the center. This mode of growth lead us to the sequence $N = 1, 9, 27, 39, 51, 59, \ldots$, etc. Unfortunately in this case it is not possible to write a closed relationships like Eqs. (1) and (2). We show in Fig. 1 two clusters with a similar number of atoms, 341 and 339, and with cubic and rounded shape, respectively. As shown in this figure, the surface sites may belong to various shells. The numbers on the clusters denote the $n$th neighbors to the central atom.

The atomic magnetic moments in clusters depend on their spatial location in the system. In principle one should calculate in a self-consistent manner the electronic structure using real-space methods. These calculations have been performed in relatively small clusters.[14] Unfortunatelly, as the cluster size increases above the 100-atom regime, the computational task becomes more and more complex and imposible to handle with the current computation facilities. Therefore, to study the magnetic properties of large atomic clusters, it is necessary to make some approximations and to use simpler models.

It is well established that the magnetic properties of transition metals may be described, as a first approximation, by Fridel's model.[17] This simple rectangular $d$-band approximation, leads to the following expression for the magnetic moment as a function of the local coordination number

$$\mu_i = \begin{cases} \mu_b \sqrt{Z_b/Z_i}, & \text{if } Z_i \geq Z_b(\mu_b/\mu_{\text{dim}})^2; \\ \mu_{\text{dim}}, & \text{otherwise}; \end{cases} \quad (3)$$

where $Z_i$ is the coordination number of the atom in the $i$-th position, $Z_b$ the coordination number in the bulk, and $\mu_b$ and $\mu_{\text{dim}}$ are the magnetic moments in the bulk and of dimers,[8] respectively. In the case of Fe, $\mu_b = 1.72\mu_B$ and $\mu_{\text{dim}} = 2.33\mu_B$, where $\mu_B$ is the Bohr magneton.

According to Eq. (3), all the surface atoms will hold a larger magnetic moment and the average magnetic moment $\bar{\mu}_N$ of the cluster is

$$\bar{\mu}_N = \frac{1}{N} \sum_{i=1}^{N} \mu_i. \quad (4)$$

Since only the atoms at the surface of the cluster have a magnetic moment that differs from the bulk value the average magnetic moment can also be written as

$$\bar{\mu}_N = \frac{1}{N} \left\{ (N - N_\sigma)\mu_b + \sum_I N_I \mu_I \right\}, \quad (5)$$

where $N_\sigma$ is the number of surface atoms, $N_I$ is the number of surface atoms of the type $I$ (with coordination number $Z_I$), $\mu_I$ is the magnetic moment of the $I$-type of atom and the sum is done over all the different types of surface atoms. For example, the surface of a perfect cubic bcc-cluster is formed by atoms on square faces, edges and vertices [see Fig. 1(a)]. Their total coordination number is 4, 2, and 1, respectively.

Equation (5) can be also rewritten in terms of the total and the partial dispersions. The total dispersion $\mathcal{D}$ is defined as the ratio of the number of surface atoms to the total number of atoms in the cluster ($N_\sigma/N$). Similarly, the partial dispersions $\mathcal{D}_I$ are defined as the ratio of the number of surface sites of a given type $I$ to the total number of atoms in the cluster ($N_I/N$). Thus, the average magnetic moment per atom can be rewritten as

$$\bar{\mu}_N = (1 - \mathcal{D})\mu_b + \sum_I \mathcal{D}_I \mu_I. \quad (6)$$

One can write analytic expressions for the total and partial dispersions for cubic $bcc$-clusters,[16] that provide also analytic equations for the average magnetic moment as a function of the cluster size. Similar expressions can be obtained for cubo-octahedral, icosahedral and simple-cubic clusters.

Recent experimental results for magnetic properties of Fe clusters in the range of 25 to 700 atoms have been published.[5] One of the puzzling findings is that the average magnetic moment presents an oscillatory behavior as a function of the cluster size. This behavior may be the result of various effects: (i) asymmetric cluster shapes, *i.e.*, it is possible that the clusters present elongate or flat shapes but with the same $bcc$-structure.[8] (ii) The clusters grow following a *rounded* process, *i.e.*, instead of forming well defined cubes, which due to energetic reasons may be less favorable, they form $bcc$-spherical-like structures. (iii) The coexistence of clusters with various geometrical structures at a given temperature.

To estimate the effect of the presence of asymmetric clusters, we consider elongated (prolate-like) and flat shape (oblate-like) simple cubic clusters. For the sake of simplicity we assume that the cluster is large enough to neglect edge and vertex contributions. In such a case it is very easy to prove that the deviation of the average magnetic moment with respect to the bulk value is

$$\frac{\bar{\mu}_N}{\mu_b} - 1 = 2\left(\frac{\mu_{\text{sup}}}{\mu_b} - 1\right)\left(\frac{1}{n_1} + \frac{1}{n_2} + \frac{1}{n_3}\right), \qquad (7)$$

where $n_i$ ($i = 1, 2, 3$) indicate the linear dimensions of the cubic cluster. Now let us fix the number of atoms in the cluster ($N = n_1 n_2 n_3$) and for the sake of simplicity let us take $n_1 = n_2$. An asymmetry paramater $f$ is defined as $(n_3/n_1)$. The case $f < 1$ ($f > 1$) corresponds to oblate (prolate) shape. Thus, Eq. (7) can be written as

$$\frac{\bar{\mu}_N}{\mu_b} - 1 = \frac{6}{\sqrt[3]{N}}\left(\frac{\mu_{\text{sup}}}{\mu_b} - 1\right)\left(\frac{2}{3}f^{1/3} + \frac{1}{3}f^{-2/3}\right). \qquad (8)$$

The last term in the right-hand side of (8) contains the dependence on the asymmetric shape of the cluster. It reduces to one in the symmetric case ($f = 1$). By performing an expansion of this coefficient around the symmetric case one obtains

$$\frac{2}{3}f^{1/3} + \frac{1}{3}f^{-2/3} \approx 1 + \frac{1}{9}(f-1)^2 - \frac{10}{81}(f-1)^3 \ldots . \qquad (9)$$

For a fixed number of atoms ($N$), the asymmetric clusters present higher values of $\bar{\mu}_N$ than the symmetric ones, and that $\bar{\mu}_N(\text{oblate}) > \bar{\mu}_N(\text{prolate}) > \bar{\mu}_N(\text{symmetric})$. One also notices that for slightly asymmetric clusters ($f \approx 1$) the correction is of second order. This result suggests that asymmetry may not be responsible of the anomalous behavior of the magnetic moment in the Fe clusters. A large asymmetry is necessary in order to get a significant change in the $\bar{\mu}_N$. For instance, to obtain a change of the 20% in $\bar{\mu}_N$ it is necessary the presence of asymmetric clusters with $f = 0.29$ for oblate-like clusters and $f = 4.14$ for prolate-like clusters. Although, this analysis holds for simple cubic lattice clusters, the results are qualitatively similar for cubic $bcc$-clusters.

The second possibility to explain the experimental results for the behavior of $\bar{\mu}_N$ in Fe clusters is that the nanostructures grow in a way that favors the formation of *spherically* symmetric structures. We calculated $\bar{\mu}_N$ for all the $bcc$-spherical clusters with sizes between 27 and 1013 atoms. We construct the cluster by adding the shells of neighbors to the central atom in a consecutive manner. For example the 27-atom

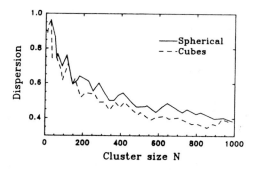

**Figure 2.** The total dispersion as a function of cluster size in bcc-clusters: the solid line corresponds to rounded clusters and the dashed curve to clusters with cubic shape grown step by step (see text).

cluster contains all the first (8), second (6) and third (12) neighbors to the central atom. All of them are located at the surface and only the central atom has saturated the total coordination number (8). The next shell of neighbors (fourth) contains 24 sites. When all those sites are occupied, the second neighbors become inner atoms and from the total of 51 only 44 (the 1st, 3rd, and 4th neighbors) form the cluster surface. This process can be continued and the results for the total dispersion as a function of the number of atoms for the *spherical bcc*-clusters is presented in Fig. 2 (solid line). We plot here also the dispersion for perfect cubes grown also shell by shell (dashed line). The analytic expression that results from Eqs. (1) and (2) and that corresponds to perfect cubes reproduces the local minima in this curve. One notices that the spherical mode of growth leads to a larger dispersion.

One can also count how many different shells form the surface of the cluster. For example the surface of the 27, 51, and 59 atoms clusters are formed by 3 different shells. When one adds the next shell (6th-neighbor) to form the 65-atom complex, the total number of different shells grows to four. The oscillations in the total dispersion as a function of $N$ are associated to the change of the total number of surface shells. When this number increases the dispersion grows as compared to the one of the previous cluster.

The oscillations of this parameter as a function of atoms brings in a natural way the same behavior in the average magnetic moment. The results for $\bar{\mu}_N$ for cubic- and spheric-like clusters are presented in Fig. 3(a). One observes that for the same number of atoms the cubic clusters posses a larger $\bar{\mu}_N$. This result is expected since the sharper the structure, the smaller the coordination number of the surface atoms.

The results for $\bar{\mu}_N$ are presented in Fig. 3b). Here, we also include the experimental results.[5] The experimental error in the determination of the magnitude of $\mu_N$ lies within 2 and 6%, being larger for the smaller aggregates. From this figure one sees that the experimental oscillatory behavior is reproduced by our calculation. However some differences are still present. (i) In the experimental results the bulk behavior is reached faster than in the model. (ii) The amplitude of the experimental oscillation is larger than the calculated. (iii) The experimental results present picks that are not reproduced by our model.

A possible explanation of the difference between theory and experiment is that small bcc-clusters may coexist with nanostructures with other geometrical structures

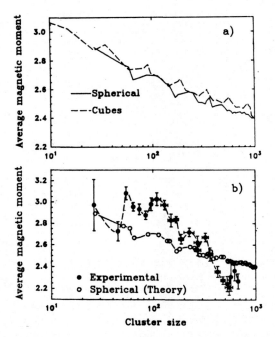

**Figure 3.** clusters with spherically symmetric *bcc*-structure as a function of cluster size. The experimental results are denoted by closed circles.[5]

and/or the very likely situation in which clusters with defects and asymmetries may be present. However, this proposal is subject to experimental verification and the further theoretical calculations.

In conclusion, we applied a simple model for the study of the average magnetic moment for Fe-clusters. Our model underestimate the values of the magnetic moments for small clusters and overestimate the values of them in the case of large clusters. This suggests that the values of the magnetic moments $\mu_i$ may depend not only on the local environment but also on the cluster size and that the growing procesess may differ from the one proposed.

We analyzed some of the possible explanations of the observed *oscillations* in the average magnetic moment. Our model shows that very strong asymmetries (oblate and prolate shapes) would be necessary to explain that particular behavior. Due to the fact that clusters with those geometries would require a large surface energy that mechanism may be excluded. We found that assuming rounded *bcc*-clusters one can produce an oscillatory behavior shown in the dependence of the average magnetic moment as a function of cluster size. The remanent discrepancies between our model and the experimental findings may be attributed to the coexistence of various geometrical structures. A possible way to improve this model is to consider some dependence of the $\mu_i$ with the different positions. This is a cumbersome task due to the large number of non-equivalent sites in a cluster. On the other hand a more realistic calculation for these type of systems must consider electronic structure calculations, as well as possible relaxation of the cluster structures. Our results represent a first step in the understanding of the rich behavior of nanostructured materials.

## Acknowledgments

We acknowledges to the Consejo Nacional de Ciencia y Tecnología (México) Grants No. 1774-E9210 and No. 4920-E9406.

## References

1. L. M. Falicov et al, J. Mater. Res. **5**, 1299 (1990) and references quoted therein.
2. D. M. Cox et al., Phys. Rev. B **32**, 7290 (1985).
3. W. A. de Herr, P. Milani, and A. Chatelain, Phys. Rev. Lett. **65**, 488 (1990).
4. J. P. Bucher et al., Phys. Rev. Lett. **66**, 3052 (1991).
5. I. M. L. Billas, J. A. Becker, A. Chatelain, and W.A. de Herr, Phys. Rev. Lett. **71**, 4067 (1993).
6. S. A. Nepijko and R. Wiesendanger, Europhys Lett. **31**, 567 (1995).
7. F. Liu, S.N. Khanna, and P. Jena, Phys. Rev. Lett. **67**, 742 (1991).
8. J. Zhao et al., Phys. Lett. A **205**, 308 (1995).
9. G. M. Pastor, J. Dorantes-Dávila, and K. H. Bennemann, Phys. Rev. B **40**, 7642 (1989).
10. F. Aguilera-Granja and J.L. Morán-López, Solid State Commun. **74**, 155 (1990) and references therein for the thin film.
11. J. Merikowski, J. Timonen, M. Manninen, and P. Jena, Phys. Rev. Lett. **66**, 938 (1991).
12. P. J. Jensen and K. H. Bennemann, Z. Phys. D **21**, 349 (1991).
13. P. V. Hendriksen, S. Linderoth, and P. A. Lindgård, Phys. Rev. B **48**, 7259 (1993).
14. A. Vega, J. Dorantes-Dávila, L. C. Balbas, and G. M. Pastor, Phys. Rev. B **47**, 4742 (1993).
15. F. Aguilera-Granja, J. L. Morán-López, and J. M. Montejano-Carrizales, Surf. Sci. **326**, 150 (1995).
16. J. M. Montejano-Carrizales and J. L. Morán-López, Nanostructured Materials **1**, 397 (1992); J. M. Montejano-Carrizales, F. Aguilera-Granja, and J. L. Morán-López, Nanostructured Materials, in press.
17. G. M. Pastor, J. Dorantes-Dávila, and K. H. Bennemann, Phys. Rev. B **50**, 15 424 (1988).

# Magnetic Anisotropy of 3d-Transition Metal Clusters, Chains, and Thin Films

J. Dorantes-Dávila,[1] and G. M. Pastor[2]

[1] Instituto de Física
Universidad Autónoma de San Luis Potosí
Alvaro Obregón 64, 78000 San Luis Potosí, S. L. P.
MEXICO

[2] Laboratoire de Physique Quantique
Université Paul Sabatier
118 route de Narbonne F-31062 Toulouse Cedex
FRANCE

## Abstract

Recent theoretical results on the magneto-anisotropic properties of low-dimensional transition-metal systems are reviewed. The magnetic anisotropy energy and related electronic properties of finite clusters, one-dimensional chains and ultra-thin films are discussed in the framework of a self-consistent tight-binding approach.

## I. Introduction

With the miniaturization of magnetic technology, the search of new low-dimensional magnetic materials involving clusters, thin films and complex overlayer structures has become increasingly important in the past years.[1] Intensive experimental search stresses the interest on theoretical studies aiming a microscopic understanding of the observed specific behaviors and providing with predictions of yet unknown phenomena. The magnetic anisotropy energy (MAE) is one of the most important properties of these magnetic materials. It determines the low-temperature orientation of the magnetization as well as the stability of the magnetization direction in mono-domains systems, characteristics which are both crucial in technological applications such as magnetic recording or memory devices. From this point of view, low-dimensional ferromagnetic transition-metals (TM) systems are attracting considerable attention, since they present MAEs which are 2–3 orders of magnitude larger than in the corresponding crystalline solids.[2-10] At the same time, changing the structure and composition of these materials should allow many possibilities of manipulating the direction of the magnetization. In the following, we discuss recent theoretical results on the magnetic-anisotropic behavior of several low-dimensional TM systems. Emphasis is given to small clusters, one-dimensional chains and low-symmetry thin films.

## II. Theory

One of the essential features of low-dimensional itinerant magnetism is the central role played by the redistributions of the spin-polarized electronic density associated to the changes in the geometrical arrangement of the atoms. An efficient way of taking these effects properly into account is provided by the self-consistent tight-binding (SCTB) approach proposed in Ref. 11. The SCTB method has been recently extended to include the spin-orbit (SO) interactions which give the dominant contribution to the MAE.[12] The electronic Hamiltonian from which the MAEs are derived is given by the sum of three terms. Besides the inter-atomic hopping term $H_0$ and the Coulomb interaction term $H_c$ treated in the unrestricted Hartree-Fock approximation[11] one includes the SO coupling term given by

$$H_{SO} = -\xi \sum_{i,\alpha\sigma,\beta\sigma'} (\vec{L}_i \cdot \vec{S}_i)_{\alpha\sigma,\beta\sigma'} \hat{c}^+_{i\alpha\sigma} \hat{c}_{i\beta\sigma'}. \tag{1}$$

where $(\vec{L}_i \cdot \vec{S}_i)_{\alpha\sigma,\beta\sigma'}$ refers to the intra-atomic matrix elements of $\vec{L} \cdot \vec{S}$ between the orbitals $\alpha$, $\beta$ of spin $\sigma$, $\sigma'$ at atom $i$, which couple the up and down spin-manifolds and which depend on the relative orientation between the average magnetization $\langle \vec{S} \rangle$ and the film structure. For each orientation of the magnetization the local densities of electronic states (DOS) $\rho_{i\alpha\sigma}(\varepsilon)$ are determined self-consistently, thus including the coupled contributions of $H_0$, $H_c$ and $H_{SO}$ on the same footing.

The electronic energy $E_\delta$ is calculated from

$$E_\delta = \sum_{i\alpha\sigma} \int_{-\infty}^{\varepsilon_F} \varepsilon \rho_{i\alpha\sigma}(\varepsilon) \, d\varepsilon - E_{dc}, \tag{2}$$

where $\delta$ refers to the magnetization direction (e.g., $\delta \equiv x, y, z$) and $E_{dc}$ to the double counting correction. The MAE is given by the change $\Delta E$ in the electronic energy $E_\delta$ associated to a change in the orientation of the magnetization $\langle \vec{S} \rangle$ with respect to the lattice. For example, $\Delta E = E_x - E_y$ measures the relative stability of the magnetization along the $x$ and $y$ axis (see, for instance, the inset of Fig. 1). The magnetization direction can be chosen without restrictions. In particular, we shall consider the magnetization direction perpendicular to the film plane, along the $z$ axis, and several direction within the $(x, y)$ plane as illustrated, for example, in the inset of Fig. 1. In this way both the perpendicular and in-plane MAEs are determined.[13]

## III. Results and discussion

The calculations to be discussed below involve Fe, Co, and Ni as magnetic elements and Pd as metallic substrate. The parameters are determined as follows. The two center $d$-electron hopping integrals are given by the canonical expression in terms of the corresponding bulk $d$-band width. In the case of Fe, Co, and Ni, the intra-atomic

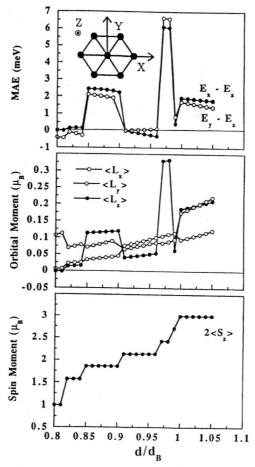

**Figure 1.** Magnetic anisotropy energy (MAE), orbital magnetic moment and spin magnetic moment of Fe$_7$ as a function of the bond length $d$ [$d_B$ = bulk nearest-neighbor (NN) distance]. The cluster structure and the considered directions of the magnetization, $x$, $y$ and $z$, are illustrated in the inset.

Coulomb integral $J$ is fitted to the bulk magnetization at $T = 0$. In the case of Pd, $J$ is obtained from local spin density (LSD) calculations (Stoner theory[14]) by taking into account the 20% reduction resulting from to correlation effects beyond the LSD approximation.[15] The SO-coupling constants $\xi$ are taken from Ref. 5.

### III.1 Clusters

The magnetic anisotropy of small 3d TM clusters deserves a special attention not only from a purely theoretical point of view but also because of its implications in cluster-beam Stern-Gerlach deflection experiments.[16,17] The MAE is one of the key parameters characterizing the dynamics of rotating clusters in an external magnetic field[18-20] where superparamagnetic behavior as well as resonance-like couplings between the rotational frequency and the Zeeman splittings are observed.[16,20]

The MAE $\Delta E$, the orbital angular momentum $\langle L_\delta \rangle$ along the magnetization direction $\delta = x, y, z$, and the average spin projection $\langle S_z \rangle$ of an Fe$_7$ cluster are shown in Fig. 1. This example is representative of a much larger number of studied sizes and structures.[6] The results are given as a function of the bond length $d$, which allows to determine the role of cluster relaxation and to infer the possible coupling of the magnetization direction to vibrations and distortions. The variations of $d/d_B$ ($d_B$ = bulk bond-length) correspond to a uniform relaxation within the given cluster geometry and quantify how $\Delta E$, $\langle L_\delta \rangle$ and $\langle S_z \rangle$ depend on the local environment of the atoms. For all studied clusters the modulus of the spin magnetization $|\langle \vec{S} \rangle|$ depends very weakly on the direction of the magnetization (typically, $|\langle S_z \rangle - \langle S_x \rangle| \sim 10^{-3}$–$10^{-4}$). Since the changes in $|\langle \vec{S} \rangle|$ for the different orientations would be indistinguishable in the scale of the plots, only the results for $\langle S_z \rangle$ are shown. Comparing Figs. 1(a) and (c) it is clear that the variations of the MAE as a function of $d/d_B$ are related to the variations of $\langle S_z \rangle$ and to the resulting changes in the electronic spectrum. For large values of $d/d_B$ the spin magnetic moments are saturated, i.e., $\langle S_z \rangle \simeq \frac{1}{2}(10 - N_d) = \frac{3}{2}$. When $d/d_B$ decreases, discrete changes in the spin polarization, $\Delta \langle S_z \rangle$, occur and non-saturated spin magnetizations are obtained.[11] For constant values of $\langle S_z \rangle$, i.e., for $d/d_B$ between two spin flips, the MAE and $\langle L_\delta \rangle$ vary continuously since the electronic spectrum and the local magnetic moments are continuous functions of $d/d_B$. This is not the case when a spin-flip occurs since here a strong and discontinuous redistribution of the spin-polarized electronic density takes place. Important changes in the energy-level structure around the Fermi energy $\varepsilon_F$ occur which modify the details of the SO mixing between these states. The resulting changes in the electronic energy depend, of course, on the explicit form of $H_{SO}$ and therefore on the direction of the magnetization (see Fig. 1). Consequently, very significant and discontinuous variations of the MAE are observed which may even lead to a change of sign of the MAE as the spin moment $\langle S_z \rangle$ decreases. Notice that the Fe$_7$ cluster presents a remarkable in-plane anisotropy $E_x - E_y \simeq 0.4$ eV. A similar situation is found for other clusters and band-fillings.[6] These results indicates that uniaxial anisotropy models[19-20] are not generally applicable to clusters.

More extensive calculations including other Fe$_N$ clusters and structures revealed the following general trends: (i) The MAE is much larger in small clusters than in the corresponding crystalline solids. In fact, $\Delta E$ is often even larger than in thin films; for instance, values of $\Delta E \sim 4$–5 meV are frequent.[6] These conclusions are in agreement with experiments on free clusters[19,21] and supported Fe nanoparticles.[22] (ii) $\Delta E$ is very sensitive to the geometrical structure of the cluster, much more than the magnetic moments. Indeed, changes of sign in $\Delta E$ are found as a function of $d$ even in situations where the magnetic moments are saturated and consequently do not depend on the cluster structure. (iii) The in-plane MAEs are considerably important in general. In some cases they are even larger than the usually considered off-plane anisotropy ($\sim 0.1$–1 meV). The in-plane MAE is of course larger for low-symmetry structures and decreases, though not monotonically, as the angle between non-equivalent $x$ and $y$ directions decreases.

## III.2 Chains

Diffusion controlled agregation and atomic manipulation with a scanning tunneling microscopy[23,24] allow to tailor one-dimensional and quasi one-dimensional magnetic devices experimentally. This should open the possibility of a variety of interesting magneto-anisotropic properties which remain, at least for the most part, to be discov-

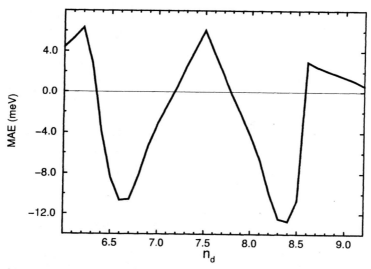

**Figure 2.** Magnetic anisotropy energy (MAE) per atom, $E_x - E_z$, of a monoatomic chain as a function of the $d$-band filling $n_d$, where $x$ refers to the magnetization direction along the chain and $z$ refers to a perpendicular direction. The results are derived from fully self-consistent calculations performed independently for each magnetization direction.

ered. In the following we discuss theoretical results on the MAEs of one dimensional TM wires, in particular concerning its dependence on $d$-band filling, on the length, width and transversal structure of the chain and on chain-substrate interactions.

The MAE of the infinite linear chain is shown in Fig. 2 as a function of the $d$-band filling $n_d$. Notice the remarkably large MAEs which in most cases are far more important than the values found in thin films[7-10] or compact clusters.[6] For example, for band-fillings close to that of Co ($n_d = 8.3$) we find $\Delta E \simeq -12$ meV while in bidimensional films one usually obtains $\Delta E \simeq 0.1-1$ meV. The actual easy and hard axes alternate among the $x$ and $z$ directions as a function of $n_d$, leading to qualitatively different behaviors for different TMs.

Finite-length chains may show specific magnetic properties, which can be significantly different from those of infinite chains. Varying the length $L$ of the chain is nowadays possible experimentally by changing the deposition or agregation conditions.[24] Therefore, it would be very useful to understand how the magnetic properties change as a function of $L$. Recently, we have investigated this problem giving emphasis on the magneto-anisotropic behavior.[25] Our results show that the MAE tends to increase with decreasing the length $L$. Finite-length monoatomic chains yield the largest MAEs of TM systems, that we are aware of. The $L$ dependence of $\Delta E$ changes with the considered elements, *i.e.*, with the $d$-band filling. Significant finite-length effects are observed only for rather short chains ($L \leq 10$ atoms). For $L \geq 20$ atoms the MAE is already very similar to the $L = \infty$ result.

In the case of multichains, the transversal structure plays a significant role on the MAE. We have considered square $bcc(001)$-like structures and triangular $fcc(111)$-like structures. The MAE shows even-odd alternations as a function of the number of atoms $N$ in the transversal direction, and decreases with increasing width of the chain. For example, for Fe and Co wires ($n_d = 7$ and $n_d = 8.3$) the easy axis is along the wire in the monoatomic case ($N = 1$) and it is perpendicular to the wire for the bichain ($N = 2$). Concerning the $d$-band filling dependence, we observe that $\Delta E$ oscillates as a function of the $n_d$ as in the monoatomic chain. The magneto-anisotropic behavior is very sensitive to the transversal structure or the inter-chain packing. For Co, for example, the direction of the magnetization is perpendicular to the plane of the multichain for square packing, while it is within the plane in case of triangular packing. Large in-plane anisotropies $E_x - E_y$ are very common. For instance, for a square tri-chain of Fe or Co we obtain $E_x - E_y \simeq 3$ meV.

The assumption of a free standing geometry limits the possibility of a direct comparison with many experiments, where the interactions between the magnetic ad-atoms and a metallic substrate can play a decisive role.[3,4] In order to quantify substrate effects on the MAE of linear chains, we performed calculations on the Co(1D)/Pd(110), a situation of particular experimental interest.[24] We find that the Pd substrate induces a change in the direction of the magnetization $\vec{M}$. In the free standing geometry $\vec{M}$ is in-line with $E_x - E_z = -12$ meV, while in Co(1D)/Pd(110) we find that $\vec{M}$ is perpendicular to substrate surface with $E_\delta - E_z \simeq 0.15$–$0.31$ meV ($\delta = x, y$). A main contribution to the magnetization direction change is the Co/Pd *interface* MAE $\Delta E_{Co/Pd}$ which favors the off-plane orientation of $\vec{M}$ [$\Delta E_{Co/Pd} \simeq 0.3$ meV (Ref. 26)]. Moreover, notice that one of the main one-dimensional features of Co(1D), namely, the remarkably large MAE, disappears upon deposition on Pd. Finally, let us point out that Co(1D)/Pd(110) presents significant in-plane MAE ($E_x - E_y \simeq 0.15$ meV) (Ref. 25).

### III.3 *bcc*(110) Thin Films

In the same way that the symmetry reduction upon going from 3 to 2 dimensions yields a remarkable increase of the *off-plane* magnetic anisotropy, one expects that a reduction of the symmetry *within the film* should result in a variety of *in-plane* anisotropic properties. In the following we discuss the *in-plane* anisotropy of $bcc(110)$ TM ultra-thin films as a function of relevant external variables such as the $d$-band filling, the strength of the interactions and the film thickness.[7]

In figure 3 results are given for the MAE $\Delta E$ of the $bcc(110)$ monolayer as a function of $d$-band filling $n_d$. $\Delta E = E_\delta - E_z$ where $\delta = x, y$ and $xy$ refers to the different in-plane magnetization directions indicated in the inset figure and $z$ refers to the direction perpendicular to the film. The isolated points are derived from fully self-consistent calculations performed independently for each magnetization direction. Positive values of the MAE correspond to a perpendicular easy axis. Notice, that the $bcc(110)$ monolayer presents a remarkable in-plane anisotropy $E_x - E_y$. In particular for $n_d = 7$, $E_x - E_y \simeq 0.4$ meV is even larger than any of the usually considered off-plane anisotropies $E_\delta - E_z$ ($\delta = x, y, xy$). The curves correspond to non-selfconsistent calculations performed by assuming that the exchange splitting $\Delta \varepsilon_x = \varepsilon_{d\downarrow} - \varepsilon_{d\uparrow}$ is independent of $n_d$. In these cases, $\Delta \varepsilon_x$ is equal to the value obtained in a self-consistent calculation on Fe neglecting SO coupling ($\xi = 0$). Comparison with the self-consistent results for Co and Ni is satisfactory. However, the converse is not true, i.e., using $\Delta \varepsilon_x(Co)$ or $\Delta \varepsilon_x(Ni)$ yields very poor results for Fe (Ref. 7).

The MAEs of the $bcc(110)$-bilayer are generally smaller than those of the monolayer.[7] However, the anisotropy within the film plane remains quantitatively at least as important as the off-plane anisotropy. Varying the band filling one finds, as in the case of the monolayer, that the easy and hard axes alternate among $x$, $y$ and $z$, and that transition between uniaxial and multi-axial MAE surfaces are present.

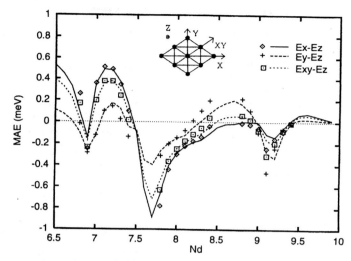

**Figure 3.** Magnetic anisotropy energy (MAE) per atom, $E_\delta - E_z$, of the $bcc(110)$ monolayer as a function of the $d$-band filling $n_d$, where $\delta = x$, $y$ and $xy$ refers to the different in-plane magnetization directions indicated in the inset figure and $z$ refers to the direction perpendicular to the film. The isolated points are derived from fully self-consistent calculations performed independently for each magnetization direction. The curves correspond to calculations using a constant exchange splitting $\Delta\varepsilon_x = \varepsilon_{d\downarrow} - \varepsilon_{d\uparrow} = 1.88$ eV, which is equal to the value obtained in self-consistent calculations neglecting SO coupling ($\xi = 0$) for $n_d = 7.0$.

While the results for the bilayer confirm qualitatively the remarkable effects already observed for the monolayer, it should be also noted that for specific elements (*i.e.*, band fillings) changing the film thickness results in very important changes in the magneto-anisotropic behavior. A very strong dependence of MAEs as a function of film thickness has been also observed in experiment.[27]

## IV. Summary

In conclusion, results for the magnetic anisotropy energy (MAE) and related electronic properties of low-dimensional $3d$ transition-metal systems were determined by calculating self-consistently the effects of the spin-orbit coupling on the spin-polarized charge distribution and on the electronic spectrum for different orientations of the magnetization. The MAE of small clusters shows a complicated, non-perturbative behavior as a function of cluster size, structure, bond-length and $d$-band filling. In agreement with experiment, the MAE is found to be considerably larger in small clusters than in the corresponding crystals, often even larger than in thin films. Moreover, the in-plane anisotropy can be of the same order of magnitude as the off-plane anisotropy. One-dimensional and quasi one-dimensional systems present a rich magneto-anisotropic behavior as a function of $d$-band filling and as a function of the length, width and transversal structure of the chain. Remarkably large values of the MAE are obtained for monoatomic chains, which are much larger than the corresponding results for two- and three-dimensional systems. Wire-substrate interactions were discussed in the case of Co(1D)/Pd(110). The magneto-anisotropic properties of $bcc(110)$ monolayers and bilayers were analized briefly. For these films, we find an important in-plane magnetic anisotropy, which is very sensitive to the $d$-band filling, the intra-atomic exchange splitting and the film thickness.

## Acknowledgments

Helpful discussions with Prof. H. Dreyssé are are greatfully acknowledged. This work has been supported in part by CONACyT (Mexico). The *Laboratoire de Physique Quantique* (Toulouse) is *Unité Mixte de Recherche* associated to the CNRS.

## References

1. See, for instance, the papers on *Magnetoelectronics*, Physics Today, April 1995, p. 24ff.
2. See, for instance, *Ultra-thin Magnetic Structures I: An Introduction to Electronic, Magnetic, and Structural Properties*, edited by J. A. C. Bland and B. Heinrich, (Springer Verlag, Berlin, 1994); M. Tischer et al., *Phys. Rev. Lett.* **75**, 1602 (1995); G. Bochi et al., *ibid.* **75**, 1839 (1995). M. Tischer et al., *Phys. Rev. Lett.* **75**, 1602 (1995); G. Bochi et al., *Phys. Rev. Lett.* **75**, 1839 (1995).
3. A. Rabe, N. Memmel, A. Steltenpohl, and Th. Fauster, *Phys. Rev. Lett.* **73**, 2728 (1994).
4. J. Kohlhepp and U. Gradmann, *J. Magn. Magn. Mater.* **139**, 347 (1995).
5. P. Bruno, *Phys. Rev. B* **39** 865 (1989); *Magnetismus von Festkörpern und Grenzflächen*, Ferienkurse des Forschungszentrums Jülich (KFA Jülich, 1993), Ch. 24.
6. G.M. Pastor, J. Dorantes-Dávila, S. Pick, and H. Dreyssé, *Phys. Rev. Lett.* **75**, 326 (1995).
7. J. Dorantes-Dávila and G. M. Pastor, *Phys. Rev. Lett.* **77**, 4450 (1996).
8. J. Dorantes-Dávila H. Dreyssé, and G. M. Pastor, *Phys. Rev. B* **55**, (1997) in press.
9. D. S. Wang, R. Wu and A. J. Freeman, *Phys. Rev. Lett.* **70**, 869 (1993); *Phys. Rev. B* **47**, 14 932 (1993).
10. S. Pick, J. Dorantes-Dávila, G.M. Pastor, and H. Dreyssé, *Phys. Rev. B* **50**, 993 (1994); S. Pick and H. Dreyssé, *ibid.* **46**, 5802 (1992); *ibid.* **48**, 13 588 (1993).
11. G. M. Pastor, J. Dorantes-Dávila, and K. H. Bennemann, *Phys. Rev. B* **40**, 7642 (1989); J. Dorantes-Dávila, H. Dreyssé and G. M. Pastor, *ibid.* **46**, 10 432 (1992).
12. For films, surfaces or multilayers dipole-dipole interactions cannot be neglected. In first approximation, they can be described by the classical expression

$$E_{\text{DD}} = \frac{1}{2N} \sum_{i \neq j} \frac{1}{R_{ij}^3} \left[ \vec{M}_i \cdot \vec{M}_j - \frac{3(\vec{R}_{ij} \cdot \vec{M}_i)(\vec{R}_{ij} \cdot \vec{M}_j)}{R_{ij}^2} \right].$$

13. For further details see Refs. 6 and 7.
14. N. E. Christensen, O. Gunnarsson, O. Jepsen, and O. K. Andersen, *J. de Phys.* (Paris) **49**, C8-17 (1988); O. K. Andersen, O. Jepsen, and D. Glötzel, *Highlights of Condensed Matter Theory*, edited by F. Bassani, F. Fumi, and M. P. Tosi (North Holland, Amsterdam, 1985).
15. G. Stollhoff, A. M. Oles, and V. Heine, *Phys. Rev. B* **41**, 7028 (1990).
16. W. A. de Heer, P. Milani, and A. Châtelain, *Phys. Rev. Lett.* **65**, 488 (1990); I. M. L. Billas, J. A. Becker, A. Châtelain, and W. A. de Heer, *Phys. Rev. Lett.* **71**, 4067 (1993).

17. J. P. Bucher, D. G. Douglas, and L. A. Bloomfield, *Phys. Rev. Lett.* **66**, 3052 (1991); D. C. Douglass, A. J. Cox, J. P. Bucher, and L. A. Bloomfield, *Phys. Rev. B* **47**, 12874 (1993).
18. S. N. Khanna and S. Linderoth, *Phys. Rev. Lett.* **67**, 742 (1991).
19. P. J. Jensen, S. Mukherjee, and K. H. Bennemann, *Z. Phys. D* **21**, 349 (1991); *Z. Phys. D* **26**, 246 (1993).
20. A. Maiti and L. M. Falicov, *Phys. Rev. B* **48**, 13596 (1993).
21. J. Becker and W. A. de Heer, *Ber. Bunsenges* **96**, 1237 (1992); I. M. Billas, J. A. Becker, and W. A. de Heer, *Z. Phys. D* **26**, 325 (1993).
22. F. Bødker, S. Mørup, and S. Linderoth, *Phys. Rev. Lett.* **72**, 282 (1994).
23. D. M. Eigler and E. K. Schweizer, *Nature* **344**, 524 (1990).
24. H. Roeder *et al.*, *Nature* **366**, 141 (1993); J. P. Bucher *et al.*, *Europhys. Lett.* **27**, 473 (1994).
25. J. Dorantes-Dávila and G. M. Pastor, submitted to *Phys. Rev. Lett.*, (1997).
26. J. Dorantes-Dávila, H. Dreyssé, and G. M. Pastor, to be published.
27. H. Frietsche, H. J. Elmers, and U. Gradmann, *J. Magn. Magn. Mater.* **135**, 343 (1994).

# Noncollinear Magnetic Structures in Small Compact Clusters

M. A. Ojeda-López,[1] J. Dorantes-Dávila,[1] and G. M. Pastor[2]

[1] Instituto de Física
Universidad Autónoma de San Luis Potosí
Alvaro Obregón 64, 78000 San Luis Potosí, S. L. P.
MEXICO

[2] Laboratoire de Physique Quantique
Université Paul Sabatier
118 route de Narbonne F-31062 Toulouse Cedex
FRANCE

## Abstract

We investigate the properties of complex magnetic structures in small compact clusters including noncollinear arrangements of local magnetic moments, spin density waves and charge density waves. The Hubbard Hamiltonian is solved within the unrestricted Hartree-Fock (UHF) approximation without imposing any symmetry constraints neither to the size or orientation of the local magnetic moments nor to local charge densities. For icosahedral and *fcc*-like clusters having $N = 13$ atoms, and for band-fillings close to half-band, the most stable solutions of UHF equations are discussed as a function of the Coulomb interaction strength $U/t$. The role of noncollinear arrangements of spins in the ground-state magnetic behavior is analyzed. The high connectivity of these compact clusters and the tendency to antiferromagnetic order between nearest-neighbors spins close to half-band filling results in magnetic frustrations and in remarkable noncollinear-spin solutions.

## I. Introduction

The study of three-dimensional spin-like arrangements of local magnetic moments is considerably important for understanding the magnetic properties of complex systems. As a result of the well known sensitivity of itinerant magnetism to the geometrical arrangement of the atoms, one expects that spiral-like–spin-density-waves (SDWs) or even more complex magnetic structure may occur.[1] The investigation of magnetic phenomena of this kind requires a symmetry-unrestricted theoretical approach in which no *a priori* assumptions are made concerning relative orientation of the local magnetic moments.

Small clusters are particularly interesting in this context. On one side $Cr_N$ clusters, and not to small $(Cr_xFe_{1-x})_N$ alloy clusters, have shown to present large local magnetic moments and antiferromagnetic (AF) order.[2] On the other hand clusters are known to modify their structure at the surface in order to maximize the local coordination number. This leads to the presence of AF correlations in compact structures and thus to magnetic frustrations which may easily lead to complex arrangements of the local magnetic moments. Quite generally, one would also like to understand the effects that the presence of a finite surface has on spiral-like SDWs.

However, very little is known so far about noncollinear magnetic behavior in finite systems. Noncollinear spin states have been investigated using the Hubbard model and a $C_{60}$-like structure with one electron per site.[3] A magnetic structure was found which matches the five fold symmetry of pentagonal rings and minimizes antiferromagnetic frustrations as the classical Heisenberg model.[4] The more general and much more complex doped case where hole delocalizations tends to destroy AF correlations and where AF states may compete with a ferromagnetic ground-state remains to be studied.

In this communication, we discuss the ground-state magnetic properties of clusters having $N = 13$ atomic sites, emphasizing the role of noncollinear magnetic states. For this purpose we consider the single-band Hubbard model in the fully unrestricted Hartree-Fock approximation. The theoretical framework is described in the next section. Our results are presented in Sect. III. Here, the icosahedral and *fcc*-like $N = 13$ clusters are taken as representative examples.

## II. Theoretical Method

We consider the single-band Hubbard Hamiltonian,[5] which in the usual notation can be written as

$$H = -t \sum_{\langle l,m \rangle, \sigma} \hat{c}_{l\sigma}^{+} \hat{c}_{m\sigma} + U \sum_{l} \hat{n}_{l\uparrow} \hat{n}_{l\downarrow}. \qquad (1)$$

The creation (annihilation) operator for an electron at site $l$ with spin $\sigma$ is referred by $\hat{c}_{l\sigma}^{+}$ ($\hat{c}_{l\sigma}$) and the corresponding number operator by $\hat{n}_{l\sigma}$. In this model the dynamics of the valence electrons is ruled by the interplay between the kinetic and Coulomb terms. The relative importance between kinetic energy and Coulomb energy can be characterized by the dimensionless parameter $U/t$. In spite of its simplicity, this Hamiltonian has played, together with related models, a major role in guiding our understanding of the many-body properties of metals, particularly concerning low-dimensional magnetism.[6] In this work we intend to use it to examine the properties of noncollinear magnetism in small compact clusters.

In the unrestricted Hartree-Fock (UHF) approximation, the ground state is a single Slater determinant

$$|UHF\rangle = \left[ \prod_k \hat{a}_k^\dagger \right] |vac\rangle, \qquad (2)$$

where the single-particle states

$$\hat{a}_k^\dagger = \sum_{l,\sigma=\uparrow,\downarrow} A_{l\sigma}^k \hat{c}_{l\sigma}^\dagger, \qquad (3)$$

are linear combinations of $|l,\sigma=\uparrow\rangle$ and $|l,\sigma=\downarrow\rangle$ states. Notice that we allow for the most general noncollinear spin arrangements since $\hat{a}_k^\dagger$ may involve, both, $\uparrow$ and $\downarrow$ spin components. The eventually *complex* coefficients $A_{i\sigma}^k$ are determined by minimizing the energy expectation value $E_\text{UHF} = \langle\text{UHF}|H|\text{UHF}\rangle$, which in terms of the density matrix

$$\rho_{l\sigma,m\sigma'} \equiv \langle\text{UHF}|\hat{c}_{l\sigma}^\dagger \hat{c}_{m\sigma'}|\text{UHF}\rangle = \sum_{k=1}^{k=\nu} \bar{A}_{l\sigma}^k A_{m\sigma'}^k, \qquad (4)$$

is given as

$$E_\text{UHF} = -t \sum_{\langle l,m\rangle,\sigma} \rho_{l\sigma,m\sigma} + U \sum_l \rho_{l\uparrow,l\uparrow}\,\rho_{l\downarrow,l\downarrow} - |\rho_{l\uparrow,l\downarrow}|^2. \qquad (5)$$

The energy minimization, together with normalization constraints on the wave function, leads to the usual self-consistent equations

$$-t \sum_m A_{m\sigma}^k + U \left( A_{l\sigma}^k \rho_{l\bar\sigma,l\bar\sigma} - A_{l\bar\sigma}^k \rho_{l\bar\sigma,l\sigma} \right) = \varepsilon_k A_{l\sigma}^k. \qquad (6)$$

For a given solution, the average of the local electronic density $\langle n_l\rangle$ and spin vectors $\langle \vec{S}_l\rangle$ [$\vec{S}_l = (S_l^x, S_l^y, S_l^z)$] are determined from

$$\langle n_l\rangle = \rho_{l\uparrow,l\uparrow} + \rho_{l\downarrow,l\downarrow}, \qquad (7)$$

and

$$\langle \vec{S}_l\rangle = \tfrac{1}{2}\left[ (\rho_{l\uparrow,l\downarrow} + \rho_{l\downarrow,l\uparrow}),\, -i(\rho_{l\uparrow,l\downarrow} - \rho_{l\downarrow,l\uparrow}),\, (\rho_{l\uparrow,l\uparrow} - \rho_{l\downarrow,l\downarrow}) \right]. \qquad (8)$$

Notice that if $\rho_{l\sigma,l\bar\sigma} = 0\ \forall l$ the local magnetic moments $\langle \vec{S}_l\rangle$ are collinear. The UHF energy can be rewritten as

$$E_\text{UHF} = -t \sum_{\langle l,m\rangle,\sigma} \rho_{l\sigma,m\sigma} + \tfrac{1}{4}U \sum_l \langle n_l\rangle^2 - U \sum_l |\langle \vec{S}_l\rangle|^2. \qquad (9)$$

One readily sees that the Hartree-Fock Coulomb energy $E_C^\text{HF}$, given by the 2nd and 3rd terms, favors a uniform density distribution and the formation of local moments $\langle \vec{S}_l\rangle$. However, due to the local character of Hubbard's Coulomb interaction, the relative orientation of different $\langle \vec{S}_l\rangle$ does not affect $E_C^\text{HF}$. It is therefore the optimization of the kinetic energy which determines the formation of complex magnetic structures (with $|\langle \vec{S}_l\rangle \cdot \langle \vec{S}_m\rangle| \times |\langle \vec{S}_l\rangle| \times |\langle \vec{S}_m\rangle| \neq 1$) as well non-uniform density distributions $\langle n_l\rangle$.

## III. Results

In this Section we present and discuss results for the magnetic order in small compact clusters as obtained by solving the self-consistent UHF Eq. (6) for the Hubbard Hamiltonian as a function of the Coulomb interaction strength $U/t$. Band-fillings close to half band are considered for which electron correlations are most important ($\nu = N, N \pm 1$ electrons). In order to analyze the magnetic arrangements resulting from the interplay between AF correlations and frustrations in compact, non-bipartite

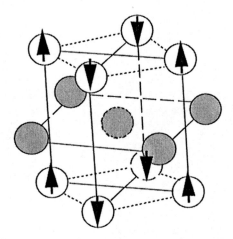

**Figure 1.** Illustration for the spin and charge distributions in a *fcc*-like cluster having $N = 13$ atoms and one hole ($\nu = N - 1$ electrons) as obtained using the Hubbard model in the UHF approximation and $U/t < 3.5$. The atoms in grey have zero local magnetic moments and an excess charge ($\langle n_i \rangle = 1.1154$ for $U/t = 0.5$). The arrows indicate the direction of the non-vanishing local magnetic moments $\langle \vec{S}_l \rangle$ which in this case are all parallel to the $z$-axis ($|\langle \vec{S}_l \rangle| = 0.26644\mu_B$ for $U/t = 0.5$).

structures, two representative cluster geometries for $N = 13$ atoms are considered: An *fcc*-like cluster, formed by a central atom and a surrounding shell of its first nearest neighbors (NN) as in the *fcc* lattice, and an icosahedral cluster formed by a central atom and 12 NN occupying the vertices of an icosahedron. An illustration of the topology of the surface of these clusters may be found in Figs. 2 and 3, respectively.

Let us consider first the *fcc*-like cluster with a single hole, *i.e.*, $\nu = N - 1$ electrons. For $U/t < 4.9$, the lowest energy solution to the UHF equations show collinear magnetic moments ($\langle \vec{S}_l \rangle \parallel \hat{z}$). The spatial symmetry is also broken, *i.e.*, the charge- and spin-density are inhomogeneous. For $U/t < 3.5$ the atoms belonging to the central plane present an enhanced electron density and vanishing local magnetic moments (for $U/t = 0.5$, $\langle n_l \rangle \simeq 1.1154$ and $\langle \vec{S}_l \rangle = 0$). The atoms of the upper and lower planes are magnetized showing AF order with every $\langle \vec{S}_l \rangle$ antiparallel to its NNs. The spin-density distribution for $U/t < 3.5$ is shown schematically in Fig. 1. The fact that the local magnetic moments vanish at the central plane can be interpreted as a consequence of magnetic frustrations. The formation of local magnetic moments at these atoms is delayed because the magnetic moments of their NNs sum up to zero. Also notice that there are no frustrations at the upper and lower squares. For $U/t > 3.5$ the surface atoms at the central plane develop non-zero local magnetic moments, in order to reduce double occupations within the UHF state. The arrangement of the local magnetic moments is still collinear. All square faces have perfect antiferromagnetic order, while the triangular faces show important frustrations. In the triangles of NNs, two local moments point in the same direction and the other one is opposite to them. An equivalent magnetic order is found in an isolated triangle for very small values of $U/t$ (Ref. 7). For $U/t > 4.9$ the spin-density distribution changes once more. Now the arrangement of the local magnetic moments is noncollinear with all surface spin having the same modulus. The central site remains nonpolarized. In fact the magnetic order is the same as the one obtained for an *fcc*-like cluster without the

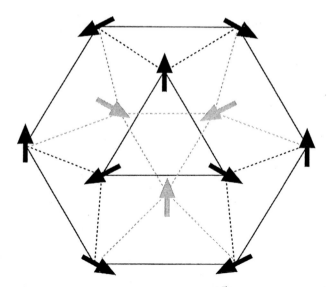

**Figure 2.** Illustration of the local magnetic moments $\langle \vec{S}_l \rangle$ at the surface of a 13-atom *fcc*-like cluster with a single hole ($\nu = N - 1$ electrons). The results are obtained using the Hubbard Hamiltonian in the UHF approximation and $U/t \geq 5$. All $\langle \vec{S}_l \rangle$ are parallel to the plane of the figure. Notice that the uppermost and lowermost triangular faces (as well as all other triangular faces after appropriate homogeneous rotations) have essentially the same spin arrangement with $\langle \vec{S}_l \rangle$ pointing along the medians.

central site ($N = 12$). As shown in Fig. 2, the moments at the triangular faces point along the medians, in the same way as classical antiferromagnetic spins on a triangle. This magnetic order minimizes frustrations and is also found in UHF calculations for the isolated triangle ($\nu = 3$ and $U/t > 0$).[7] It is worth noting that in spite of the three-dimensional structure of the cluster, all magnetic moments lay on the same plane, *i.e.*, the spin distribution is two-dimensional.

In the case of the icosahedron ($N = 13$ and $\nu = 12$) we obtain a noncollinear arrangement of the local magnetic moments at the surface already for very small $U/t$. This is probably due to the five-fold degeneracy of the single-particle spectrum at $U = 0$. The magnetic order for $0.8 \leq U/t \leq 5$ is illustrated in Fig. 3. At the central atom (not shown in the figure) the magnetic moment is zero. In this range of values of $U/t$ ($U/t < 5$) the moments are coplanar, as already observed in the *fcc* structure. The tendency to avoid frustrations within NN triangular rings plays a dominant role in determining the overall arrangement of the local moments. However, one observes in Fig. 3 that some triangles have "optimal" antiferromagnetic order with the spin polarizations along the medians, while in other triangles two bond are frustrated. For larger values of $U/t$, a less frustrated solution, which is favored by the antiferromagnetic correlations, yields the lowest UHF energy. This is actually the case for $U/t > 5$, where we find a truly three-dimensional arrangement of the surface moments. An illustration of the magnetic order is given in Fig. 4. The magnetic moment at the central atom is zero, so that the same arrangement of surface spins is obtained even if the central site is lacking. The topology of the spins is very interesting. For example, if the 12 surface $\langle \vec{S}_l \rangle$ are brought to a common origin one observes that they also form an icosahedron. Concerning the magnetic moments $\langle \vec{S}_l \rangle$ on pentagonal

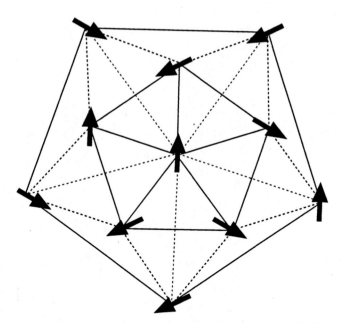

**Figure 3.** Illustration of the magnetic order at the surface of an icosahedral 13-atom cluster carrying a single hole ($\nu = N - 1$ electrons). The results are obtained by using the UHF approximation to the Hubbard model with $0.8 < U/t \leq 5$. All local magnetic moments $\langle \vec{S}_l \rangle$, represented by the arrows, are parallel to the plane of the figure. Notice that the position of the atoms in the lower pentagonal rings have been expanded to easy the visualization.

rings we find a small component perpendicular to the ring which is antiparallel to the moment at the atom capping the ring. The projections of $\langle \vec{S}_l \rangle$ within the ring plane present the same order as the one found for the isolated pentagon (see Fig. 4).

For half-band filling (*i.e.*, $\nu = 13$) we obtain that the *fcc* structure is more stable than the icosahedral structure for $0 \leq U/t \leq 3.7$. In this range of $U/t$ the magnetic moments are collinear. The size of spin polarizations $|\langle \vec{S}_l \rangle|$ is not uniform at the cluster surface. The atoms belonging to the central (001) plane (shown in grey in Fig. 5) present much larger moments than the atoms at the upper and lower (001) planes. For example for $U/t = 0.5$, $|\langle \vec{S}_l \rangle| = 0.2598\mu_B$ at the the central plane, while the moments at the other surface atoms are $|\langle \vec{S}_l \rangle| = 0.0046\mu_B$. In contrast to the single-hole case ($\nu = N-1$) now the central atom has a non-vanishing spin polarization ($|\langle \vec{S}_1 \rangle| = 0.0020\mu_B$ for $U/t = 0.5$). Notice that the magnetic order within the upper and lower (001) planes is ferromagnetic-like (with small magnetic moments) and that the surface moments belonging to different (001) planes couple antiferromagnetically. Thus, the magnetic moments at the surface of the central plane are not frustrated since all its NNs are antiparallel to them. Antiferromagnetic correlations appear to be important in the ground state already for small values of $U/t$. Unavoidable frustrations are found among the smallest magnetic moments (see Fig. 5). The nature of the lowest-energy UHF solution changes for $U/t > 5$. In this case the magnetic order is similar to the one found in this $U/t$ range for $\nu = N - 1$, which is illustrated in Fig. 2. As in the single-hole case, the direction of the spin polarizations at the triangular faces is along the medians. However, for $\nu = N$ the central site has non-

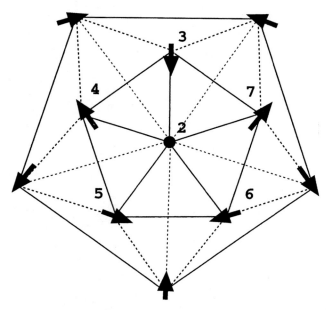

**Figure 4.** Illustration of the magnetic order at the surface of an icosahedral 13-atom cluster as in Fig. 3. The number of electrons is $\nu = N - 1$ (single hole) and the Coulomb repulsion is $U/t > 5$. The magnetic moment at site 2 is perpendicular to the surface (along the radial direction). The $\langle \vec{S}_l \rangle$ at sites $l = 3$–$7$ have an antiparallel projection to $\langle \vec{S}_2 \rangle$ ($\langle \vec{S}_l \rangle \cdot \langle \vec{S}_2 \rangle / |\langle \vec{S}_l \rangle||\langle \vec{S}_2 \rangle| = -0.4472$).

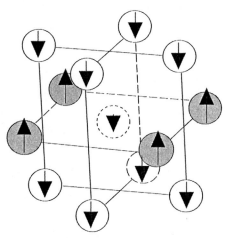

**Figure 5.** Illustration of the magnetic order in a *fcc*-like 13-atom cluster at half-band filling ($\nu = N$) and $U/t < 3.7$. All local magnetic moments are parallel to the $z$ axis.

zero magnetic moment $\langle \vec{S}_1 \rangle$. The surface $\langle \vec{S}_l \rangle$ tend to couple antiparallel to $\langle \vec{S}_1 \rangle$ which leads to a truly three-dimensional magnetic structure. Also for the icosahedron we find a close relationship between the $\nu = N$ and the $\nu = N - 1$ results, in particular concerning the arrangements of the local moments at the cluster surface.

Let us finally recall that in non-bipartite structures as the ones considered here, the behavior corresponding to one excess electron $\nu = N + 1$ cannot be inferred from the one-hole results. For the $fcc$-line 13-atom cluster with $\nu = 14$ electrons we obtain that the lowest energy UHF solution is the fully polarized ferromagnetic state for $U/t \geq 22.4$. This is in agreement with Nagaoka's theorem and with exact Lanczos diagonalizations.[6] On the other side, for $U/t \leq 3.7$ the solution is paramagnetic. In the intermediate range $3.7 < U/t < 22.4$, many different solutions are found to the UHF equations, each one of these yields the lowest energy over a small range of $U/t$. A more detail account of our calculations will be reported elsewhere.

## Acknowledgments

This work has been supported by CONACyT (Mexico). The Laboratoire de Physique Quantique is Unité Mixte de Recherche, associated to the CNRS.

## References

1. David R. Penn, *Phys. Rev.* **142**, 350 (1966); Ytunoda, *J. Phys. Condens. Matter* **1**, 10 427 (1989).
2. A. Vega, J. Dorantes-Dávila, G. M. Pastor, and L. C. Balbás, *Z. Phys. D* **19**, 263 (1991).
3. F. Willaime and L. M. Falicov, *J. Chem. Phys.* **98**, 6369 (1993); L. Bergomi, J. P. Blaizot, Th. Jolicoeur, and E. Dagotto, *Phys. Rev. B* **47**, 5539 (1993).
4. D. Coffey and S. A. Trugman, *Phys. Rev. Lett.* **69**, 176 (1992).
5. J. Hubbard, *Proc. R. Soc. London, Ser. A* **276**, 238 (1963); *ibid.* **281**, 401 (1964); J. Kanamori, *Prog. Theor. Phys.* **30**, 275 (1963); M. C. Gutzwiller, *Phys. Rev. Lett.* **10**, 159 (1963).
6. G. M. Pastor, R. Hirsch, and B. Mühlschlegel, *Phys. Rev. Lett.* **72**, 3879 (1994); *Phys. Rev. B* **53**, 10 382 (1996).
7. M. A. Ojeda-López, *Rev. Mex. Fís.* **43**, 280 (1997); M. A. Ojeda-López, J. Dorantes-Dávila, G. M. Pastor, *J. Appl. Phys.* **81**, 4170 (1997).

# Phase Transitions in Ising Square Antiferromagnets: A Controversial System

E. López-Sandoval,[1] F. Aguilera-Granja,[2,*] and J. L. Morán-López[1]

[1] Instituto de Física
Universidad Autónoma de San Luis Potosí
Alvaro Obregón 64, 78000 San Luis Potosí, S. L. P.
MEXICO

[2] Departamento de Física Teórica
Universidad de Valladolid
47011 Valladolid
SPAIN

## Abstract

The phase transitions occurring in the Ising square antiferromagnet with first- ($J_1$) and second- ($J_2$) nearest-neighbors interactions are studied as a function of the $V (\equiv J_2/J_1)$ parameter using the Monte Carlo simulation method. Our simulation results based on the analysis of the magnetization at the critical temperature indicate that the system undergoes a second-order phase transition for $V < 0.5$ and a first-order phase transition for $0.5 < V < 1.35$. Our results confirm previous calculations performed within the 9-point approximation of the cluster variation method.

## I. Introduction

Although the Ising square antiferromagnetic system with nearest- ($J_1$) and next-nearest-neighbor interactions ($J_2$) is a relatively old problem, there are still some facts not well know in this simple model, as it has been pointed out recently.[1] In particular, the behavior of the system close to $V = J_2/J_1 = 0.5$ seems to be a very controversial point, since various behaviors have been suggested: (i) non-universal behavior and continuous change in the critical exponents as a function of the $V$ parameter in the vicinity of $V > 0.5$ (Ref. 2), (ii) one dimensional-like behavior for $V = 0.5$ (Ref. 3), and (iii) a tricritical point at $V = 0.5$, that is, the order of the transition changes from second order (for $V < 0.5$) to first order for $0.5 < V \leq 1.142$ (Refs. 1 and 4). A more detailed discussion on this problem can be found in the Ref. 1 and works quoted therein.

Here, we are interested in studying the Ising square antiferromagnetic system with first- and second-neighbor interactions, in particular the nature of the transitions in the range $0.5 \leq V \approx 1.3$. Although this has been studied before within the cluster variation method (CVM) in different approximations,[1,4] this has been criticized due to the fact that the CVM is a mean field method. Here, we use the Monte Carlo method to study this range of $V$ values by looking at the long-range order (LRO) parameter or magnetization, to establish the nature of the transition for $V > 0.5$.

In the Sect. II we present the model. The results are discussed in Sect. III. Our conclusions and a summary are presented in Sect. IV.

## II. Model

We consider the two dimensional spin-$\frac{1}{2}$ Ising square lattice model with first- ($J_1$) and second-neighbor ($J_2$) interactions. The Hamiltonian that describes the system is

$$\mathcal{H} = J_1 {\sum}' \sigma_i \sigma_j + J_2 {\sum}'' \sigma_i \sigma_k, \qquad (1)$$

where $\sigma_\ell = \pm 1$ and the sums ${\sum}'$ and ${\sum}''$ are over nearest- and next-nearest-neighbors, respectively. In particular in this work, we restrict ourselves to the case in which both interactions are antiferromagnetic ($J_1, J_2 > 0$). The ground state in terms of the $V$ parameter is well known.[5] For $V < 0.5$ the system develops a simple antiferromagnetic order (AF) consisting of two ferromagnetic interpenetrating sublattices with antiferromagnetic coupling. For $V > 0.5$, the system develops a superantiferromagnetic order (SAF). This phase can be described as an alternative arrangement of single ferromagnetic rows of opposite oriented spins. For the point $V = 0.5$, the magnetic order disappears due to the highly degenerate state.[1-5]

To describe the thermodynamics of the system we use a standard important-sampling Monte Carlo method (MC) for a system of $L \times L$ lattice points with periodic boundary condictions (PBC) applied to eliminate surface effects. Only even values of $L$ were used so that no "misfit seams" were present. Succesive states were generated using a single-site spin-flip mechanism where the transition probability is

$$W(\sigma_i \to -\sigma_i) = W_0 \exp\left(\frac{-\Delta E_i}{k_B T}\right), \qquad (2)$$

where $W_0$ is a normalization factor and $\Delta E_i$ is the energy difference involved in the transition and is calculated from (1). The thermodynamic average for the order parameter was determined by calculating the weighted statistical mechanical average over the limited number of configurations sampled. At least $10^4$ Monte Carlo steps (MCS) were taken for each one of the lattice points. Each data point for a given configuration is averaged over a given number of events for every one of the different $V$ values. A full description of the MC method used here can be found in Ref. 6.

To describe the two antiferromagenetic phases quoted above, the square lattice in the system has to subdivided into four square equivalent interpenetrating sublattices: $\alpha, \beta, \gamma$, and $\delta$. The magnetization at each sublattice is

$$M_\nu = \frac{4}{N} \sum_{i \in \nu} \langle \sigma_i \rangle, \qquad (3)$$

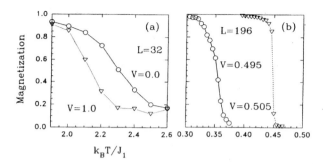

**Figure 1.** Magnetization for different system sizes and $V$ values as indicated in the figure.

where $N$ is the total number of lattice sites $(L \times L)$ and $\nu$ is the lattice type. The magnetization in the AF and SAF states are given in terms of the sublattice magnetization by

$$M_{AF} = \tfrac{1}{4}[M_\alpha + M_\gamma - (M_\beta + M_\delta)], \qquad (4a)$$

$$M_{SAF} = \tfrac{1}{4}[M_\alpha + M_\delta - (M_\beta + M_\gamma)]. \qquad (4b)$$

Before we present the results, we would like to mention that we can reproduce with our MC algorithm the phase diagram reported by Landau[5] for the whole range of $V$ values.

## III. Results

In the Fig. 1(a), we present the magnetization for $V = 0.0$ and $V = 1.0$ for a system with $L = 32$. The magnetization in both cases presents a very smooth behavior. The critical temperature $T_c$ was determinated from the infection point of a polynomial function interpolated within the simulated points. In the case $V = 0.0$ our simulation gives $k_B T_c/J_1 = 2.29$ in comparison with the exact value 2.269, and for $V = 1.0$ $k_B T_c/J_1$ is 2.1 in comparison with the one from the extrapolated from the finite size scaling 2.07. In both cases the difference is less than 1.5%. The smooth behavior of the magnetization in the case of $V = 0.0$ indicates a second-order phase transition, as it is well known.[5] However in the case of $V = 1.0$ a more careful analysis is required as we see below. Notice that for temperatures above the critical temperature it is observed a remaing tail in the magnetization due to the finite size of the cell ($L \times L$ regardless of the PBC). This tail vanishes only in the limit $L \to \infty$ (Refs. 2, 5, and 6).

To find out the nature of the transitions close to 0.5, we performed simulations of the magnetization for $V = 0.495$ and $0.505$ and to minimize the tail effect we increase the system size up to $L = 196$. The results are presented in Fig. 1(b), the first thing to be noticed is that close to $V = 0.5$ the magnetization conserves the value to 1.0 until very close to $T_c(V)$, this fact was pointed out before by Landau,[5] however, in his calculation he did not get so close to $V = 0.5$ as we did. The critical temperatures for $V = 0.495$ and $0.505$ using the above definition are 0.356 and 0.447, respectively.

Now a key point is: How to define a first or second order phase transition based on numerical results of magnetization? This is not an easy question in the case of numerical calculations, and any criteria taken has an arbitrariness in their definition. In the case of the Ising-like models a possible criterion could be the width of the

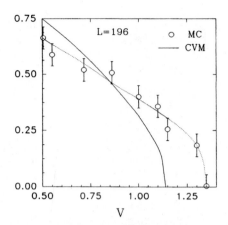

**Figure 2.** Discontinuity of the magnetization $[\Delta M = M(T_{c-}) - M(T_{c+})]$ calculated with the nine-point CVM approximation by Morán-López et al.,[1] in a continuous line and the MC simulation results on circles (dashed line is an aid to the eye).

temperature range $\tau$ associated with a decrease $\mathcal{D}$ in the LRO from 0.85% to $1/\sqrt{L}$. By this simple criterion we take into account the remaining magnetization (tail effects). Now we need a reference value to decide when a transition is first or second order. This can be taken from the Onsager exact solution for the magnetization; in this case $\tau$ is $0.06T_c(V=0)$. With this criterion, we can say that a second-order transition occurs if $\tau \approx 0.06T_c(V)$ or larger than that, and a *weak first order* if $\tau$ is smaller than that value. Now, with this criterion let us examine the cases $V = 0.495$ and $0.505$ of our MC simulation. The results for $\tau$ are 0.1 and 0.01 of $T_c$, respectively. Then the phase transition at $V = 0.495$ is of second order while for $V = 0.505$, it can be considered as a *weak first-order* phase transition.

We can get more information about the nature of the transition by looking the discontinuity in the magnetization $[\Delta M = M(T_{c-}) - M(T_{c+})]$; being $T_{c\pm} = T_c(1 \pm \epsilon)$ and $\epsilon$ an infinitesimal value. $\Delta M$ is different from zero for a first-order phase transition and zero in the case of a second-order phase transition. However, before we present the results and compare with the CVM results,[1] we have to mention that in order to perform the comparison in the case of the MC results it is necessary to consider the remaing tail due to the finite size. To correct that, it is reasonable to subtract the factor $1/\sqrt{L}$ to $\Delta M$ (Ref. 7). The results are presented in Fig. 2, the continuous line is the result of the nine-point approximation of CVM while the circles are our MC results. Notice that both MC and CVM present the same qualitative behavior. Both predict a range where the system presents a first-order phase transition. It is important to mention that this first-order phase transition was suggested first using the nine-point CVM approximation before,[1] and that if the correction due to the remainig tail is not considered the conclusions still hold.

## IV. Summary and Conclusions

In this work we studied the behavior presented for a two dimensional Ising square antiferromagnetic lattice with first- and second-neighbor interactions. We calculated in particular the magnetization of the system in the range $0.5 \leq V \leq 1.35$. Our results indicate a possible *weak first-order* phase transition in the range $0.5 \leq V \approx 1.3$. The results of this simulation confirm some calculations based on the different CVM approximations that predict a first-order phase transition for $0.5 < V \approx 1.14$ (Refs. 1 and 4). Since this is may not be the final answer to the behavior of this system, it is required more extensive simulations with larger lattice sizes to avoid tails effects that could mask the order of the transition.

## Acknowledgments

This work has been supported by DGICYT Grants PB95-0720-C02-01 and SAB95-0390. One of us (FAG) also acknowledges support from Consejo Nacional de Ciencia y Tecnología, México (Grant 961015).

## References

* Permanent address: Instituto de Física, Universidad Autónoma de San Luis Potosí, 78000 San Luis Potosí, S.L.P., Mexico
1. J. L. Morán-López, F. Aguilera-Granja and J. M. Sanchez, *Phys. Rev. B* **48**, 3519 (1993); *J. Phys.: Condensed Matter* **6**, 9759 (1994).
2. K. Binder and D. P. Landau., *Phys. Rev. B* **31**, 5946 (1985).
3. M. D. Grynberg and B. Tanatar, *Phys. Rev. B* **45**, 2876 (1992).
4. F. Aguilera-Granja and J. L. Morán-López., *J. Phys.: Condensed Matter* **5**, A195 (1993).
5. D. P. Landau, *Phys. Rev. B* **21**, 1285 (1980).
6. K. Binder and D. W. Heermann, *Monte Carlo Simulation in Statistical Physics* (Springer-Verlag, Berlin, 1990).
7. D. P. Landau, *Phys. Rev. B* **7**, 2997 (1976).

# Interfacial Interdiffusion and Magnetic Properties of Transition Metal Based Materials

M. Freyss,[1] D. Stoeffler,[1] S. Miethaner,[2] G. Bayreuther,[2] and H. Dreyssé[1]

[1] Institut de Physique et Chimie des Matériaux de Strasbourg
Université Louis Pasteur
23 rue du Loess, F-67037 Strasbourg
FRANCE

[2] Institut für Experimentelle and Angewandte Physik
Universität Regensburg
93040 Regensburg
GERMANY

## Abstract

Experimental results and calculations of the Cr/Fe system magnetic order are reported. The role of the interfacial diffusion of Cr in Fe substrate during the Cr growth on Fe(001) is particulary investigated. The magnetic moment distribution is computed in the real-space by means of a self-consistent Hubbard-type tight-binding Hamiltonian. A simulation of the variation of the total sample magnetization during Cr growth shows that the more Cr and Fe are interdiffused at the interface, the more important is the decrease of the magnetization. The effects of this interdiffusion are found to be more important if the growth mode is not layer by layer. The local magnetic moments are strongly affected by this interdiffusion. Experimental studies are explained within this framework. For Fe/Cr/Fe trilayers, experimental evidences of spin-flip in the upper Fe film are shown. Limitations and extensions of the model are discussed.

## I. Introduction

The possibility to taylor artificial materials at an atomic scale has opened the way to the preparation of new materials with specific properties. Magnetic properties are particularly interesting for technological applications.[1] The Fe/Cr system is probably the most studied system, theoretically and experimentally.[2] The antiferromagnetic (AF) coupling through the Cr spacer shown by Grünberg et al.[3] and the giant magnetoresistance by the group of Fert (Ref. 4) have strongly contributed to this interest. The Fe/Cr interface appears as a "paradigm," since a lot of problems encountered in the artificial magnetic materials are present.[2]

In this contribution we shall focus on two particular aspects where the interaction between experimental work and theoretical approach is particularly useful. Theoretical description of low-dimensional magnetism has been restricted for a long time to idealized model systems. Only recently first-principles band structure formalisms have been derived and used in the real space.[5] Semi-empirical approaches have gained reliability and shown their ability to describe complex systems.[6] Experimental studies indicate clearly that the growth processes are complex. We will illustrate these difficulties by considering the growth of Cr on a Fe(001) substrate. In such a case our calculations display a possible Cr-Fe interdiffusion at the interface. In a second part we discuss the magnetic arrangement of a Fe/Cr/Fe trilayer. We will show that the orientation of the surface atoms magnetic moments is related to a Cr-Fe interdiffusion at the first interface (between Cr and the Fe substrate) and to possible topological antiferromagnetism.

## II. Cr Growth on Fe(001)

Experimental studies of the Cr growth on Fe(001) have shown contrasting results. On one side, a very large variation of the sample moment by $5\mu_B$ per two-dimensional unit cell has been found.[7] On the other side, oscillatory behavior of the magnetization for low Cr thickness (typically up to 2 ML) and for large thickness more or less no variation have been shown.[8] Obviously the growth conditions of the films are the key quantities. The temperature of preparation is a crucial parameter. As it can be expected by thermodynamic arguments, an increase of the temperature could induce interdiffusion of Cr into Fe. Recently various experimental works have clearly demonstrated that supposing a perfectly flat interface during the Cr growth on Fe(001) is only a rough approximation. This assumption has the great advantage to make computations more tractable. However, the "real" systems are much more complex. For instance by means of scanning tunneling microscopy (STM) and spectroscopy, Davies et al.,[9] measured the concentraction of Cr on the surface layer during its growth. After the deposition of 1 ML of Cr, only 10% of the Cr remain in the surface layer, indicating that most of Cr atoms have diffused into the Fe substrate. Furthermore, it is only after the deposition of 2–3 ML of Cr that the surface layer contains more Cr than Fe. A similar trend has been found by Pfandzelter et al.[10] using proton- and electron-induced Auger-electron spectroscopy. The Cr concentration profile is, from the surface layer to the bulk: 45%, 55%, and 0% for 1 ML of Cr deposited, and 70%, 100%, and 30% for 2 ML of Cr deposited. These results are different from those of Venus and Heinrich[11] who show a gradual decrease of the Cr concentration from the surface to the bulk.

In this work we report recent calculations of local magnetic moment distribution during the Cr growth on a Fe(001) substrate taking into account interfacial diffusion. An interdiffusion over two layers is supposed at the Fe interface (Fig. 1), simulating the $Cr_n/Fe_xCr_{1-x}/Fe_{1-x}Cr_x/Fe(001)$ system. When no interdiffusion occurs ($x = 0$, in Fig. 1), the direction of the local magnetic moment of all atoms is clearly defined by the strong AF coupling between Cr and nearest-neighbors Cr or Fe atoms. For

# Interfacial Interdiffusion and Magnetic Properties of ...

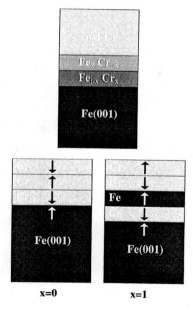

**Figure 1.** Schematic view of studied systems: without interdiffusion ($x = 0$); complete monolayer Fe-Cr interchange at the Fe(001) interface ($x = 1$) and two layers alloy at the Fe(001) interface. All layers are found to display parallel local magnetic moments for $n = 0$ and $n = 1$. The direction of the layer magnetization is given by the arrow. Since no spin-orbit coupling is consider in this work, the direction of the magnetic moment is arbitrary.

a complete exchange of one Fe layer with a Cr layer, previous work[12] has shown that the exchanged Fe layer displays a small positive magnetic moment (according to the sign convention of Fig. 1) whereas the Cr atoms have clearly oriented local magnetic moments, once again imposed by the AF Cr-Cr and Cr-Fe coupling. In these two cases, the Fe/Cr interface has been supposed to be abrupt and no magnetic frustrations are present. This is no more the case when partial interdiffusion occurs and the situation becomes more delicate. In our calculations we have assumed ordered two-layers magnetic structures (for this reason, the values of $x$ correspond to the simple 2D arrangements). This model should be able to give indications of the modifications induced by interdiffusion.

In figure 2 numerical solutions for $Cr_2/Cr_{2/3}Fe_{1/3}/Cr_{1/3}Fe_{2/3}/Fe(100)$ and $Cr_2/Cr_{1/2}Fe_{1/2}/Cr_{1/2}Fe_{1/2}/Fe(001)$ are reported. Calculations have been performed in the real space by solving a self-consistent Hubbard type tight-binding Hamiltonian.[6,12–14] In figure 2, only the most stable configurations are given. For Cr layers with a thickness larger than 2 ML, usually, only two numerical solutions are found. For thinner Cr layers, a larger number of solutions can be found according to the value of $x$ (Ref. 15). Figure 2 is a good example of the complex magnetic moment distributions in such frustrated systems. The magnetic moments of the surface Cr atoms are of opposite sign for those two values of $x$. For $x = \frac{1}{3}$, the Fe atoms, in larger amount than Cr in the first mixed layer, impose negative Cr moments in the second mixed layer. Then those negative Cr moments impose positive moments on the first pure Cr layer and due to strong AF Cr-Cr coupling, a negative surface atoms moments. For $x = \frac{1}{2}$, the Fe atoms of the second mixed layer force pure Cr layer to

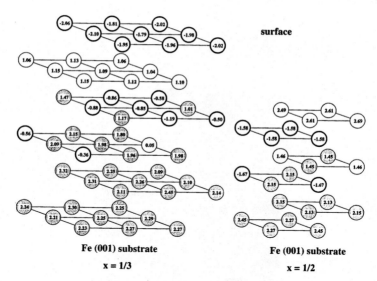

**Figure 2.** Ground state and metastable solution for $Cr_2/Cr_{2/3}Fe_{1/3}/Cr_{1/3}Fe_{2/3}/Fe(100)$ and $Cr_2/Cr_{1/2}Fe_{1/2}/Cr_{1/2}Fe_{1/2}/Fe(001)$. A distinction is made between the two directions of the Cr magnetic moments by means of the linewidth of the circles. In the case $x = \frac{1}{3}$, the vertical scale has been expanded for more visibility. Only the unit cell of each layer is shown, as well as only the the two first layers of the semi-infinite Fe substrate.

be negative. Then the surface moments are positive. It is also interesting to note the different values of the Cr moments for those two values of $x$. For instance, the Cr surface atoms for $x = \frac{1}{2}$ display a large moment, as shown in pure Cr(001) surface.[16] It is a completely different situation for $x = \frac{1}{3}$ where the surface atoms bear a much smaller average magnetic moment. It should also be noticed that "frustrated" atoms have smaller magnetic moments as already observed.[6] A simple rule emerges from the calculations: The configuration with less "frustrated" bound is, generally, the most stable. However, it must be noticed here that the energy differences can be small and, *a priori* coexistence of two magnetic phases cannot be excluded at finite temperature. In any case the fully self-consistent calculations must be performed, general trends can only give very rough order of magnitude moments.[15]

The total magnetization induced by the Cr films is computed within a model where the key ingredients are $x$ (the degree of interdiffusion at the interface, as defined previously) and $\theta_i$, the coverage of the $i$th layer *versus* time. In this simple model[15] the system consists on $Cr_p/Fe_xCr_{1-x}/Cr_xFe_{1-x}/Fe(001)$ arrangements where $p$ is given by the set of $\theta_i$. The local magnetic moments are again computed self-consistently for each different configuration. Two different regimes have been considered. In a first case a nearly layer-by-layer growth mode is assumed. As it can be seen from Fig. 3, the magnetization displays a "periodic" behavior, which can be expected directly due to the strong AF coupling between adjacent Cr layers. For a large interdiffusion ($x = 0.89$), few layers are necessary to get this behavior and now the mean value (around $-4.5\mu_B$) is large as compared to the case $x = 0$ ($\simeq 0.5\mu_B$). This situation is different when a strong non-layer-by-layer growth mode is considered. In figure 4 we report the magnetization induced by the Cr deposition for different values of the interfacial diffusion. The oscillations are damped, as compared to a layer-by-layer growth mode.

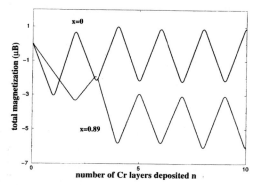

**Figure 3.** Simulation of the variation of the total magnetization during Cr growth for two interfacial alloy concentration $x$, as defined in Fig. 1, supposing a quasi-layer-by-layer growth mode.

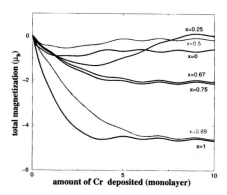

**Figure 4.** Simulation of the variation magnetization during Cr growth *versus* the interfacial alloy concentration $x$, as a defined in Fig. 1 supposing a strongly non-layer-by-layer growth mode.

General features appear from Fig. 4. In the first stage of the growth (typically for a coverage lower than 3 ML), whatever the value of $x$ is, the magnetization decreases. For a larger coverage, two cases occur. When $x$ is larger than 0.5, the magnetization continues to decrease and tends to an asymptotic value, which increases, in absolute value, as a function of $x$. This larger decrease of magnetization is related to the strong reduction of the Fe moments at the upper mixed layers (of the order of $1\mu_B$ or less). When $x$ is smaller than 0.5, the magnetization increases for coverage larger than 3 ML and finally tends to the asymptotic value. This increase is particularly important for $x = 0.25$. This atypical behavior can be explained by the fact that around this value of $x$, a large change in the local magnetic moments of the alloy region occurs when the thickness of Cr increases.

In a recent work, Miethaner and Bayreuther[17] have studied strongly faceted surfaces with a large step density by using particular growth conditions (*e.g.*, growth temperature 300 K) and after depositing several Au/Fe/Cr sandwich layers. After the deposition of a gold layer of 30 Å a Fe films of 12.6 Å was deposited followed by

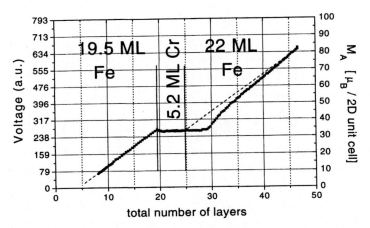

**Figure 5.** Total spontaneous magnetization of a Fe/Cr/Fe trilayer grown on Au (001) measured by *in situ* AGM.[18]

8.4 Å of Cr, 12.2 Å of Fe and a further Au layer (all grown at 300 K). It has been observed that the magnetic moment of the sample does not change upon deposition of Cr on top of the first Fe film in contrast to previous work.[7] Furthermore by using different growth conditions.[18] LEED, RHEED, and *in situ* STM were used for structural characterization of the films. Magnetization curves were measured continuously during film growth using an UHV alternating gradient magnetometer (AGM). The spontaneous moment is obtained by extrapolating from saturation back to $H = 0$. In figure 5 the total sample magnetization is reported for a substrate with intermediate step density during the film deposition. In the first stage the linear increase of the moment during Fe growth, as expected, shows a rate of $2.2\mu_B$ per Fe atom. Then follows a distinct moment decrease during Cr deposited which saturates after about 2 ML of Cr at $1.4\mu_B$ per 2D unit cell. The asymptotic value of the total sample moment variation reached during the Cr growth for, relatively, thin Cr films (typically less than 10 ML), topological antiferromagnetism for Cr magnetic arrangement is the driving force. As shown in model calculations,[6,19] the ferromagnetic order of the Fe substrate imposes domains of Cr with opposite direction of the local moments. If the number of terraces is large (or the density high enough), the average Cr moment should vanish. It is the case in the work where strongly faceted Fe surfaces have been created.[17] In the experimental work reported in Fig. 5, it is no more the case and the asymptotic value is far to vanish: interdifusion of Cr at the Fe interface occurs. A stronger interdiffusion should also have been present in the work of Turner and Bayreuther.[7]

## III. The Fe/Cr/Fe Trilayer

SEMPA experiments[20] have shown that in Fe/Cr/Fe trilayer the AF coupling of the Fe layer through the Cr spacer can display a long period, as expected theoretically by Fermi surface topology arguments[21] and a short period also predicted by band structure calculations.[22,23] The growth temperature of the sample leads to different microscopic arrangements and to two different oscillation periods. The short period of 2 monolayers (ML) can be easily understood: As it is well known bulk Cr presents

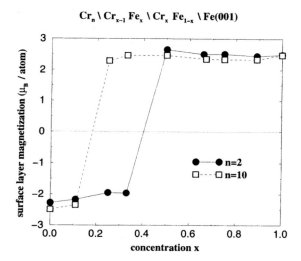

**Figure 6.** Values of the average surface atoms magnetic moment as a function of the concentration $x$ when two and eleven Cr layers are deposited with two interfacial mixed layers. The sign of the surface atoms moment characterizes the phase of the antiferromagnetic layer-by-layer structure of Cr.

a spin density wave (SDW) of about 2.1 ML at room temperature. In the previously mentionned SEMPA work, this was clearly recognizable. A good approximation to describe bcc Cr, especially for very thin film, is to suppose that Cr has a perfect antiferromagnetic order. Theoretical studies[24,25] indicated that AF order is prefered for middle serie transition metals whereas ferromagnetic (F) order is the rule for transition metals at the end of the series and thus an AF coupling at the Fe/Cr (001) interface is expected. This later point has been confirmed by various experimental works.[8,26] The short period of 2 ML is a direct consequence of the AF first neighbor Cr-Cr and Fe-Cr interactions.

However, a subtle point was not understood in the SEMPA experiments: The polarization of the top Fe layer was opposite to the expected value according the parity of the number of Cr layers. This aspect was nearly forgotten until a couple of experimental works addressed the question of possible interdiffusion near the Cr/Fe interface as mentionned in the previous Section. To understand this behavior, we have performed a systematic study of the electronic structure and the magnetic moment distribution $versus$ $x$. When Fe is deposited on Cr, no interdiffusion of Fe in Cr occurs.[11] Than for Fe films grown on perfectly flat interface, all Fe are ferromagnetically coupled and the direction of the surface atom moment, which can be observed experimentally, is opposite to the direction of the top Cr layer. Thus the systematical study of $Cr_n/Fe_xCr_{1-x}/Cr_xFe_{1-x}/Fe(001)$ system can address to this problem. Considering three layers of Cr (i.e., a film of two layers of Cr on top of a two-layer alloy), the average value of the surface Cr atoms shows a drastic change around $x = 0.50$ (see Fig. 6). This result indicates that if more Cr atoms than Fe atoms are at the interface of the Fe substrate, the phase of the Cr magnetic arrangement changes. This conjecture has been checked for large Cr thickness. Surprisingly (see Fig. 6), for 11 layers of Cr, this transition occurs for smaller interdiffusion (around $x = 0.3$), suggesting that a small interdiffusion is able to reverse the magnetization of the Cr layers.

In the experiments performed at NIST,[20] a Fe whisker has been used and the width of the terraces was particulary large. In the work reported in Fig. 5, there is a much larger density of terraces at the Cr/Fe interface, inducing surely no abrupt interface at the second Fe/Cr interface. Once again, one can expect that topological antiferromagnetism plays an importan role. In such a case, the Fe film presents, for very small thickness, many domains with opposite direction of the local moments. When the Fe coverage increases, previous theoretical work[6] has suggested progressive spin-flips. On Figure 5, the magnetic moment of the second Fe layer increases at a higher rate than for $2.2\mu_B$ per Fe atom (dashed line in Fig. 5). It indicates clearly the transition from the topological antiferromagnetism to a simple Fe ferromagnetic order. As it can be seen in Fig. 5, it takes around 20 ML before the Fe ferromagnetic order is fully recovered.

## IV. Conclusion and Outlook

In this communication we have reported recent results on the Fe/Cr magnetic interface, combining experimental studies and numerical simulations. The growth of Cr on a Fe(001) substrate has been particularly investigated. Within a simple model of growth, the variation of the total sample magnetization has been determined for two situations: layer-by-layer growth model and strongly non-layer-by-layer growth mode. As shown in recent experimental works, the effect of the Cr diffusion into Fe substrate has been taken into account by considering a two-layer alloy near the Fe substrate. The rate of interdiffusion in such a model is found to play an important role when the growth mode is far from being layer by layer. This interdiffusion affects strongly the magnetic order of the Cr film. The exchange of one quarter of a monolayer of Fe and Cr is enough to reverse the layer-by-layer antiferromagnetic arrangement of a 10 ML thick Cr film. For Cr on Fe(001), the asymptotic value of the total sample magnetization is a good indication of the interdiffusion at the Fe interface. For Fe deposited on Cr/Fe(001), the recent experimental works shown in this paper give clear evidences of topological antiferromagnetism of Fe.

In our calculations, a perfect 1D symmetry has been supposed. A next step is to consider kink effects. A real-space tight-binding framework should be able to answer this question. Obviously, fully self-consistent calculation of these systems will require large computational facilities. Two other aspects should be considered in a next future in order to describe the growth mode of magnetic films at an atomic scale. First a perfectly order alloy has been supposed at the interface. A completely disordered alloy be described by means of the coherent potential approximation (CPA).[27] In fact chemical and magnetic order should be considered simultaneously but a complete description has not yet been done. Another limitation in our work is the collinearity imposed to all magnetic moments. We could imagine that noncollinear will minimize the energy lost by magnetic frustrations. Unfortunately calculations of noncollinearity magnetic order for nonperfect systems require too large computing time.[15] The use of multiprocessor architecture computers is absolutely necessary for such but only significant improvements on the algorithms will allow to address this question.

## Acknowledgments

This work has been realized with a partial support of the french-german collaboration program PROCOPE and the European Network TMR "Interface Magnetism." Fruitful discussions with C. Demangeat are acknowledged.

# References

1. See for instance J. A. C. Bland and B. Heinrich, *Ultrathin Magnetic Structures I and II* (Springer, Berlin, 1994).
2. For instance more than 30 papers have been presented in the "Fe/Cr Interface Magnetism Workshop" held in Strasbourg, 2–3 june 1996.
3. P. Grünberg et al., *Phys. Rev. Lett.* **57**, 2442 (1986).
4. M. N. Baibich et al., *Phys. Rev. Lett.* **61**, 2472 (1988); A. Barthelemy, A. Fert, R. Morel, and L. Streren, *Physics World*, 34 (november, 1994).
5. For a recent review see H. Dreyssé and C. Demangeat, *Surf. Sci. Rep.* **28**, 65 (1997).
6. A. Vega, D. Stoeffler, H. Dreyssé, and C. Demangeat, *Europhy. Lett.* **31**, 561 (1995).
7. C. Turtur and G. Bayreuther, *Phys. Rev. Lett.* **72**, 1557 (1994).
8. P. Fruchs, K. Totland, and M. Landolt, *Phys. Rev. B* **53**, 3293 (1995).
9. A. Davies, J. A. Stroscio, D. T. Pierce, and R. J. Celotta, *Phys. Rev. Lett.* **76**, 4175 (1996).
10. R. Pfandzelter, T. Igel, and H. Winter, *Phys. Rev. B* **54**, 4496 (1996).
11. D. Venus and B. Heinrich, *Phys. Rev. B* **53**, R1733 (1996).
12. D. Stoeffler, A. Vega, H. Dreyssé, and C. Demangeat, *Mat. Res. Soc. Symp. Proc.* **384**, 247 (1995).
13. D. Stoeffler, Ph.D. Tesis, Strasbourg, 1992 (unpublished); D. Stoeffler, F. Gautier, *J. Magn. Magn. Mater.* **147**, 260 (1995).
14. M. Freyss, D. Stoeffler and H. Dreyssé, *Phys. Rev. B* **54**, R12 677 (1996).
15. M. Freyss, D. Stoeffler, and H. Dreyssé, *Phys. Rev. B* **56**, 6047 (1997).
16. A. Vega et al., *Phys. Rev. B* **49**, 12 797 (1994).
17. S. Miethaner and G. Bayreuther, *J. Magn. Magn. Mater.* **148**, 42 (1995).
18. S. Miethaner and G. Bayreuther, to be published.
19. D. Stoeffler and F. Gautier, *J. Magn. Magn. Mater.* **156**, 114 (1996).
20. J. Unguris, R. J. Cellota, and D. T. Pierce, *Phys. Rev. Lett.* **67**, 140 (1991).
21. P. Bruno and C. Chappert, *Phys. Rev. Lett.* **67**, 1602 (1991).
22. D. Stoeffler and F. Gautier, *Prog. Theor. Phys.* Suppl. **101**, 139 (1990); D. Stoeffler and F. Gautier, *Magn. Magn. Mater.* **121**, 259 (1993).
23. F. Herman, J. Sticht, and M. van Schilegaarde, *J. Appl. Phys.* **69**, 4783 (1991).
24. V. Heine and J. H. Samson, *J. Phys. F* **13**, 2155 (1983).
25. D. Stoeffler and H. Dreyssé, *Solid State Commun.* **79**, 645 (1991).
26. F. U. Hillerbrecht et al., *Europhy. Lett.* **19**, 711 (1992).
27. I. Turek et al., *Phys. Rev. Lett.* **74**, 2551 (1995).

# Magnetic Surface Enhancement and the Curie Temperature in Ising Thin Films

S. Meza-Aguilar,[1] F. Aguilera-Granja,[2,*] and J. L. Morán-López[3]

[1] Institut de Physique el Chimie des Matériaux de Strasburg
GEMM, UMR46, CNRS-ULP, 23 rue du Loess
67037, Starsburg Cedex
FRANCE

[2] Departamento de Física Teórica
Universidad de Valladolid
47011 Valladolid
SPAIN

[3] Instituto de Física
Universidad Autónoma de San Luis Potosí
Alvaro Obregón 64, 78000 San Luis Potosí, S.L.P.
MEXICO

## Abstract

We study the magnetic surface enhancement and the Curie temperature in ferromagnetic Ising thin films as a function of the surface orientations and structure. We use a simplified version of the cluster variation method in the pair approximation that only considers as inequivalent those sites on the surface, while all the internal sites are considered as equivalent regardless of the position. We calculate the critical value of the surface interaction $J_{sc}$, beyond which the surface dominates the magnetic properties of the system. Our results are in agreement with those of the reaction field approximation and the renormalization group theory for the case of simple cubic lattice thin films.

## I. Introduction

It is well established experimentally that there is an enhancement in the magnetic moments of atoms at the surface of thin films. This has been reported for Fe, Co, and

Gd thin films and surfaces.[1,2] There is also experimental evidence that the Curie temperature of thin films of some materials also presents an enhancement respect to the bulk Curie temperature.[1,2] These surface phenomena are very sensitive to the topology of the surface and not only to the reduced coordination number.[1,2] The surface orientation is an important parameter in determining the surface behavior, since the vicinity and the number of neighbors are different for different directions. Therefore, a dependence on the surface direction is also expected. The surface exchange interaction is other important parameter in the determination of the surface properties. It is important to study its effects on the surface behavior.

Among the theoretical efforts to explain the surface enhancement in the magnetic moment and the Curie temperature at the surfaces in macroscopic systems and thin films, there are some phenomenological models that assume a different exchange interaction at the surface $J_s$ than in the bulk $J$ (Refs. 3 and 6). In these models the main goal is the calculation of the critical value of the surface effective interaction $J_{sc}$, beyond which the surface dominates the magnetic properties of the system. However, in most of the cases the theoretical models only study the (100) surface of a simple cubic lattice, and within the mean field approximation in which an analytical expression for the critical surface interaction $J_{sc}$ is obtained. Only a few cases in other structures and surface orientations are considered.[3,4] The reader may find an excellent review of the different values reported for the critical surface interaction $J_{sc}$ for the different approximations in the paper by Tsallis.[6]

Here, we are interested in the surface enhancement and the Curie temperature in ferromagnetic Ising thin films as a function of the different geometrical properties. We calculate the Curie temperature and the critical value of the surface interaction $J_{sc}$ for thin films with various crystal structures and orientations. To that aim we assume an effective magnetic moment at each site of the system, and only two types of sites: surface and internal sites. The probability distribution that describes these two kinds of points is calculated within a simplified version of the pair approximation of the cluster variation method (CVM).[7] The model worked out here is similar to the one presented by Hellenthal[8] in the sixties. However in our case we considered correlations between first nearest-neighbors and the position dependence in the case of surface sites.

The layout of the paper is as follows. In Sect. II we outline the model and the results are presented and discussed in Sect. III. Our conclusions and a summary are presented in Sect. IV.

## II. Model

To describe the magnetic properties of the thin films we adopt the Ising model with spin-$\frac{1}{2}$. In this model we consider only first nearest-neighbor ferromagnetic interactions. We assume that there are only two kinds of sites; surface sites and internal sites. All the internal sites, those below the surface, are equivalent among themselves regardless of the position. In the case of pairs of sites there are three different types; surface pairs, when both sites are at the surface, intermediate pair when a site is in the surface and the other one in the interior and internal pairs when both sites of the pair are in the interior, of the system. The spin probabilities are denoted by $x_i$ with $i = 1$ (2) for spin up (down). The probability of finding a pair of first nearest-neighbor spins $i$–$j$ is denoted by $y_{ij}$. Since we take into account the position dependence we need and additional super-index in the probabilities: in this way we write $x_i^{(n)}$ and

## Table I

Geometrical characteristics of thin films in terms of the coordination number within the plane $z_0$ and in two adjacent planes $z_1$, the number of atoms in the plane $\mathcal{N}$ and the number of parallel planes $N$. These expresions hold only for $N \geq 3$.

| | |
|---|---|
| $\alpha$ | $N\mathcal{N}$ |
| $\alpha_s$ | $2\mathcal{N}$ |
| $\alpha_i$ | $(N-2)\mathcal{N}$ |
| $\gamma$ | $\frac{1}{2}\mathcal{N}[z_0 N + 2z_1(N-1)]$ |
| $\gamma_s$ | $z_0 \mathcal{N}$ |
| $\gamma_m$ | $2z_1 \mathcal{N}$ |
| $\gamma_i$ | $\frac{1}{2}\mathcal{N}[z_0(N-2) + 2z_1(N-3)]$ |

$y_{ij}^{(n)}$ for sites and pairs, respectively. The super-index value for the surface is $(s)$, for the internal $(i)$ and for the intermediate $(m)$.

In this approximation the geometrical characteristics of the system are given by the number of sites ($\alpha$) and pairs ($\gamma$). The number of surface sites is denoted $\alpha_s$ and the number of internal sites by $\alpha_i$. The number of pairs at the surface $\gamma_s$, the number of internal pairs $\gamma_i$ and the number of intermediate pairs $\gamma_m$. Notice that $\alpha = \alpha_s + \alpha_i$ and $\gamma = \gamma_s + \gamma_m + \gamma_i$. In this model the geometrical characteristics and the surface orientation are described in terms of the coordination number within the plane $z_0$, the coordination number between planes $z_1$, the number of parallel planes in the film $N$, and $\mathcal{N}$ the number of atoms in each plane. Their values are given in Table I. These values only hold for films with $N \geq 3$; the case for $N = 2$ is different. In this case $\gamma = (z_0 + z_1)\mathcal{N}$, $\gamma_m = z_1 \mathcal{N}$ and $\gamma_i = 0$. The other geometrical characteristics are given in Table I.

The set of variables that describe the system are the pair probabilities. They are not independent since they satisfy normalization constraints for the three different kinds of pairs

$$\sum_{ij} y_{ij}^{(n)} = 1, \qquad (1)$$

being $n$ the kind of pairs ($n = s, i, m$). In the case of the single site probabilities they are given as follows

$$x_i^{(s)} = \sum_j y_{ij}^{(s)} = \sum_j y_{ij}^{(m)}, \qquad (2a)$$

$$x_i^{(i)} = \sum_j y_{ij}^{(i)} = \sum_j y_{ji}^{(m)}. \qquad (2b)$$

Notice that the $x_i^{(s)}$ and $x_i^{(i)}$ can be written in two different ways and both have to be equal. These conditions are named the consistency constraints. Considering the symmetry $y_{ij}^{(k)} = y_{ji}^{(k)}$ for $k = s$ and $i$, and Eqs. (1) and (2) the number of independent variables in this model is only five. The magnetic behavior is described in terms of a long-range order parameter (LRO)

$$\eta^{(\ell)} = x_1^{(\ell)} - x_2^{(\ell)}, \qquad (3)$$

## Table II

Dependent variables in terms of the independent ones. The elements of the table correspond to the coefficient of the linear combination of the set of independent variables $\{1, \eta^{(s)}, \eta^{(i)}, y_{12}^{(s)}, y_{21}^{(m)}, y_{12}^{(i)}\}$.

| | 1 | $\eta^{(s)}$ | $\eta^{(i)}$ | $y_{12}^{(s)}$ | $y_{21}^{(m)}$ | $y_{12}^{(i)}$ |
|---|---|---|---|---|---|---|
| $x_1^{(s)}$ | 1/2 | 1/2 | | | | |
| $x_2^{(s)}$ | 1/2 | −1/2 | | | | |
| $y_{11}^{(s)}$ | 1/2 | 1/2 | | −1 | | |
| $y_{22}^{(s)}$ | 1/2 | −1/2 | | −1 | | |
| $y_{11}^{(m)}$ | 1/2 | | | 1/2 | −1 | |
| $y_{12}^{(m)}$ | | 1/2 | −1/2 | | 1 | |
| $y_{22}^{(m)}$ | 1/2 | | −1/2 | | −1 | |
| $x_1^{(i)}$ | 1/2 | | 1/2 | | | |
| $x_2^{(i)}$ | 1/2 | | −1/2 | | | |
| $y_{11}^{(i)}$ | 1/2 | | 1/2 | | | −1 |
| $y_{22}^{(i)}$ | 1/2 | | −1/2 | | | −1 |

where $\ell$ is $s$ ($i$) for the surface (for the internal points). These LRO can be used as two of the independent variables. The other independent variables can be chosen as $y_{12}^{(s)}$, $y_{12}^{(i)}$, and $y_{21}^{(m)}$. These can be interpreted as short-range order parameters. Using Eqs. (1)–(3) we can eliminate the dependent variables in terms of the independent ones as shown in Table II. As an example of the meaning of this table we can write

$$y_{12}^{(m)} = \tfrac{1}{2}\left[\eta^{(s)} - \eta^{(i)}\right] + y_{21}^{(m)}. \tag{4}$$

For the internal energy part, the parameters used to describe a ferromagnetic system are

$$\varepsilon_{ij}^{(n)} = \begin{cases} -J_n, & \text{for parallel pair of spins } (i = j); \\ J_n, & \text{for antiparallel pair of spins } (i \neq j); \end{cases} \tag{5}$$

where $J_n > 0$, $n$ is the position, and the subindex ($i$-$j$) the spin variables. The internal energy in terms of the pair probabilities and considering only first-nearest-neighbor interactions is then given by

$$E = -\gamma_s J_s \left[y_{11}^{(s)} + y_{22}^{(s)} - 2y_{12}^{(s)}\right] - \gamma_m J_m \left[y_{11}^{(m)} + y_{22}^{(m)} - y_{12}^{(m)} - y_{21}^{(m)}\right] \\ - \gamma_i J_i \left[y_{11}^{(i)} + y_{22}^{(i)} - 2y_{12}^{(i)}\right], \tag{6}$$

where $\gamma_l$ is the total number of pairs of $l$ type. For the sake of simplicity, assuming that only the surface interaction is different and calling $J_m = J_i = J$, Eq. (6) can be rewritten in terms of the independent variables as follows

$$E = \gamma_s J_s \left[4y_{12}^{(s)} - 1\right] + \gamma_m J\left[4y_{21}^{(m)} - 1 + \eta^{(s)} - \eta^{(i)}\right] + \gamma_i J\left[4y_{12}^{(i)} - 1\right]. \tag{7}$$

In the case of the entropy, it is very easy to prove that in the pair approximation is given by[7]

$$\frac{S}{k_B} = (2\gamma_s + \gamma_m - \alpha_s)\sum_i \mathcal{L}(x_i^{(s)}) + (2\gamma_i + \gamma_m - \alpha_i)\sum_i \mathcal{L}(x_i^{(i)})$$
$$- \gamma_s \sum_{ij} \mathcal{L}(y_{ij}^{(s)}) - \gamma_m \sum_{ij} \mathcal{L}(y_{ij}^{(m)}) - \gamma_i \sum_{ij} \mathcal{L}(y_{ij}^{(i)}), \tag{8}$$

where $\mathcal{L}(v) = v \ln v - v$. Using Eqs. (7) and (8) we can write the free energy ($\mathcal{F} = E - TS$) in terms of the independent variables or order parameters $\{\eta^{(s)}, \eta^{(i)}, y_{12}^{(s)}, y_{21}^{(m)}, y_{12}^{(i)}\}$. The minimization of $\mathcal{F}$ with respect to the independent variables leads us to the following set of equations

$$\frac{\partial \mathcal{F}}{\partial \eta^{(s)}} = 0: \quad \gamma_s \ln \frac{y_{11}^{(s)}}{y_{22}^{(s)}} + \gamma_m \ln \frac{y_{12}^{(m)}}{y_{22}^{(m)}} - p \ln \frac{x_1^{(s)}}{x_2^{(s)}} + 2\gamma_m J\beta = 0, \tag{9a}$$

$$\frac{\partial \mathcal{F}}{\partial \eta^{(i)}} = 0: \quad \gamma_i \ln \frac{y_{11}^{(i)}}{y_{22}^{(i)}} + \gamma_m \ln \frac{y_{11}^{(m)}}{y_{12}^{(m)}} - q \ln \frac{x_1^{(i)}}{x_2^{(i)}} - 2\gamma_m J\beta = 0, \tag{9b}$$

$$\frac{\partial \mathcal{F}}{\partial y_{12}^{(s)}} = 0: \quad y_{11}^{(s)} y_{22}^{(s)} - \left(y_{12}^{(s)}\right)^2 \exp(4J_s\beta) = 0, \tag{9c}$$

$$\frac{\partial \mathcal{F}}{\partial y_{12}^{(m)}} = 0: \quad y_{11}^{(m)} y_{22}^{(m)} - y_{12}^{(m)} y_{21}^{(m)} \exp(4J\beta) = 0, \tag{9d}$$

$$\frac{\partial \mathcal{F}}{\partial y_{12}^{(i)}} = 0: \quad y_{11}^{(i)} y_{22}^{(i)} - \left(y_{12}^{(i)}\right)^2 \exp(4J\beta) = 0, \tag{9e}$$

where $\beta = 1/k_B T$, $p = 2\gamma_s + \gamma_m - \alpha_s$ and $q = 2\gamma_i + \gamma_m - \alpha_i$, respectively. To find the order-disorder or Curie temperature (where the system change from ferromagnetic to paramagnetic state), we can solve numerically the set of Eqs. (9) for the independent variables $\{\eta^{(s)}, \eta^{(i)}, y_{12}^{(s)}, y_{21}^{(m)}, y_{12}^{(i)}\}$. We can also find the dependence of the order parameters with the pair interactions and film thickness. However instead of that, we calculate directly the Curie temperature by evaluating the determinant of the Hessian matrix at the paramagnetic state ($\eta = 0$), that is

$$\left(\mathcal{F}_{\eta^{(s)}\eta^{(s)}} \mathcal{F}_{\eta^{(i)}\eta^{(i)}} - \mathcal{F}_{\eta^{(s)}\eta^{(i)}} \mathcal{F}_{\eta^{(i)}\eta^{(s)}}\right)\big|_{T_c} = 0, \tag{10}$$

where $\mathcal{F}_{vu}$ are the second order derivatives of the free energy respect to $v$ and $u$. Evaluating numerically Eq. (10) we can find the Curie temperature and the critical value of the surface interaction for a given geometry and thin film thickness.

Notice that in this approximation all the geometrical properties ($\alpha_l$ and $\gamma_l$) can be written just in terms of the coordination numbers, the number of atoms, and type of atoms (surface or interior). In general these parameters do not define in unique form a structure due to the fact that different structures may have the same number of sites and pairs. In order to distinguish them we have to go farther in building polygons and polyhedra (triangles, squares, tetrahedrons, ... etc.) in the structure and to use a more complicated CVM approximation that have to be tailored for each one of the structures of interest.[7] However the approximation used here, although has some limitations, allows us to study many structures.

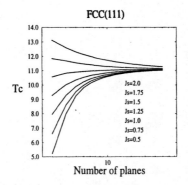

**Figure 1.** Critical temperature for a $fcc(111)$ thin film for various surface interactions as a function of the number of atomic plains. The saturation to the bulk temperature value may be reached in 20 atomic planes approximately if the surface interaction is equal to the bulk value ($J_s = J$) and it takes much longer if $J_s > J$.

## III. Results

For the sake of simplicity all the results presented here are in units of $k_B/J$. In figure 1, we present the results for the Curie or critical temperature $T_c$ for a $fcc(111)$ thin film as a function of the thickness (number of planes) and the surface interaction ($J_s$). The results are as expected; $T_c$ is a monotonous function of the thickness and reaches the asymptotic value for films thicker than 20 planes when the surface interaction $J_s = J$. The asymptotic limit for the critical temperature value in this approximation for a thick film ($N \to \infty$) is $k_B T_c = 2J/\ln[z/(z-2)]$ being $z$ ($= z_0 + 2z_1$) the bulk coordination number. This monotonous increasing behavior in $T_c$ changes if the surface exchange interaction increases, and in some cases it is possible that $T_c$ becomes larger than the bulk one if the surface exacerbation of the $J_s$ is large enough. In figure 1, we can notice that the change from monotonous increasing in the temperature to monotonous decreasing is around $J_s = 1.5$, where the temperature dependence is almost independent of the thickness and has approximately the bulk value in the whole range. This behavior of the surface indicates the existence of a critical value for the surface interaction beyond which the surface dominates the magnetic behavior of the system. The surface enhancement has been reported in Gd (0001) that in this approximation is equivalent to the $fcc(111)$.[2]

In figure 2, we present $T_c$ as a function of the surface interaction for $sc(100)$, $bcc(110)$, and $fcc(111)$ thin films. In all the cases we observe that the dependence of $T_c$ is very similar, *i.e.*, there is a flat like behavior for small $J_s$ values and a linear increasing region with the surface interaction for large values. Notice that in these figures there is a *common point*. This point separates the region where the bulk dominates over the surface. The *common point* corresponds to the critical value of the surface interaction $J_{sc}$. It is worth to notice that for values of $J_s$ smaller than the critical value, $T_c$ increase as the thin film becomes thicker, while for values larger than the critical $J_{sc}$ it is the other way around. In this range the thin film has higher critical temperature than the thicker ones for a fixed surface interaction. The different critical values for different thin films and surface orientations calculated here are presented in Table III. In the same table are presented the corresponding values obtained within the mean field (MF or Bragg-Williams) approximation for thin films.[3]

## Table III
Critical values for surface coupling obtained in this work and corresponding ones predicted by the mean field approximation.

|            | $J_{sc}$ | $J_{sc}(MF)$ |
|------------|----------|--------------|
| $sc(100)$  | 1.321    | 1.25         |
| $sc(110)$  | 2.711    | 2.0          |
| $bcc(110)$ | 1.638    | 1.5          |
| $fcc(100)$ | 2.223    | 2.0          |
| $fcc(110)$ | 4.882    | 3.5          |
| $fcc(111)$ | 1.585    | 1.5          |

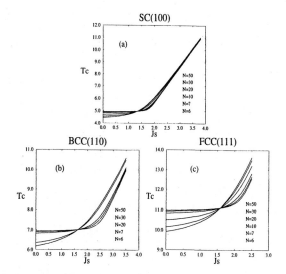

**Figure 2.** Critical temperature as a function of the surface interaction for $sc(100)$, $bcc(110)$, and $fcc(111)$ thin films of different thickness in (a), (b), and (c), respectively.

In the literature there are many reports on the surface critical value for three surfaces the $sc(100)$, $fcc(100)$, and the $fcc(111)$.[4,6] In all the cases there is a wide range of values of critical surface interaction values that goes from 1.25 for the MF to 1.5 for the Monte Carlo simulations in the $sc(100)$.[3,5,6] For $fcc(100)$ it goes from 2.0 in the MF to 2.25 in the tetrahedron approximation of the CVM,[4] and in the case of $fcc(111)$, 1.5 for MF to 1.715 for the octahedron-tetrahedron approximation of the CVM.[4] In the case of the $sc(100)$ we found 1.321, that is in good agreement with the value calculated recently with the reaction field approximation by Ilkovic[5] that gives $J_{sc} = 1.316$, the calculated within the effective field theory[10] 1.3068, and the renormalization group theory 1.307 (see Ref. 11). In the case of the $fcc(100)$ we found 2.223 that is in a very good agreement with the CVM tetrahedron approximation 2.25. It is important to mention that out of the mean field approximation where an analytical expression is possible ($J_s = 1 + z_1/z_0$) we do not know of any theoretical calculation for thin films with other structures and surfaces orientations.

## IV. Summary and Conclusions

We studied the Curie temperature and the surface enhancement in ferromagnetic Ising thin films as a function of the lattice structure and different surface orientations. We used a simplified version of the pair approximation of the cluster variational method that only considers as inequivalent those sites on the surface while all the internal sites are considered as equivalent regardless of the position. We found a monotonous increase in the Curie temperature as a function of the thickness of films for $J_s \leq J$, however this situation changes if $J_s$ exceeds a critical value $J_{sc}$. We calculated the critical value for the surface interaction, beyond which the surface dominates the ferromagnetic properties of the system. We compare with some values in the literature and we found that the Bragg-Williams approximation underestimate these values for all the cases. Good agreement was found for the critical values of $J_{sc}$ for $sc(100)$ and the $fcc(100)$ with those in the literature calculated by different methods.[4,5,9-11]

## Acknowledgments

This work has been supported by DGICYT Grants PB95-0720-C02-01 and SAB95-0390. One of us (FAG) also acknowledges support from Consejo Nacional de Ciencia y Tecnología, México (Grant 961015).

## References

* Permanent address: Instituto de Física, Universidad Autónoma de San Luis Potosí, 78000 San Luis Potosí, S. L. P., México
1. L. M. Falicov et al., *J. Mater. Res.* **5**, 1299 (1990) and references therein.
2. D. Weller, S. F. Alvarado, W. Gudat, Schröder, and M. Campagna, *Phys. Rev. Lett.* **54**, 1555 (1985).
3. F. Aguilera-Granja and J. L. Morán-López, *Phys. Rev. B* **31**, 7146 (1985); *Solid State Commun.* **74**, 155 (1990).
4. J. M. Sánchez and J. L. Morán-López, *Springer Proc. Phys.* **14**, (Springer, Berlin, Heidelberg, 1986); *Springer Proc. Phys.* **50**, (Springer, Berlin, Heidelberg, 1990).
5. D. P. Landau and K. Binder *J. Magn. Magn. Mater.* **104-107**, 841 (1992).
6. C. Tsallis, *Magnetism, Magnetic Materials and Their Applications*, edited by F. Leccabue and J. L. Sánchez-Llamazares, (IOP Publishing LTD, 1992).
7. R. Kikuchi, *Phys. Rev.* **81**, 988 (1951); *Prog. Theor. Phys.* Suppl. **115**, 165 (1994).
8. W. Hellenthal, *Z. Phys.* **170**, 303 (1962).
9. V. Ilkovic, *Surf. Sci.* **365**, 168 (1996).
10. T. Kaneyoshi, *Phys. Rev. B* **39**, 557 (1989).
11. T. W. Burkhard and E. Eisenrieger, *Phys. Rev. B* **16**, 3213 (1977).

# Conductivity Oscillations of Magnetic Multilayers

Miguel Kiwi,[1] Ana María Llois,[2] Ricardo Ramírez,[1] and Mariana Weissmann[2]

[1] Facultad de Física
Pontificia Universidad Católica de Chile
Casilla 306, Santiago 22
CHILE

[2] Departamento de Física
Comisión Nacional de Energía Atómica
Avda. del Libertador 8250, 1429 Buenos Aires
ARGENTINA

## Abstract

The electrical conductivity of the superlattice systems Ni/Co, Ni/Cu, and Pd/Ag, as a function of layer thickness, is investigated theoretically. Experimentally an oscillatory dependence was found for the first two, while the latter exhibits a monotonous behavior. First, a band theory based calculation is carried out, which fails to yield significant conductivity variations, but which points to the presence of $d$-character quantum well states close to $E_F$. This, led us to put forward and explore a model Hamiltonian that incorporates a scattering mechanism of these carriers against $d$-character quantum well states, which are present in only one of the superlattice materials. This latter model yields results in agreement with the magnitude of the experimentally observed oscillations.

## I. Introduction

Magnetic multilayers have attracted a lot of attention lately, both of experimentalist and theoreticians. This interest is related to the oscillatory long-range exchange coupling between magnetic layers, separated by nonmagnetic spacers,[1,2] that they exhibit and the associated giant magnetoresistance.[3] In addition to the interesting basic physics problems their understanding poses, the latter has important technological applications. In addition, recent measurements of the resistivity and anisotropic magnetoresistance of Ni/Co and Ni/Cu multilayer systems[4,5] show a clear-cut oscillatory dependence on layer thickness, which seems to be the first observation of superlattice effects on transport properties.

On the other hand, the possibility of achieving the ballistic limit, in which the bound on the electrical conductance is not due to lattice imperfections, but rather to the presence of a constriction, has been intensively investigated over the years[6,7] since Landauer[8] first suggested it in 1957. Thus, in this context it is of special interest that the experiments of Ref. 4, which report the oscillatory dependence of the conductivity on layer thickness, were performed very near (if not at) the ballistic limit. Moreover, since the amplitude of the oscillations decreases when the number of repeated layers is reduced, there is good reason to attribute them to the superlattice structure of the system.

## II. Band Theory Calculation

To test this hypothesis we investigate the transport properties of Co-Ni superlattices grown in the (111) direction as a function of the number of layers of each material. Our aim is to understand how the presence of interfaces modifies the transport properties of the system, as compared with bulk Ni or Co. That is, we concentrate on the consequences that the superlattice band structure has on the electric conductivity, but we neglect other scattering mechanisms. For example, we assume an infinite periodic system, with no disorder in the bulk or at the interfaces. We use the semiclassical approximation (Boltzmann's equation) and therefore only makes use of the electronic energy bands, but not the eigenvectors required by the Kubo-Greenwood formula, based on linear response theory. Moreover, each spin component is assumed to contribute independently to the conduction.

Presently, available experimental results are obtained in the diffusive limit and thus are more adequately interpreted by the diffusive conductivity $\sigma$, than by the ballistic conductance $G$. However, the recent production of the superlattices with mean free path of the order of magnitude of their thickness,[4] as well as other mesoscopic systems, make the ballistic conductance $G$ an important physical quantity. This is reflected in several recent, but quit dissimilar, calculations of the magnetoresistance of magnetic multilayers.[9-12] In fact, there are significant conceptual differences between the ballistic conductance calculation of Schep et al.,[9] and the conductivity evaluation of Oguchi[10] and Zhan et al.[11] The former procedure[9] applies to a system where "the total resistance is dominated by a classical point contact with a diameter $A$ shorter that the mean free path and larger that the electron wavelength. The conductance is finite due to the finite cross section $A$." However, for the conductivity the scattering mechanism is due to the presence of impurities and essentially two approaches have been used in the calculations. The first one[13,14] focuses mainly on the spatial, angular and spin dependence of the relaxation time $\tau$ and uses a free electron band structure. The alternative procedure, used by Oguchi,[10] Zhan et al.[11] and in this Section, focuses on the importance of band structure and Fermi surface topology and assumes a constant $\tau$, independent of spin, number of interfaces and angular direction of the wave vector.

Ballistic and diffusive conductivity therefore correspond to two experimentally different transport regimes. However, different these idealizations are, in the limit of

## Table I

Charge transfer and magnetization in the interface vicinity.
The symbol denotes the interface layer.

| Layer | Occupation (e) | Magnetic moment ($\mu_B$) |
|---|---|---|
| Ni(I) | 10.036 | 0.579 |
| Ni(I − 1) | 10.001 | 0.572 |
| Ni(bulk) | 10.000 | 0.613 |
| Co(I) | 8.937 | 1.680 |
| Co(I − 1) | 9.027 | 1.558 |
| Co(bulk) | 9.000 | 1.595 |

constant relaxation time, they lead to remarkably similar calculation algorithms, and thus we compute both the ballistic conductances $G$ and the diffusive conductivity $\sigma$.

In previous papers[15,16] we have shown that a Hubbard tight-binding Hamiltonian, solved in the unrestricted Hartree-Fock approximation and parametrized to fit bulk equilibrium values of transition metal characteristics, gives suitable information on surface and superlattices, such as densities of states, band structures, magnetization and charge transfers. We therefore use this procedure to calculate self-consistently the energy bands, taking into account charge transfers and including a Madelung correlation for the repeated dipole layers. Orbitals of $s$-, $p$- and $d$-character are considered. The tight-binding parameters are obtained from Anderson[17] for the pure materials and the geometric mean is used for the Co-Ni hopping integrals. Only nearest neighbor interactions and perfect sharp interfaces are considered.

For a superlattice of $n$ Ni layers and $m$ Co layers, denoted as $n$Ni/$m$Co, the smallest real space unit cell was chosen, which has $(n + m)$-atoms. For $n + m \leq 8$ the Hamiltonian is solved self-consistently in reciprocal space, using 500 $k$-points in the reduced Brillouin zone. For a large number of layers, as in the experimental samples, we assume that atoms far away from the interfaces have bulk Ni or Co occupations. Near the surfaces we assume that the occupations are the same as those calculated for $n + m = 4$; this implies a charge transfer from Co to Ni, leading to a smaller Ni magnetic moment, in the two layers closest to the interface, and to an oscillating magnetic moment of the Co layers. In fact, the Co atoms at the interface lose electrons, increasing their magnetic moment, while the opposite happens to those in the subsequent layer. All this is summarized in Table I.

When $n+m > 8$ the Fermi energy is not calculated self-consistently and a weighted average of the Fermi energies of both bulk materials is adopted for the superlattice Fermi level. A similar approximation scheme was used in previous papers to avoid excessive computations.[9–11] Using these two approximations, for the occupations and the Fermi energy, the electronic energy bands can be obtained in a single iteration, using as many points in the Brillouin zone as needed. In particular, for most calculations 4800 $k$-points in each hexagonal plane perpendicular to the (111) direction were used. The number of such planes depends on the cell size, so as to cover uniformly the Brillouin zone. The larger the unit cell, the smaller the number of planes, and of course the larger the number of foldings.

The conductivity tensor is given as a sum of contributions of each energy band, labeled by the band index $n$ and the spin index $s$,

$$\sigma = \sum_{n,s} \sigma(n,s), \qquad (1)$$

where

$$\sigma(n,s) \propto \int \tau(n,s,k)\, v^2(n,s,k)\, \delta[\epsilon_{n,s}(k) - E_F]\, d^3k, \qquad (2)$$

and where $\tau$ is the impurity scattering relaxation time, which we assume to be constant, and thus a multiplicative factor in front of the above integral; $v$ is the group velocity of the quasi-particles, or equivalently the derivative of the band energy with respect to $k$. When multiplied by the $\delta$-function the latter becomes the Fermi velocity. The formula above is valid only for zero temperature; to extend the treatment to $T \neq 0$ the derivative of the Fermi function should be included in the integrand, in substitution of the $\delta$-function.

Analogously, the expression for the ballistic conductance $G$ is given by,

$$G = \sum_{n,s} G(n,s), \qquad (3)$$

where

$$G(n,s) \propto A \int |v(n,s,k)|\, \delta[\epsilon_{n,s}(k) - E_F]\, d^3k, \qquad (4)$$

where $A$ is the constriction cross section.[9]

In fact, the ballistic conductance is a tensor of which we have calculated only two diagonal elements, one along the superlattice growth direction and the other perpendicular to it. This implies considering the components of the velocity only along these directions. Moreover, the diagonal elements of this tensor, $\mathbf{G}_{ii}$, are proportional to the Fermi surface projection normal to the $i$-direction.[9]

## III. Band Theory Results

As a first step, to establish a reference standard for the superlattice transport coefficients, we have calculated $G$ and $\sigma$ of pure bulk Ni and pure bulk Co inferred from the band structure obtained from large unit cells, similar in size to those of the superlattices. This way we set bounds to errors due to the discretization of $k$-space and Brillouin zone folding. For example, Table II shows results for fcc Ni and Co considered as multilayers grown in the (111) direction, with 6 atoms per unit cell. The values of $G$ and $\sigma$ are normalized respectively to $G_{\text{IP}}^{\text{tot}}(\text{Ni}) = 1$ and $\sigma_{\text{IP}}^{\text{tot}}(\text{Ni}) = 1$, where "tot" denotes the sum of majority and minority contributions. Majority spin bands at the Fermi level are mostly of $sp$-character, therefore the density of states is small and so is $G$. However, they cross the Fermi energy with a large slope, yielding a large Fermi velocity and $\sigma$. Minority spin bands have a much larger density of states at the Fermi energy and they are of $d$-character. $G$ is large, but the crossing slope is small, so that $\sigma$ is about the same as for the majority band. 1200 $k$-values are calculated on the displayed portion of the $(k_x, k_y)$-plane and the results for 15 different $k_z$ planes are subsequently superimposed and therefore, the number of $k$-points used in 72 000.

Here we report calculations for superlattices of the type 9Co/$n$Ni, that is, with a fixed number of Co layers about 20 Å wide and a variable number of Ni layers. These values were taken from the experimental work of Ref. 4. As perfect epitaxy is assumed

## Table II

Ballistic conductance $G$ and conductivity $\sigma$ calculated for the indicated superlattices. Each spin contribution is displayed independently, and all values are normalized respectively to $G_{\text{IP}}^{\text{tot}}(\text{Ni}) = 1$ and $\sigma_{\text{IP}}^{\text{tot}}(\text{Ni}) = 1$. The majority ($\uparrow$) spin contribution is almost independent of the number $n$ of Ni layers; thus, only a few cases are shown. In the calculation of the minority spin contribution for $n \leq 10$ we use 28 800 $k$-points, while for $n \geq 9$ we employ only 19 200. For $n = 9$ and $n = 10$ both results are displayed (the ones denoted by * correspond to 19 200 $k$-points) in order to exibit the errors due to discretization.

|           | $G_{\text{IP}}\uparrow$ | $\sigma_{\text{IP}}\uparrow$ | $G_{\text{PP}}\uparrow$ | $\sigma_{\text{PP}}\uparrow$ |
|-----------|------|------|------|------|
| 6 Ni      | 0.287 | 0.522 | 0.239 | 0.448 |
| 6 Co      | 0.277 | 0.540 | 0.235 | 0.500 |
| 9Co/4Ni   | 0.280 | 0.552 | 0.211 | 0.428 |
| 9Co/5Ni   | 0.280 | 0.549 | 0.211 | 0.420 |
| 9Co/6Ni   | 0.280 | 0.544 | 0.211 | 0.417 |
| 9Co/7Ni   | 0.280 | 0.542 | 0.208 | 0.402 |
| 9Co/8Ni   | 0.280 | 0.539 | 0.208 | 0.407 |
|           | $G_{\text{IP}}\uparrow$ | $\sigma_{\text{IP}}\uparrow$ | $G_{\text{PP}}\uparrow$ | $\sigma_{\text{PP}}\uparrow$ |
| 6 Ni      | 0.713 | 0.477 | 0.633 | 0.415 |
| 6 Co      | 0.502 | 0.370 | 0.505 | 0.344 |
| 9Co/4Ni   | 0.426 | 0.223 | 0.235 | 0.082 |
| 9Co/5Ni   | 0.432 | 0.231 | 0.239 | 0.087 |
| 9Co/6Ni   | 0.429 | 0.229 | 0.208 | 0.070 |
| 9Co/7Ni   | 0.446 | 0.236 | 0.225 | 0.083 |
| 9Co/8Ni   | 0.453 | 0.243 | 0.232 | 0.080 |
| 9Co/9Ni   | 0.464 | 0.249 | 0.235 | 0.077 |
| *9Co/9Ni  | 0.464 | 0.249 | 0.214 | 0.063 |
| 9Co/10Ni  | 0.477 | 0.251 | 0.249 | 0.080 |
| *9Co/10Ni | 0.477 | 0.251 | 0.228 | 0.066 |
| 9Co/11Ni  | 0.474 | 0.262 | 0.190 | 0.055 |
| 9Co/12Ni  | 0.481 | 0.266 | 0.190 | 0.056 |
| 9Co/13Ni  | 0.477 | 0.271 | 0.173 | 0.054 |
| 9Co/14Ni  | 0.481 | 0.270 | 0.183 | 0.057 |

in the superlattice, the same grid of the previous cases is used in the $(k_x, k_y)$-plane. Along $k_z$, the (111) direction, the number of partitions is selected so as to cover the whole Brillouin zone as uniformy as possible. This implies a change from 6 to 4 planes around $n = 9$ or $n = 10$, and therefore a slight change in accuracy. Table II shows the results obtained for $n$ varying from 4 to 14 (8 to 28 Å of Ni).

The most interesting results refer to the minority spin, which is predominantly of $d$-character. The effect of the interfaces is such that both $G$ and $\sigma$ are always smaller in the superlattices than in the pure elements. As far as the precision is concerned, for the particular case $n = 6$, the number of $k$-points was increased to 250 000 in the reduced zone and the values obtained for in plane (IP) and perpendicular to the plane conduction (PP) were: (i) $G_{IP} = 0.426$ and $\sigma_{IP} = 0.236$; and (ii) $G_{PP} = 0.220$ and $\sigma = 0.081$. Comparing with the corresponding $n = 6$ values in Table II one observes that the accuracy of these calculations, for the in plane transport parameters, is better and also less dependent on the $k_z$ increment (see $n = 9$ and $n = 10$ of Table II) than the perpendicular to the plane values. The effect of interfaces is of course stronger on $G_{PP}$ and $\sigma_{PP}$. Moreover, we checked that slight changes in the value of $E_F$ induce only minor variations, of less than 10%, on the results provided in Table II, so that the qualitative conclusions outlined above remain valid.

$G_{IP}$ and $\sigma_{IP}$ increase with the number of Ni layers per unit cell, since the corresponding values for pure Ni are larger than those for pure Co. In the examples we calculated the observed increase is almost linear. $G_{PP}$ and $\sigma_{PP}$ are almost independent of the number of the Ni layers $n$, but this may be related to larger errors in the calculation.

The majority spin contribution is very similar in all the superlattices considered and also similar to the calculated values for the pure bulk materials. As mentioned before for the pure materials, the number of bands crossing the Fermi energy is smaller for the majority band, but the slope is larger, so that: $G_{IP}\uparrow < G_{IP}\downarrow$, while $\sigma_{IP}\uparrow > \sigma_{IP}\downarrow$. On the other hand $G_{PP}\uparrow \simeq G_{PP}\downarrow$ and $\sigma_{PP}\uparrow \gg \sigma_{PP}\downarrow$, where $\uparrow (\downarrow)$ stands for majority (minority) band. Therefore, the qualitative results obtained depend on the type of transport calculation performed and it is clear that Fermi velocity plays an important role.

Since $\sigma$ is the important quantity describing the present experimental situation, in which the mean free path is smaller than the sample size (diffusive regime), the majority spin conduction is the dominant term, provided that $\tau(\uparrow) = \tau(\downarrow)$. Under these circumstances, interfaces play a lesser role than in the ballistic case.

A word of warning of the conclusions that may be drawn when this problem is reduced to a one dimensional system, is in order. For example, let us focus on the $\Lambda$-L line of the Brillouin zone, which is perpendicular to the (111) growth direction. Using the calculations of the previous section we find that the Fermi energy lies in a gap of majority spin bands for all 9Co/$n$Ni superlattices. Thus a primitive one-dimensional model would lead us to the conclusion that those bands do not contribute to the conductivity.

On the other hand, we derive the important result that the minority bands close to the Fermi energy $E_F$ are almost nondispersive; as a function of $n$ they sometimes cross the Fermi energy, while in the other instances $E_F$ lies within a gap. However, within the approximations of this model calculation, we do not find oscillations of the transport properties, as a function of the number of layers of each material, of the magnitude reported by Gallego et al.[4] The cause of this discrepancy is certainly not to be found in the accuracy of our calculations, and thus originates in one or more of the ingredients that were not included in our model, such as: The dependence of the relaxation time $\tau$ on the number of layers of each type and/or on spin, the finite size of the system and the imperfections of the interfaces (such as roughness, layer width fluctuations, and the presence of nonmagnetic impurities).

## IV. Model Hamiltonian Calculation

In what follows, and based on the structure calculations reported above, we here put forward a theoretical model that accounts for the transport properties of the superlattices.

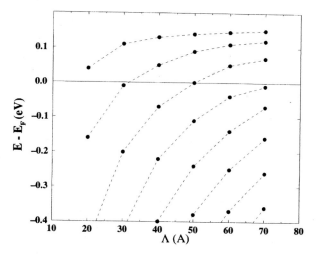

**Figure 1.** Quantum well levels at $\Gamma$ for the superlattices Co(0.6 $\Lambda$)/Ni(0.4 $\Lambda$) as a function of $\Lambda$. The dashed lines link QWS with the same number of nodes.

The conductivity of metallic systems depends only on the quantum states with eigen-energies close to the Fermi level $E_F$, and for superlattices these states shift as a function of the number of layers of each element. To study the dependence of the conductivity on the thickess of each superlattice material, we have considered fcc superlattices grown in the (111) direction, in particular Ni/Co and Pd/Ag. In each of these three cases, only one of the pure elements of the pair has a d band that crosses the Fermi level in the $\Gamma$-L direction. It is therefore pertinent to pay special attention to the $\Gamma$-L direction of the Brillouin zone of these superlattices, where we expect folding effects to be important.

But, an interesting property that does depend on the thickness of the Ni and Co layers was noticed as an outcome of the work of Ref. 18: Quantum well states (QWS) present in the Ni layers, of d-character and minority spin, appear in the Brillouin zone along the $\Gamma$-L line and its vicinity. In addition, for some particular layer thicknesses they have energies very close to $E_F$. A similar result was recently reported for Cu/Co superlattices grown in the (100) direction.[19] We believe that the presence of these non-dispersive states near $E_F$ bear relation to the experimentally observed conductivity oscillations. Arguments of this type have been used to explain the influence on the conductivity of the different dilute transition metal impurities in Cu (Ref. 20): when the impurity d-level lies close to the Cu Fermi level, scattering of the conduction electrons is enhanced and hence an increase of the resistivity is to be expected.

In figure 5 of Ref. 18 we showed the calculated energy levels at the $\Lambda$ point, for superlattices having 9 Co layers (18 Å) and a variable number of Ni layers. This geometry was chosen to compare with the experiments of Ref. 4. We found a periodic approach towards $E_F$ of the energy levels, with a periodicity of nearly 4 Ni layers, which is quite similar to that of the experimentally observed resistivity maxima. The figure resembles those due to the discretization of the perpendicular wave vector in a thin film, as for example shown in Refs. 21 and 22, and in fact this effect occurs both in slabs and superlattices.

Actually, to obtain the energy of these QWS, as a function of layers of each material, it is not necessary to perform a fully self-consistent calculation, such as in Ref. 18.

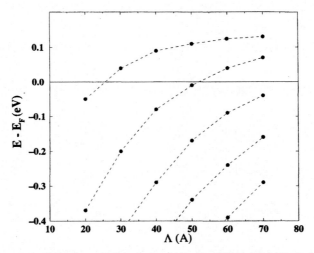

**Figure 2.** Quantum well levels at $\Gamma$ for the superlattices Ni(0.6 $\Lambda$)/Cu(0.4 $\Lambda$) as a function of $\Lambda$. The dashed lines link QWS with the same number of nodes.

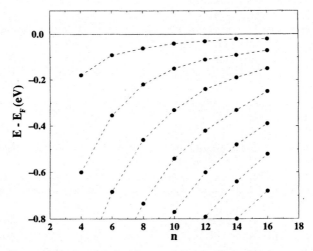

**Figure 3.** Quantum well levels at $\Gamma$ for the superlattices 9Ag/nPd as a function of $n$. The deshed lines link QWS with the same number of nodes.

In fact they can be obtained, without a significant loss in precision, using a simple tight-binding method with the parameters of both bulk materials, but shifting the diagonal elements so as to align the two Fermi levels. The results shown in Figs. 1 and 2, for Ni/Co and Ni/Cu, respectively, were calculated within this approximation. In both cases are QWS of minority spin and $d$-character, in the Ni layers, whose energies approach periodically the Fermi energy. Inspection of the wave-vectors shows that the state of highest energy is nodeless, the second has one node, and so on. Experiments[5] show resistivity oscillations in these two cases, with approximately the same periodicity found through our calculations.

In the case of Pd and Ag there is a large interfacial misfit, since their lattice constants differ by 5%. Moreover, Pd becomes magnetic when strained to adopt the lattice constant of Ag. However, when we performed a full self-consistent calculations for superlattices with only a few layers of Pd, they turned out to be nonmagnetic. Consequently, we assumed that the magnetic splitting remains small for superlattices with additional Pd layers, and the Pd/Ag results we report here are nonmagnetic. Figure 3 shows the QWS for Pd/Ag which are of d-character and located in the Pd ayers. The main difference with the Ni/Cu cases is that all lie below $E_F$ and approach monotonously this energy level as the thickness of Pd increases. Some states of sp-character also appear close to $E_F$ in this case, but they are not QWS. The experiments reported in Ref. 5 show an increasing resistivity with Pd thickness, that saturates at about 25 Å (or 12 monolayers), with no oscillations.

Consequently, in the three pairs of elements we have examined in detail: Ni/Co, Ni/Cu and Pd/Ag, there definitely is a correlation between the existence of QWS close to $E_F$ and the maxima of the measured resistivity.

How can we reconcile our previous non-oscillating conductivity results for the Ni/Co superlattices,[18] with the oscillations observed experimentally and the periodic appearance of QWS near the Fermi energy? Evidently, we have to modify some of the hypothesis of the previous calculations and consider the different scattering mechanisms separately. In particular, we may envision a spin-flip process that scatter and sp-character conduction electrons of the majority band into a quantum well d-character minority state. Actually, spin relaxation effects in NiCu, which increase as the Ni concentrations grows, have been observed experimentally by Hsu et al.[23] Moreover their measurements, in Co/NiCu multilayers, established that spin-flip scattering does occur in NiCu alloys.

As the d-states are nondispersive they do not contribute to the conductivity; on the contrary, if their energies are close to $E_F$ they should decrease the total conductivity. In the Boltzmann picture this implies that $\tau$ is not constant and that each scattering mechanism contributes separately to the resistivity with a different $\tau$ value. Similar mechanisms were already proposed by other authors. Among them we mention Suzuki and Taga,[24] who estimated the influence of d-band states on the minority conduction channel of Co/Cu superlattices, and Campbell et al.,[13,25] who studied the influence of spin-flip processes on the resistivity of Fe and Ni, based alloys.

Following the above arguments we study an additional contribution to the resistivity due to sd scattering (where the d-states are QWS). In the first two systems under consideration, Ni/Co and Ni/Cu, the scattering process involves a spin-flip, while for Pd/Ag this is not necessary the case. In addition, we believe that the essence of the two channel model is preserved, since electrons are only rarely scattered by this mechanism, but one has to keep in mind that there is a small spd hybridzation of the genuine multilayer bands, and thus sd scattering only constitutes a simplified notation for a more complex reality. We consider the following model Hamiltonian, which incorporates the interaction between a conduction s-electron and quantum well, mainly d-state:

$$H = \sum_{\vec{k}} \varepsilon_{\vec{k}\uparrow} c^\dagger_{\vec{k}\uparrow} c_{\vec{k}\uparrow} + E_d d^\dagger_\downarrow d_\downarrow + \sum_{\vec{k}} \left( V_{k\uparrow d\downarrow} c^\dagger_{\vec{k}\uparrow} d_\downarrow + V_{d_\downarrow k\uparrow} d^\dagger_\downarrow c_{\vec{k}\uparrow} \right), \tag{5}$$

where the first term is the quasi-free electron Hamiltonian of the majority spin (mainly d-character) minority spin, and the third term is the interaction between the previous two.

In pure bulk magnetic metals the current is transported almost evenly by both spin channels. But, from Table II and assuming the system to be in the diffusive regime, we conclude that in superlattices the majority band is responsible for about 80% of the total conductivity $\sigma$. We also found that $\sigma$ is only weakly anisotropic, as the effect of the interfaces is rather negligible on the majority carries. Of course the genuine bands are hybridized and it is the small $d$-character of the majority states that allows for a nonzero spin-orbit coupling with the QWS, which are pure $d$ minority states. For the time being we have assumed that a nondispersive state, with energy $E_d$, exists throughout the Brillouin zone. Realistic superlattice band structure calculations show that this is actually the case for a macroscopic fraction of $k$-space. Thus, as already pointed out in Ref. 19, if $E_d$ is close to the Fermi energy then "a significant region of the superlattice Fermi surface has three dimensional character." This provides a justification for the oversimplified hypothesis we put forward.

The Kubo formula for the conductivity is related to the diagonal element of the retarded Green's function $G_{kk}^{\text{ret}}(\omega)$, as follows:[26]

$$\sigma = \frac{4e^2\hbar^3}{3m^2} \int \frac{d^3k\, k^2}{(2\pi)^3} \int \frac{d\omega}{2\pi} \left[\text{Im}\, G_{kk}^{\text{ret}}(\omega)\right]^2 \delta(\omega - E_F). \tag{6}$$

The Hamiltonian of Eq. (5) can be solved exactly and the relevant Green's function reads

$$G_{kk}^{\text{ret}}(\omega) = \frac{1}{\omega - \varepsilon_k + i0^+}\left(1 + \frac{|V_{k\uparrow d\downarrow}|^2}{\omega - E_d + i\Gamma} \frac{1}{\omega - \varepsilon_k + i0^+}\right), \tag{7}$$

with

$$\Gamma = -\text{Im} \sum_{\vec{k}} \frac{|V_{k\uparrow d\downarrow}|^2}{\omega - \varepsilon_k + i0^+} \approx \pi \mathcal{D}_s(E_F) \langle |V_{k\uparrow d\downarrow}|^2 \rangle_{\text{FS}}, \tag{8}$$

where $\mathcal{D}_s(E_F)$ is the density of states of the majority $s$ band at the Fermi surface. The square of the interaction matrix element $\langle |V_{k\uparrow d\downarrow}|^2 \rangle_{\text{FS}}$, averaged over the Fermi surface, from now on will be denoted as $V^2$ and assumed to be constant.

However, the width of the unperturbed Green's function [i.e., the first term of Eq. (7)] is nonzero due to impurity scattering. This finite width, which we denote as $\gamma$ and which simply replaces $0^+ \to \gamma$ in Eq. (7), can be estimated from the experimental conductivity of pure Ni at low temperatures. In fact, the expression for $\sigma$ can readily be evaluated (ignoring vertex correlations) in the $V = 0$ limit, to yield the usual result $\sigma_0 = e^2 n_0 \tau/m$, if $\tau = \hbar/2\gamma$. Here $n_0 = p_F^3/3\pi^2\hbar^3$ is the total carrier density, $p_F$ the Fermi momentum, $e$ and $m$ the electron charge and mass.

For $V \neq 0$ the integrals in Eq. (6) can also be evaluated analytically, in the physically interesting range of $\gamma \gg \Gamma$, that is, when the $sd$ scattering is small compared to impurity scattering. To new contributions to the conductivity $\sigma$ appear, in addition to $\sigma_0$. They are proportional to $V^2$ and $V^4$, and can be written in the following way:

$$\frac{\sigma}{\sigma_0} = 1 - \frac{\Gamma}{\pi\gamma^2\mathcal{D}(E_F)} \frac{1-y^2}{(1+y^2)^3}\left(4 - \frac{1}{\pi\gamma^2\mathcal{D}(E_F)} \frac{1-y^2}{1+y^2}\right), \tag{9}$$

where $y = (E_d - E_F)/\gamma$.

Of course, when $y = 0$ the physically interesting maximal resistivity case is realized, while for $y \gg 1$ the additional contributions to $\sigma$ vanish. Thus, oscillations of $\sigma$ as a function of $E_d - E_F$, or equivalently as a consequence of variations of the multilayer thickness, are generated. Moreover, an order of magnitude estimate for the extra ($y = 0$) contributions to $\sigma$ can be carried out quite straightforwardly. The conductivity of pure Ni, or pure Co, implies a value of $\gamma \sim 0.1$ eV. On the other hand, the results of our full band-structure calculations for the superlattices yield $\mathcal{D}_s(E_F) \approx 1$ state/(eV·unit cell). Consequently, the experimentally observed conductivity reduction can be explained on the basis of a value of the QWS width $\Gamma \sim 0.01$ eV, which seems quite reasonable. It is important to realize that in spite of the fact that the impurity scattering width is much larger than the QWS width ($\gamma \gg \Gamma$), the reduction of $\sigma$ due to the QWS can be quite significant. In fact, $\Gamma/\gamma = 0.1$ yields a 25% reduction of $\sigma$; for $\Gamma/\gamma = 0.2$ a 50% reduction is achieved. We believe this to be the main consequence that can be derived from our model Hamiltonian.

## V. Conclusion

In conclusion, after a band theory calculation proved insufficient to explain the observed transport properties of magnetic multilayer system, we put forward a model Hamiltonian which seems to provide a sound basis for the description of the resistivity, as a function of layer thickness, of three superlattice systems: Ni/Co, Ni/Cu, and Pd/Ag. The conductivity oscillations can be traced to the presence of quantum well states with energies close to $E_F$, which shift as a the thickness of the layers varies. Calculations of the conductivity using a constant relaxation time $\tau$ independent of wave number, band index and number of layers of each element, do not reflect the experimental results. Since in these superlattices the current is mainly carried by the $sp$-character majority spin electrons, we propose that the origin of the conductivity minima is due to scattering of these electrons against the QWS.

## Acknowledgments

We gratefully acknowledge Prof. Ivan K. Schuller and Dr. Sihong Kim for motivating our interest in the subject and for illuminating discussions. This work was supported by the Consejo Nacional de Investigaciones Científicas y Técnicas (CONICET, Argentina), the Fondo Nacional de Investigaciones Científicas y Tecnológicas (FONDECYT, Chile) under grant No. 1971212, and by ICTP (Trieste).

## References

1. A. Cebollada et al., *Phys. Rev. B* **39**, 9726 (1989).
2. S. S. P. Parkin et al., *Phys. Rev. Lett.* **64**, 2304 (1990), *Phys. Rev. Lett.* **66**, 2152 (1991).
3. M. Baibich et al., *Phys. Rev. Lett.* **61**, 2472 (1988).
4. J. M. Gallego et al., *Phys. Rev. Lett.* **74**, 4515 (1995).
5. S. Kim, D. Lederman, J. M. Gallego, and I. K. Schuller, *Phys. Rev. B* **54**, R5291 (1996); S. Kim, Ph.D. Thesis, University of California, San Diego, 1996.
6. See R. Landauer, *Z. Phys. B* **68**, 217 (1987) and references therein.
7. H. van Houten and C. Beenakker, *Phys. Today* **49**, 22 (1996).
8. R. Landauer, *IBM J. Res. Dev.* **1**, 233 (1957).
9. K. M. Schep, P. J. Kelly and G. E. W. Bauer, *Phys. Rev. Lett.* **74**, 568 (1995).
10. T. Oguchi, *Mater. Sci. and Eng. B* **31**, 311 (1995).
11. P. Zhan et al., *Phys. Rev. Lett.* **75**, 2996 (1995).

12. R. Gomez Abal et al., *Phys. Rev. B* **53**, R8844 (1996).
13. A. Fert et al., *J. Phys. F: Metal Phys.* **6**, 849 (1976).
14. H. E. Camblong, *Phys. Rev. B* **51**, 1855 (1995).
15. G. Fabricius et al., *Phys. Rev. B* **49**, 2121 (1994).
16. G. Fabricius, A. M. Llois, and M. Weissmann, *J. Phys. Cond. Matter* **6**, 5017 (1994).
17. O. K. Andersen, O. Jepsen, and D. Glötzel, *Highlights of Condensed Matter Theory* (Amsterdam, 1984).
18. M. Weissmann et al., *Phys. Rev. B* **15**, 335 (1996).
19. J. L. Pérez-Díaz and M. C. Muñoz, *Phys. Rev. Lett.* **76**, 4967 (1996).
20. I. Mertig et al., *J. Phys. F: Metal Phys.* **12**, 1689 (1982).
21. J. E. Ortega et al., *Phys. Rev. B* **47**, 1540 (1993).
22. N. V. Smith et al., *Phys. Rev. B* **49**, 332 (1994).
23. S. Y. Hsu et al., *Phys. Rev. B* **54**, 9027 (1996).
24. M. Suzuki and Y. Taga, *J. Phys. Cond. Matter* **7**, 8497 (1995).
25. I. A. Campbell et al., *Phil. Mag.* **15**, 977 (1967).
26. G. D. Mahan, *Many Particle Physics* (Plenum Press, New York, 1981), Chapter 7.

# Slave-Boson Approach to Electron Correlations and Magnetism in Low-Dimensional Systems

E. Muñoz Sandoval,[1] J. Dorantes-Dávila,[1] and G. M. Pastor[2]

[1] Instituto de Física
Universidad Autónoma de San Luis Potosí
Alvaro Obregón 64, 78000 San Luis Potosí, S.L.P.
MEXICO

[2] Laboratoire de Physique Quantique
Université Paul Sabatier
118 route de Narbonne, F-31062 Toulouse
FRANCE

## Abstract

The electronic and magnetic properties of low-dimensional systems are investigated by using a saddle-point slave-boson approximation to the Hubbard model. A local approach based on a real-space expansion of the local Green's function is presented. The changes in the electronic properties are determined as a function of the local coordination number $z$. Results are given for the magnetic moments, magnetic order, average number of double occupations and hopping renormalizations. The environment dependence of the electronic correlations is discussed.

## I. Introduction

Low-dimensional systems can be characterized by the presence of surfaces or interfaces, which restrict the propagation of particles and excitations. Examples of such systems, which lie between the atom (dimension zero) and the periodic solid (dimension three), are molecules, clusters, small particles, ideal chains, quasi one-dimensional systems, thin films, overlayers, surfaces, multilayers, and granular structures. These materials are known to present a variety of remarkable properties, which are not only most interesting from a fundamental point of view, but which have also opened numerous possibilities for technological applications.[1] Besides quantifying the overwhelming diversity of material-specific behaviors, theorists have dedicated considerable research effort to understand in simple terms which are the common features shared by these systems. Local approaches to electronic structure theory have been proven to be most successful, since they allow to relate the electronic properties to the local environment of the atoms in a physically transparent way.[2-5]

One of the most interesting and challenging problems in this context is itinerant magnetism. A large number of experimental and theoretical works have shown that the magnetic behavior of itinerant electrons depend very strongly on the system dimensions and on the local atomic environment. Let us recall, for example, the enhancement of the local magnetic moments at Fe atoms as their local coordination number is reduced (e.g., in thin films, near surfaces and in small clusters)[5-8] or the onset of magnetism in small clusters of 4d-transition metals which are nonmagnetic in the solid (e.g., $Rh_N$).[9] From a microscopic point of view, these remarkable properties are the result of a delicate balance between hybridizations effects, which favor equal filling of spin states, and Coulomb interactions, which according to Hund's rules favor the formation of local magnetic moments. Hybridizations and the associated electron delocalization tend to reduce the ground-state kinetic energy $E_K$ but at the same time they involve local charge fluctuations which increase the Coulomb-interaction energy $E_c$. The interplay between $E_K$ and $E_c$ leads to correlations in the electronic motion, which play a central role in determining the magnetic properties. In addition, one expects quite generally that electron correlations should become more important as the system dimensions are reduced since $E_K$ decreases with decreasing coordination number $z$. Moreover, the reduction of $z$ should hinder the backflow of density excitations responsible for dynamic screening of charge fluctuations.

## II. Theoretical Method

In order to investigate the trends in the electronic and magnetic properties of low-dimensional systems we consider the well-known Hubbard Hamiltonian:[10]

$$H = -t \sum_{\langle i,j \rangle, \sigma} \hat{c}_{i\sigma}^{+} \hat{c}_{j\sigma} + U \sum_{i} \hat{n}_{i\uparrow} \hat{n}_{i\downarrow} . \tag{1}$$

Equation (1) can be regarded as a minimum model for itinerant electrons, since it includes electron delocalization, described by the first term, and Coulomb repulsions associated to local charge fluctuations, described by the second term. In the saddle-point slave-boson approximation[11] the ground-state properties of the Hubbard model are derived from an effective Hamiltonian

$$\hat{H}' = \sum_{i\sigma} \varepsilon_i' \hat{n}_{i\sigma} + \sum_{i \neq j} t'_{ij\sigma} \hat{c}_{i\sigma}^{+} \hat{c}_{j\sigma} , \tag{2}$$

which describes the electrons as if they were independent quasi-particles having, as a result of Coulomb interactions and correlations, shifted energy levels $\varepsilon_{i\sigma}'$ and renormalized hopping integrals $t'_{ij\sigma} = q_{ij}^{\sigma} t_{ij}$. The hopping renormalization factor $q_{ij}^{\sigma} = \langle \hat{z}_{i\sigma}^{+} \hat{z}_{j\sigma} \rangle$ is given by

$$q_{ij}^{\sigma} = \frac{(e_i p_{i\sigma} + p_{i\bar{\sigma}} d_i)(e_j p_{j\sigma} + p_{j\bar{\sigma}} d_j)}{\sqrt{n_{i\sigma}(1 - n_{i\sigma}) n_{j\sigma}(1 - n_{j\sigma})}} . \tag{3}$$

The local variables $e_i$, $p_{i\sigma}$, and $d_i$ are the saddle-point values of boson operators and represent, respectively, the ground-state probability for the site $i$ to be empty, singly occupied with spin $\sigma$ or doubly occupied. The actual $T = 0$ values of $e_i$, $p_{i\sigma}$, and $d_i$ are determined by minimizing the electronic energy

$$E = \sum_{i\sigma} \int_{-\infty}^{\varepsilon_F} (\varepsilon - \varepsilon'_{i\sigma}) \rho_{i\sigma}(\varepsilon) \, d\varepsilon + U \sum_i d_i^2, \qquad (4)$$

under certain self-consistency constraints on the spin-polarized electronic density.[11,12] The first term in Eq. (4) represents the kinetic energy $E_K$ renormalized by correlations. Here, $\rho_{i\sigma}(\varepsilon) = -(1/\pi) \operatorname{Im} \{G_{i\sigma,i\sigma}(\varepsilon)\}$ is the local density of states (LDOS) $[G(\varepsilon) = (\varepsilon - H')^{-1}]$ and $\varepsilon_F$ the Fermi energy. The second term represents the Coulomb energy $E_c$ associated to local charge fluctuations; $d_i^2$ is the average double occupation at site $i$.

The environment dependence of the electronic properties enters in the calculation of $\rho_{i\sigma}(\varepsilon)$. A systematic expansion of $\rho_{i\sigma}(\varepsilon)$ which allows to include the contributions to the LDOS from a local point of view is provided by the Haydock-Heine-Kelly recursion scheme.[13] In order to identify the effects of the local coordination number on the electronic structure we consider in the following the third moment expansion. Assuming for simplicity that the system structure is bipartite, with sublattices denoted by $A$ and $B$, $\rho_{i\sigma}(\varepsilon)$ is given by

$$\rho_{i\sigma}(\varepsilon) = \frac{b_{i\sigma}}{\pi} \frac{\sqrt{1 - \left(\frac{\varepsilon - \sigma\Delta/2}{2b_{i\sigma}}\right)^2}}{\Delta(\varepsilon + \sigma\Delta/2) + b_{i\sigma}^2}, \qquad (5)$$

where $\Delta = \varepsilon_{A\sigma} - \varepsilon_{B\sigma}$ and $b_{i\sigma} = q_{ij}^\sigma t_{ij} \sqrt{z}$. Equation (5) corresponds to a density of states centered at $\frac{1}{2}\sigma\Delta$ and having an effective band width $4b_{i\sigma}$. The electronic energy $E$ is minimized by considering paramagnetic, ferromagnetic, and antiferromagnetic self-consistent solutions.

## III. Results and Discussion

The ground-state magnetic order—ferromagnetic (FM), antiferromagnetic (AF) or paramagnetic (PM)—has been determined as a function of $U/t$, the average coordination number $\bar{z}$ and the number of electrons per site $n$ (Ref. 12). Different regimes may be distinguished depending on the value of $U/W$, where $W = 4t\sqrt{\bar{z}}$ is the band width of the unrenormalized single-particle LDOS. For small values of $U/W$ ($0 \leq U/W \leq \frac{1}{2}$) PM order dominates for all values of $n$. For $U/W > \frac{1}{2}$ and close to half-band filling ($1 \leq n \leq 1.05$) we find AF order. Starting from the strongly correlated limit and decreasing $U/W$ we observe that the AF region extends to larger values of $n$ (from up to $n = 1.0$ for $U/W = \infty$ to up $n = 1.25$ for $U/W = 1.31$). A change of behavior is found for $U/W < 1$, where AF order is displaced by PM order. For $n > 1.25$, the PM solution is always the most stable one irrespectively of the values of $U/W$ and $\bar{z}$. Finally, for sufficiently large $U/W$ ($U/W > 7$) we find FM order. The FM region is located between the AF and PM domains. Starting from a point where all three phases have the same energy ($n = 1.12$, $U/W = 7.15$) the range of electron densities where the ferromagnetism dominates increases monotonically with increasing $U$. In agreement with Nagaoka's theorem,[14] for $n = 1$ and $U/W = \infty$, the FM and AF solutions have the same energy. In contrast to Hartree-Fock results, the FM solution is less stable than the PM one for $n > 1.38$ even if the Coulomb interaction strength $U/W$ is arbitrary large. This illustrates the ability of saddle-point slave-boson approach to suppress local charge fluctuations, an important feature in order to determine the ground-state energy of low-spin states.

It is also interesting to identify the magnetic transitions which may occur as a function of $\bar{z}$ for different values of $U/t$ and $n$. Taking into account that in low dimensional systems $1 \leq \bar{z} \leq 12$ and that the single-particle band width is given by $W = 4t\sqrt{\bar{z}}$, the following magnetic transitions are found as $\bar{z}$ increases: (i) from FM to AF order for $28.6 \leq U/t \leq \infty$ and $1 < n \leq 1.12$, (ii) from FM to PM order for $28.6 \leq U/t \leq \infty$ and $1.12 \leq n \leq 1.38$, (iii) from AF to PM order for $2.13 \leq U/t \leq 8.24$ and $1 \leq n \leq 1.12$, (iv) from PM to AF order for $5.23 \leq U/t \leq 99.1$ and $1.12 \leq n \leq 1.25$, (v) a re-entrant transition of the type PM-AF-PM for $2.38 \leq U/t \leq 99.1$ and $1.12 \leq n \leq 1.25$ and (vi) a double transition of the type FM-PM-AF for $28.6 \leq U/t \leq 218.7$ and $1.12 \leq n \leq 1.25$. In the following we discuss several electronic and magnetic properties as a function of $\bar{z}$. As a representative example we consider the band filling $n = 1.1$ which illustrates different FM, AF, and PM behaviors.

In figure 1 results are given for (a) the $U/t$-$\bar{z}$ magnetic phase diagram, (b) the local magnetic moments $\mu$, (c) the Coulomb energy $E_c$ and (d) the kinetic energy $E_K$ [see Eq. (4)]. For $U/t = 7$ we observe a transition from PM to AF order as $\bar{z}$ decreases. Notice the onset and increase of the magnetic moments as the dimension of the system decreases. This enhancement of $\mu$ is qualitatively in agreement with realistic $spd$-band Hartree-Fock calculations on Cr clusters.[15] Figs. 1(c) and 1(d) illustrate the interplay between electron delocalization and Coulomb-energy fluctuations. As $\bar{z}$ decrease the kinetic energy associated to band formation decreases. In other words, the electrons tend to be more localized. The Coulomb interaction energy and the average number of double occupations $d^2$, are reduced accordingly.

For half-band filling ($n = 1$) a qualitatively similar PM-AF transition is also observed.[12] The critical coordination number $z_c$ for the onset of AF order and the value of the magnetic moment $\mu$ decrease as we increase $n$ from $n = 1.0$ to $n = 1.1$, e.g., for $n = 1.0$ and $U/t = 7\bar{z}_c = 10.7$ and $\mu = 0.96$ while for $n = 1.1$ and $U/t = 7\bar{z}_c = 10.12$ and $\mu = 0.83$. This is a consequence of the decreasing stability of the AF solution as we move away from half-band filling. As the dimensions are reduced, the average double occupations $d^2$ are suppressed more rapidly in the AF state than in the PM state. Qualitatively this is the same for $n = 1.0$ and for $n > 1$. However, the difference between the slopes $\partial d^2/\partial \bar{z}$ in the AF and PM solutions is much less important if $n > 1$. In other words, the suppression of double occupations becomes less efficient in the AF state as $n$ increases ($n \geq 1$).

For large values of $U/t$ a qualitatively different magnetic behavior is found as a function of $\bar{z}$. For example, for $U/t = 90$ we obtain that in the three-dimensional system ($\bar{z} = 12$) there is AF order with unsaturated magnetic moments $\mu \simeq 0.7$ and a considerably renormalized kinetic energy. As shown in Fig. 1(b) $\mu$ now decreases slowly with decreasing $\bar{z}$. The Coulomb interaction energy is reduced to its minimum possible value $d^2_{\min} = n - 1 = 0.1$ already for $\bar{z} = 12$. Therefore, reducing the system dimensions has little influence on $E_c$, which decreases only very weakly as $\bar{z}$ decreases [see Fig. 1(c)]. For sufficiently small $\bar{z}$ we observe a transition from AF to FM order. The critical coordination number $\bar{z}_c$, below which FM sets in, increases with increasing $U/t$. The same holds for the magnetic moments $\mu$, in qualitatively agreement with Hartree-Fock $d$-band model calculations.[7] Notice the abrupt increase of $\mu$ at $\bar{z} = \bar{z}_c$, which equals its saturation value $\mu_{\text{sat}} = 2 - n$ for $\bar{z} < \bar{z}_c$. In contrast, the Coulomb energy is nearly unaffected by the transition since already for $\bar{z} > \bar{z}_c$ (in the AF regime) $d^2$ is very close to the minimum value $d^2_{\min} = n - 1$. As the magnetic moment

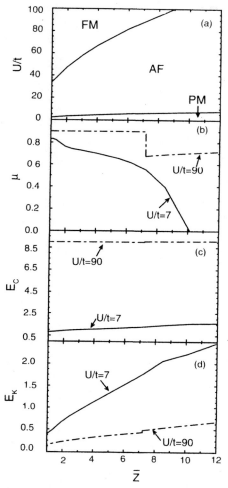

**Figure 1.** Electronic and magnetic ground-state properties of low-dimensional bipartite structures: (a) $U/t$-$\bar{z}$ magnetic-phase diagram, (b) average local magnetic moment $\mu$, (c) Coulomb energy $E_c$ and (d) kinetic energy $E_K$ [see Eq. (4)]. The results, given as a function of the average local coordination number $\bar{z}$, were derived from the Hubbard Hamiltonian by using a slave-boson mean-field method and the third-moment expansion of the local densities of states. The band filling is $n = 1.1$. In (b)–(d), representative values of the Coulomb repulsion $U/t$ are used as indicated.

develops ($\bar{z} < \bar{z}_c$) the majority and minority spins renormalization factors $q^{\uparrow}$ and $q^{\downarrow}$ split. $q^{\downarrow}$ increases with $\mu$ and eventually tends to 1 when $\mu \to 2-n$. The minority electrons give the dominant contribution to the kinetic energy $E_K$. In contrast, $q^{\uparrow}$ decreases as $\mu$ increases and the majority-electron contribution to $E_K$ becomes less and less important as the up band tends to be completely filled ($n_{\uparrow} \to 1$ as $\bar{z} \to 1$ for $n > 1$).

It is also interesting to compare our local-approach with previous calculations on infinite bidimensional systems in which rectangular or elliptical single-particle densities of states were used.[11,16-18] In general, the results are qualitatively similar. The main difference concerns the ground-state magnetic behavior for $n \to 1$ and $U \to 0$. While our calculations yield that the transition from PM to AF order occurs at a finite value of $U/t$, in Refs. 11, 16–18, the AF order sets in already for an arbitrary small $U > 0$. This discrepancy is due to the existence of perfect nesting in the bidimensional Fermi surface and not to differences in the shape of the single-particle LDOS. In fact, other calculations show that in the case of honeycomb[19] and Bethe lattices[20] the transition from PM to AF order occurs at a finite $U/t$. Finally, it should be also noted, that in our case the relative stability between different magnetic phases is determined by comparing the ground-state energies, while in Refs. 11, 16, and 17, the transition is inferred from the divergence of the staggered susceptibility. A more detailed account of the present study will be published elsewhere.[12]

## Acknowledgments

This work has been supported by CONACyT (Mexico). The Laboratoire de Physique Quantique (Toulouse) is *Unité Mixte de Recherche* associated to the CNRS.

## References

1. See, for instance, the papers on *Magnetoelectronics*, Physics Today, April 1995, p. 24ff.
2. V. Heine, in *Solid State Physics*, edited by H. Ehrenreich, F. Seitz, and D. Turnbull (Academic, New York, 1980), Vol. 35, p. 1.
3. D. Tománek, S. Mukherjee, and K. H. Bennemann, *Phys. Rev. B* **28**, 665 (1983).
4. G. M. Pastor, J. Dorantes-Dávila, and K. H. Bennemann, *Chem. Phys. Lett.* **148**, 459 (1988).
5. L. M. Falicov and G. A. Somorjai, *Proc. Natl. Acad. Sci. USA* **82**, 2207 (1985).
6. D. R. Salahub and R. P. Messmer, *Surf. Sci.* **106**, 415 (1981); K. Lee, J. Callaway, and S. Dhar, *Phys. Rev. B* **30**, 1724 (1985); K. Lee, J. Callaway, K. Wong, R. Tang, and A. Ziegler, *ibid.* **31**, 1796 (1985).
7. G. M. Pastor, J. Dorantes-Dávila, and K. H. Bennemann, *Physica B* **149**, 22 (1988); *Phys. Rev. B* **40**, 7642 (1989); J. Dorantes-Dávila, H. Dreyssé, and G. M. Pastor, *Phys. Rev. B* **46**, 10432 (1992).
8. I. M. L. Billas, J. A. Becker, A. Châtelain, and W. A. de Heer, *Phys. Rev. Lett.* **71**, 4067 (1993); I. M. L. Billas, A. Châtelain, and W. A. de Heer, *Science* **265**, 1662 (1994); J. P. Bucher, D. C. Douglas, and L. A. Bloomfield, *Phys. Rev. Lett.* **66**, 3052 (1991); *Phys. Rev. B* **45**, 6341 (1992); D. C. Douglass, A. J. Cox, J. P. Bucher, and L. A. Bloomfield, *Phys. Rev. B* **47**, 12874 (1993); FS. E. Apsel, J. W. Emert, J. Deng, and L. A. Bloomfield, *Phys. Rev. Lett.* **76**, 1441 (1996).
9. A. J. Cox, J. G. Louderback, and L. A. Bloomfield, *Phys. Rev. Lett.* **71**, 923 (1993); A. J. Cox, J. G. Louderback, S. E. Apsel, and L. A. Bloomfield, *Phys. Rev. B* **49**, 12 295 (1994).
10. J. Hubbard, *Proc. R. Soc. London Ser. A* **276**, 238 (1963); **281**, 401 (1964); J. Kanamori, *Prog. Theor. Phys.* **30**, 275 (1963); M. C. Gutzwiller, *Phys. Rev. Lett.* **10**, 159 (1963).
11. G. Kotliar and A. E. Ruckenstein, *Phys. Rev. Lett.* **57**, 1362 (1986).

12. E. Muñoz-Sandoval, J. Dorantes-Dávila, and G. M. Pastor, submitted to *Phys. Rev. B* (1997).
13. R. Haydock, in *Solid State Physics*, edited by H. Ehrenreich, F. Seitz, and D. Turnbull, (Academic, New York, 1980), Vol. 35, p. 215.
14. Y. Nagaoka, *Solid State Commun.* **3**, 409 (1965); D. J. Thouless, *Proc. Phys. Soc. London* **86**, 893 (1965); Y. Nagaoka, *Phys. Rev.* **147**, 392 (1996); H. Tasaki, *Phys. Rev.* **40**, 9192 (1989).
15. A. Vega, J. Dorantes-Dávila, G. M. Pastor, and L. C. Balbás, *Z. Phys. D* **19**, 263 (1991).
16. S. M. M. Evans, *Europhys. Lett.* **20** 53 (1992).
17. P. Denteneer and M. Blaauboer, *J. Phys. Condens. Matter* **7**, 151 (1995).
18. M. Deeg, H. Fehske, and H. Bütner, *Z. Phys. B* **91**, 31 (1993).
19. R. Frésard and K. Doll, *Proceedings of NATO Advanced Research Workshop on the Physics and Mathematical Physics of the Hubbard Model*, edited by Dionys Baeriswyl, David K. Campbell, Jose M. P. Carmelo, Francisco Guinea, and Enrique Louis NATO ASI Series, Series B: Physics **343**, 385 (Plenum Press, New York, 1995).
20. W. Zhang, M. Avignon, and K. H. Bennemann, *Phys. Rev. B* **45**, 12 478 (1992).

# Segregation and Ordering at a Ni-10 at.% Al Surface from First Principles

T. C. Schulthess[1] and R. Monnier[2]

[1] Oak Ridge National Laboratory
Oak Ridge, TN 37831-6114
USA

[2] Laboratorium für Festkörperphysik
ETH-Hönggerberg
CH-8093 Zürich
SWITZERLAND

## Abstract

We have calculated the surface energy and work function of the (111) surface of Ni-10 at.% Al, as a function of the Al concentration in the first lattice plane, by means of the coherent potential approximation (CPA) for inhomogeneous systems, implemented within the layer Korringa-Kohn-Rostoker (LKKR) multiple scattering formalism. Our treatment includes the charge correlation ignored in the standard single-site implementation of the CPA. Temperature effects are accounted within the mean field approximation for the configurational entropy. We find that at 1000 K the surface concentration of Al is doubled with respect to its bulk nominal value, in agreement with the results of recent surface ion scattering experiments. Allowing the formation of an ordered $L1_2$ plane at the surface further lowers the free energy of the system.

## I. Introduction

As a rule, the surface composition of a random binary alloy differs from that of the bulk. This has also been observed for the (111) face of Ni-10 at.% Al (Ref. 1), where at a temperature of 1000 K, the concentration of Al in the first lattice plane has been found to be 21%. What makes this system particulary interesting is that, upon increasing the temperature by 100 K, the equilibrium surface concentration of Al rises to 25% and does not change anymore up to the highest temperature of observation, *i.e.*, 1300 K. Furthemore, the surface composition remains "pinned" at

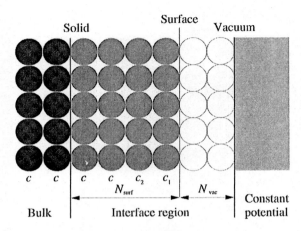

**Figure 1.** Schematic setup for surface electronic structure calculation within the LKKR-CPA method. The interface region consist of $N_{\text{vac}}$ layers of empty spheres (white circles) and $N_{\text{surf}}$ atomic layers beneath the surface plane (grey spheres left of the surface plane). The composition in the first few layers next to the surface can vary while the concentration in layers on the bulk side of the surface region are fixed to the bulk value. The interface region is sandwiched between a semi-infinite bulk (left side) and a constant vacuum potential (right side).

its high temperature value, even after the system has been cooled back to 1000 K. Although in the temperature range of the experiment, the bulk system is well inside the disordered region of the phase diagram,[2] the above findings suggest that an ordered $L1_2$ layer could be formed at the surface.

The purpose of this paper is to verify this conjecture with the help of a parameter-free, first-principles calculation. In a first step, we investigate the variation of the surface energy with the composition of the surface layer for the solid solution Ni-10 at.% Al, treating the configurational disorder within the coherent potential approximation (CPA) for inhomogeneous systems. The effect of charge correlations, which are neglected in the single-site CPA, is included by extending Johnson and Pinski's[3] scheme to surface. After the surface composition of the random alloy has been established, we investigate the possibility of chemical ordering in the surface layer.

## II. Energetics for a Chemically Disordered Surface Region

The typical setup for computing the electronic structure of a semi-infinite alloy with nominal bulk composition $A_c B_{1-c}$ in the atomic sphere approximation (ASA) is shown schematically in Fig. 1. The crystal consists of an interface region which is embedded self-consistently between a semi-infinite bulk (left side), for which the potential is determined in a separate bulk calculation, and vacuum (right side) which has a constant potential. The interface region consists of a vacuum part, which is modeled by $N_{\text{vac}}$ layers of empty sphere and a surface region which is composed of $N_{\text{surf}}$ layers for which the layers near the surface may have an average composition that differs from the bulk value. Atoms near the surface will loose electrons to the vacuum. On the bulk side several layers with bulk composition are included into the surface region to allow the charge density to converge to the bulk density when the bulk side of the interface region is approached.

The free energy of the interface region is calculated from

$$\Omega(T, \{c_i\}) = E_{\text{surf}}(T, \{c_i\}) - TS(T, \{c_i\}) - \mu(T, c) \sum_i (c_i - c), \quad (1)$$

where $E_{\text{surf}}$ is the internal energy of the layers in the interface region of Fig. 1. The entropy $S$ is determined within the mean field approximation, i.e.,

$$S(T, \{c_i\}) = -k_B \sum_i [c_i \ln c_i + (1 - c_i) \ln(1 - c_i)], \quad (2)$$

and the chemical potential is found from the concentraction derivative of the bulk energy,

$$\mu(T, c) = \left. \frac{dE_{\text{bulk}}(T, c')}{dc'} \right|_c. \quad (3)$$

Only the electronic degree of freedom is considered in the internal energy and we neglect its temperature dependence. This leaves us now with only an explicit temperature dependence through $TS(\{c_I\})$ in the free energy.

The energetic quantity of interest is the difference between the surface free energy and the bulk free energy, i.e.,

$$\Delta\Omega(T, \{c_i\}) = \Omega(T, \{c_i\}) - \Omega(T, c)$$
$$= E_{\text{surf}}(\{c_i\}) - N_{\text{surf}} E_{\text{bulk}}(c) - T\Delta S(\{c_i\}) - \mu(T, c) \sum_i (c_i - c), \quad (4)$$

which we will refer to as the *surface energy* of the system with concentraction profile $\{c_i\}$. Note, that in this last expression $E_{\text{surf}}$ is the electronic energy of $N_{\text{surf}} + N_{\text{vac}}$ interfacial layers (Fig. 1), while $E_{\text{bulk}}$ is the energy per bulk unit cell which is multiplied by the number of layers in Eq. (4).

The contribution of a single atomic sphere to the internal energy is

$$E_i = T[\rho_i] + U[\rho_i] + E_{\text{ex}}[\rho_i] + E_{M,i}, \quad (5)$$

where the charge density $\rho_i$ is taken to be zero outside the sphere and the usual functional expressions are used for the kinetic energy, $T[\rho_i]$, the potential energy, $U[\rho_i]$, and the exchange and correlation energy, $E_{\text{ex}}[\rho_i]$ (Ref. 4). Since these three quantities are local, the configuration average is straightforwardly calculated from the concentration weighted sum over the species that may occupy the atomic site. In the case of the Madelung energy,

$$E_M = \frac{1}{2} {\sum_j}' \frac{\langle Q_i Q_j \rangle}{|\mathbf{R}_i - \mathbf{R}_j|}, \quad (6)$$

the determination of the configurational average, which we denoted by $\langle \cdots \rangle$, is far from trivial and requires some further discussion. In this last expression $Q_i$ is the net charge of the sphere centered $\mathbf{R}_i$ and the prime denotes that the term $j = i$ has been excluded from the summation.

Within the single-site CPA, the charge on site $i$ is independent of the environment of the site. This approximation implies, that charges on different sites $i$ and $j$ are statistically uncorrelated, i.e.,

$$\langle Q_i Q_j \rangle = \langle Q_i \rangle \langle Q_j \rangle.$$

Due to the global charge neutrally in the bulk, $\langle Q_i \rangle = 0$, the single-site CPA implies that the Madelung energy vanishes for bulk solid solution. In the surface region, the charges are averaged over intralayer coordinates only, i.e., $\langle Q_i \rangle$ is replaced by

$$Q_I = \langle Q_i \rangle_{\text{layer},I} = c_I Q_I^A + (1 - c_I) Q_I^B,$$

where $Q_I^{A(B)}$ denote the effective charge on a site in layer $I$ occupied with an atom of species $A(B)$. Within the single-site CPA the Madelung energy of the semi-infinite crystal (Fig. 1) is computed for a 2D periodic array of effective charges $Q_I$, which are nonzero due to the broken symmetry perpendicular to the surface.

In the "exact" Madelung problem the net charge on a site depends on the environment even for a random alloy with *no* correlations in the site occupancies. Determining the configurational average in Eq. (6) is formidable task. In order to make the problem solvable within the context of a self-consistent electronic structure calculation, it is therefore necessary to include charge correlations in a simpler way.

Magri et al.,[5] showed that a very good agreement of the ordering energy with experimental values can be found when the Madelung energy is calculated for effective charges $Q_i^{\alpha=A,B}$ which are proportional to number of the atoms of opposite type in the nearest-neightbor shell. Thus, assuming that only the charge correlation within the first neighbor cluster contribute to the Madelung energy, the binary solid solution can be viewed as a random alloy of $2(Z+1)$ components, where $Z$ is the coordination number. Implementing the single-site CPA for this $2(Z+1)$ component alloy, Johnson and Pinski[3] have confirmed that the excess charge on a site is essentially proportional to the number of unlike nearest neighbors after a small system dependent offset has been added. For zero offset the average charge $Q^{A(B)}$ of an atom of type $A(B)$ is completely screened by the first neighbor shell, which implies[3] that, at the mean field level, change correlations effects can be included self-consistently in the single-site CPA through a simple shift of the effective one-electron potential for each species $\alpha$ by:

$$\Delta V_{\text{scr}}^\alpha = \frac{Q^\alpha}{R_{nn}}, \tag{7}$$

where $R_{nn}$ is the nearest-neighbor distance and $\alpha$ refers to the species. The corresponding correction to the Madelung energy,

$$E_{\text{scr}} = -\frac{1}{2} \sum_\alpha c_\alpha \frac{(Q^\alpha)^2}{R_{nn}}, \tag{8}$$

of this so-called screened CPA treatment leads to formation energies for the solid solution which are in close agreement with the charge correlated CPA treatment of the $2(Z+1)$ component alloy.[3]

Assuming that the electronic response to a charge perturbation is the same at the surface as in the bulk,[6] the screened CPA can be straightforwardly applied to the surface problem. The average excess charge on a site in layer $I$ occupied by species $\alpha$ is,

$$\Delta Q_I^\alpha = Q_I^\alpha - Q_I,$$

which when introduced into the expression corresponding to the Eq. (7) leads to the screening correction, $\Delta V_{\text{scr},I}^\alpha$, to the potential of this layer. The screening correction to the energy ot the $I$-th layer is then found from:

$$\Delta E_{\text{scr},I} = -\frac{1}{2} \sum_\alpha c_\alpha \Delta Q_I^\alpha \Delta V_{\text{scr},I}^\alpha. \tag{9}$$

## Table I

Work function and surface energy for the (111) face of pure Al and Ni. The LMTO result are from Skriver and Rosengaard[8] and the experimental values are those quoted by these authors.

|    | Exp. | LMTO | LKKR | Exp. | LMTO | LKKR |
|----|------|------|------|------|------|------|
| Al | 4.24 | 4.54 | 4.54 | 0.51 | 0.56 | 0.59 |
| Ni | 5.35 | 5.77 | 5.82 | 0.82 | 0.88 | 0.84 |

The electronic charge density is determined self-consistently using the LKKR-CPA multiple scattering approach,[7] where in the case of the screened CPA the correction $\Delta V_{scr,I}^{\alpha}$ is added to the potential during the self-consistency procedure. For surface calculation the ($l=1$, $m=0$) dipoles are included into the electrostatic part of the problem in addition to the point charge using the procedure which was introduced by Skriver and Rosengaard[8] and which Crampin[9] successfully applied to the calculation of the work functions for binary alloys with the LKKR-CPA method. A reasonable compromise between numerical accuracy and computing time is attained with $s, p$, and $d$ partial waves, 16 energy points on a semicircular contour in the complex energy plane, 45 special $k$-points in the 1/12-th section of the 2D Brillouin zone and 13 plane waves for the interlayer scattering. With this basis size and a surface region which always includes two layers of empty spheres in the vacuum and a minimun of four layers with bulk composition in addition to the layers with different composition at the surface, the numerical error in the surface energy is estimated to be less than 1 meV. As a final test, we have computed the work function and the surface energy for the (111) face of pure Al and Ni, respectively, and the results are compared with those obtained by Skriver and Rosengaard with the linear muffin-tin orbital (LMTO) method, and with experimental values quoted by these authors in Table I. The good agreement between the two sets of theoretical results, and their closeness to the experimental values, together with the success of the screened CPA in predicting energies in binary solid solution makes us confident that the LKKR-CPA approach will yield reliable results for the inhomogeneous surfaces region of a Ni-10 at.% Al random alloy.

A sensitive measure of the surface composition is offered by the work function, and in Fig. 2, we show our calculated values as a function of the Al concentration in the first lattice plane. The first observation is that the standard single-site CPA and the screened CPA produce virtually indistinguishable results. This is not entirely unexpected, since the surface dipole barrier is determined by the average charge within the planes and not by its distribution among the components of the alloy. More interesting is the fact that a single layer of pure Al at the surface of a Ni-10 at.% Al random alloy is sufficient to bring the value of the work function from 5.6 eV to 4.6 eV, which is only 0.06 eV higher than the value for the (111) face of pure Al!

Figure 3 shows the surface free energy of the alloy at 1000 K as a function of the composition of the first lattice plane. A shallow minimum is observed around 20 at.% Al, which is close to the value measured experimentally.[1] Charge correlation effects do not affect the position of the minimun, but lead to steeper rise of the surface energy at higher Al concentrations. This is due to the fact that with equal atomic sphere radii for Ni and Al, the Madelung correction enhances the excess charge on the Al sites considerably.

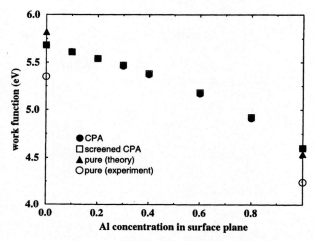

**Figure 2.** Work function of Ni-10 at.% Al as a function of Al concentration in surface plane (screened and unscreened).

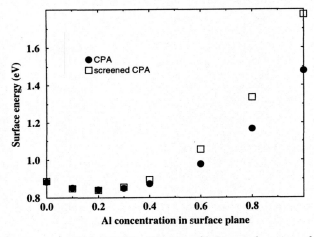

**Figure 3.** Surface energy at $T = 1000$ K of Ni-10 at.% Al as a function of Al concentration in surface plane (screened and unscreened).

## III. Ni-10 at.% Al Solid Solution with Ni$_3$Al Over-Layer

Since the surface of Ni-10 at.% Al at 1000 K is enriched with about 20 at.% Al and further heating of the sample leads to concentration that are even closer to the one of Ni$_3$Al, one is lead to the assumption that the surface is ordered in the L1$_2$ structure. Within the LKKR-CPA method and with the (111) stacking of layers, an ordered layer can be included by increasing the number of atoms per 2D unit cell from one (for complete randomness) to four. An ordered layer consists of a unit cell with one Al and three Ni atoms on defined sites, say, Al occupying the site at the origin of the unit cell while Ni occupies the other three sites. The introduction of a four-site 2D unit cell leads to a significant increase in computer time, and one has find a new compromise between the size of the plane waves basis and $k$-space integration meshes which are used in the calculation. We have found a reasonable agreement with one atom per unit cell calculations when 37 plane waves and the equivalent of 3 $k$-point in the 1/12-th section of the Brillouin zone are used (note that computer time increases with the third power of the number of plane waves). It is important to compare energies which are obtained with equal basis size. We therefore have to repeat the homogeneous calculations with a four atoms per layer 2D unit cell.

For ordered $Ni_3Al$ the value of the work function is 5.64 eV and the surface energy is 0.80 eV/atom. Both values compare well with the corresponding LMTO results[10] of 5.33 eV and 0.79 eV/atom, respectively.

Covering Ni-10 at.% Al with one layer of ordered $Ni_3Al$ reduces the surface energy at $T = 0$ K by about 0.35 eV. This values is of the order of the ordering energy of bulk $Ni_3Al$, which we find to be 0.435 eV/atom.

## IV. Conclusion

The theoretical results presented in this work for the segregation at the (111) surface of Ni-10 at.% Al demonstrate the capability of present-day *ab initio* methods for inhomogeneous alloys. The Al enrichment in the outermost lattice plane obtained at $T = 1000$ K is in good agreement with the experimental observation. Although this enrichment is purely entropy induced (the equilibrium surface composition at $T = 0$ K is the same as in the bulk), we find that, covering the surface with an ordered $Ni_3Al$ layer in the $L1_2$ structure further lowers the surface energy. We therefore conjecture that the concentration of 25% Al measured at higher temperature[1] is due to the formation of such an order layer.

## Acknowledgments

This work has been financially supported by the Swiss National Science Foundation. Oak Ridge National Laboratory is operated for the U.S. Department of Energy by Martin Marieta Energy Systems, Inc. under Contract No. DEAC05-84OR21400.

## References

1. T. C. Schulthess, E. Wetli, and M. Erbudak, *Surf. Sci.* **320**, L95 (1994).
2. T. M. Massalsky (editor), *Binary Alloy Phase Diagrams*, (ASM International, USA, 1990).
3. D. D. Johnson and F. J. Pinski, *Phys. Rev. B* **48**, 11 553 (1993).
4. R. M. Dreizler and E. K. U. Gross, *Density Functional Theory* (Springer-Verlag, Berlin, 1990).
5. R. Magri, S. H. Wei, and A. Zunger, *Phys. Rev. B* **42**, 11 338 (1990).
6. J. P. Perdew and R. Monnier, *J. Phys. F* **10**, L287 (1980).
7. J. M. MacLaren, S. Crampin, D. D. Vedensky, and J. B. Pendry, *Phys. Rev. B* **40**, 12 164 (1989); S. Crampin *et al.*, *Phys. Rev. B* **45**, 464 (1992).
8. H. L. Skiver and N. M. Rosengaard, *Phys. Rev. B* **43**, 9538 (1991).
9. S. Crampin, *J. Phys.: Condens. Matter* **5**, L443 (1993).
10. N. M. Rosengaard, private comunication.

# Study of Deep Level Defect in Polycrystalline Cadmium Sulfide Films

U. Pal, R. Silva González, F. Donado, M. L. Hernández, and
J. M. Gracia-Jiménez

*Instituto de Física*
*Universidad Autonoma de Puebla*
*Apartado postal J-48, 72570 Puebla, Pue.*
*MEXICO*

## Abstract

Photoluminescence (PL) spectroscopy is used to study the deep defect levels in CdS films grown by chemical bath deposition (CBD) and thermal evaporation (TE) techniques. The characteristics of PL bands and their evolution are studied mainly in three different types of (a) near stoichiometrically grown CBD (there after CBD-a), (b) off stoichiometrically grown CBD (there after CBD-b), and (c) near stoichiometrically grown TE samples to identify the origin of different defect levels in them. In general, TE films present more distinct features in PL spectra than the CBD films especially in the subband edge spectral region, indicating a lower concentration of defect states in them. A remarkable change in composition on thermal treatment (TT) of CBD-b [there after CBD-b (TT)] films is observed along with the usual change in phase (cubic to hexagonal). It has been observed that the green emission (GE) band is more related to the crystalline phase of the films than to their chemical composition. The evolutions of other low energy emissions, *e.g.*, so-called red emission (RE) and yellow emissions (YE) on temperature and film composition are studied.

## I. Introduction

There has been a strong interest in CdS because of the importance of this material in thin film photovoltaics[1] and luminescent devices.[2] Generally the as grown films prepared by CBD or TE techniques acquire the characteristic of metastable cubic ($\beta$-CdS or howleyite) phase and with suitable thermal treatment (TT), transform into the stable hexagonal ($\alpha$-CdS or greenockite) crystalline phase.[3,4] As the defects play a key role in determining its performance in devices, a considerable effort has been placed in the electronic and optical characterization of defects in this material.[5] The characterization of localized impurity and defect levels in CdS has been the subject of consideration since the past three decades. Several workers[3,6-10] have reported about identification and origin of different defect levels over the wide spectral range. Emission or absorption bands in the different spectral regions were connected to the presence of different interstitial of component atoms, vacancies, interstitial-vacancy complexes or to the phase transition. However, the results and the explanations are sometimes contradictory and there is still insufficient understanding of the defects in this compound.

Generally, the interface states in the CdS/CdTe solar cells arising from the lattice mismatch and difference in their thermal expansion coefficients act as trapping centers and lead to an increase in dark current density. However, the bulk defect density in CdS controls the short wave length spectral response. As the defect states in CdS are believed to be controlled by both its chemical composition and crystalline phase, the choice of CdS epilayers with suitable phase and composition is important.

In the present communication, we present a PL study of deep levels in polycrystalline CBD and TE grown CdS films with different stoichiometry and crystalline phase. To verify the effect of phase transition and chemical composition separately, the PL measurements were made on nonstoichiometric CdS films of cubic phase, near stoichiometric CdS films of cubic phase and near stoichiometric CdS films of hexagonal phase. It is observed that the evolutions of defect bands are different for the films of similar chemical composition and crystalline phase if they are prepared by different routes. It is also seen that the so-called GE band is closely related to the hexagonal crystalline phase and the bands in the intermediate spectral range (*i.e.*, YE and RE bands) are strongly related to the chemical composition and other defects of the films.

## II. Experiments

Investigations were made on the CdS films grown by CBD and TE techniques. The CBD films were grown on properly cleaned glass substrates using an aqueous bath containing freshly prepared 0.02M $CdCl_2$, 0.5M $NH_4NO_3$, 0.1M NaOH, and 0.1M $CS(NH_2)_2$ solutions. The substrates were immersed vertically in the solution kept at 333 K. The solution was continuously stirred during deposition process to obtain homogeneous distribution of the chemical components. To attain the difference in stoichiometry in the films, the ratio of $CdCl_2$ and $NH_4NO_3$ solutions was varied in the reaction mixture. For CBD-a samples, the ratio was maintained 1:8 and for the CBD-b samples the ratio was 1:2. However, the pH of the reaction mixture in both cases was maintained the same (8.8). The deposition process was repeated for six times to obtain a sufficient film thickness. The final film thicknesses were of the order of 0.5 $\mu$m. The as grown CBD-b films were seen to be of non-stoichiometric and they were subject to a thermal annealing treatment (TT) in a partially evacuated ($10^{-2}$ Torr) quartz tube at 623 K for 2 hours.

The TE films were grown on glass substrates by evaporating CdS powders (99.999%) from a tungsten boat in an evacuated ($2 \times 10^{-6}$ Torr) JEOL, JEE-400 evaporation system. The films were grown at different substrate temperatures ($T_s$) and at a constant deposition rate of 120 nm min$^{-1}$.

The composition and morphology of the films were studied by a JEOL, JSM 5400LV scanning electron microscope (SEM) with NORAN system. XRD patterns were recorded by a Phillips (PW 3710) x-ray diffractometer. The PL spectra of the as grown and annealed samples were taken by mounting them on a cold finger

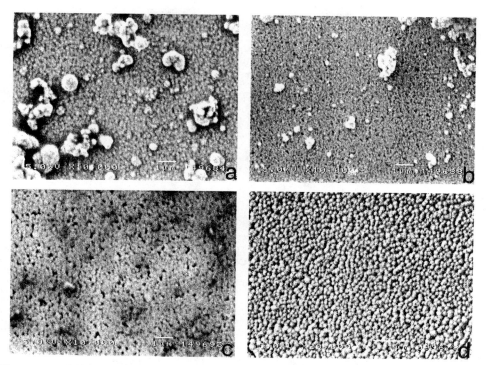

**Figure 1.** SEM micrographs of (a) CBD-a, (b) CBD-b, (c) CBD-b (TT), and (d) TE ($T_s = RT$) films.

of a Displex system and were held at temperatures ranging from 20–300 K. The samples were excited by the 488 nm line of a Spectra Physics cw Ar ion laser. The luminescence spectra were analysed by a Zeiss SPL monochromator and recorded by a lock-in technique in HP $x$-$y$ plotter.

## III. Results

As deposited CBD-a and CBD-b films were smooth, adherent and bright yellow in color. Where as, the TE films were bright orange. The XRD patterns of as grown CBD films revealed that the films are of cubic crystalline phase. On thermal annealing, the CBD-b (TT) films revealed a predominant hexagonal phase. The TE films grown at $T_s = RT$ show a predominant hexagonal phase and might have some cubic phase mixed. In the TE films grown at $T_s = 473$ K the fraction of hexagonal phase increased.

The EDAX results revealed a near stoichiometric composition with Cd and S atomic ratio 1.09 for the CBD-a films. The CBD-b films were nonstoichiometric with Cd and S ratio 0.60. On TT, the Cd and S ratio in CBD-b (TT) films became 1.09. The ratio of Cd and S in TE ($T_s = RT$) films was 1.06, whereas, the ratio was 1.02 for the films grown at $T_s = 473$ K.

The SEM micrographs of CBD-a, CBD-b, CBD-b (TT), and TE (RT) films are shown in Fig. 1. The CBD-a and TE films revealed a similar microstructural feature. However, in TE films, the crystallites were well resolved and there were no superficial

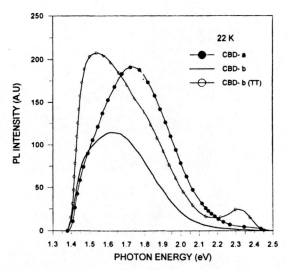

**Figure 2.** Low temperature (22K) PL spectra of CBD-a, CBD-b, and CBD-b (TT) films.

Cd rich clusters (cauliflower structures) in them. The CBD-b films revealed a different morphology from the other films. The average crystallite size in them was smaller than the CBD-a or TE films. On TT, there was no noticeable change in morphology except a slight increase in crystallite size and a better surface homogeneity.

In figure 2, the low temperature PL spectra of CBD-a, CBD-b and CBD-b (TT) films are presented. The integrated PL intensity in CBD-b films was about 50% that of CBD-a films. However, the intensities were similar for the CBD-a and CBD-b (TT) films.

There appeared no GE band in as grown CBD-b films. On TT, the GE band appeared along with a considerable increase of PL intensity. In CBD-a films a much less intense broad GE band appeared. The evolutions of so-called YE and RE bands were different in the films. The PL bands resolved more clearly in the TE films. The 22 K PL spectrum of a TE ($T_s = RT$) film is presented in Fig. 3. There appeared a broad low intense GE band.

In figure 4, the PL spectrum (22 K) of a TE film deposited at $T_s = 473$ K is shown. In this case a more intense and sharper GE band appeared. The over all PL intensity increased when compared to the intensity of TE ($T_s = RT$) films. The evolution of YE and RE bands were also different for the films deposited at different $T_s$. In the PL spectra of most of the TE films, appeared about four low energy bands apart from the GE band. In the inset of the Fig. 3, the evolution of those four bands appeared near about 1.45 eV, 1.58 eV, 1.70 eV, and 1.85 eV (at 22 K) with temperature are shown. Apart from the 1.45 eV band, the evolutions of all other bands were similar, indicating a similar origin for all of them. However, all the bands saturated below 90 K. The activation energies calculated from their evolutions above 100 K were 36 meV, 35 meV, 35 meV, and 32 meV, respectively.

From the SEM micrographs, it is clear that on TT of CBD films, the morphology becomes more homogeneous along with an increase in crystallite size value. Though in the TE ($T_s = RT$) films, the average crystallite size value is comparable to the crystallite size of CBD-a films, there appeared no superficial cauliflower structures (Cd rich) indicating a better chemical homogeneity in them.

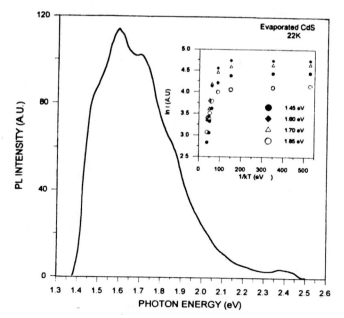

**Figure 3.** Low temperature PL spectrum of TE ($T_s = RT$) film. Inset shows the temperature evolution of the peaks near 1.45 eV, 1.60 eV, 1.70 eV, and 1.85 eV. The activation energies calculated from the evolutions above 100 K are 36 meV, 35 meV, 35 meV, and 32 meV, respectively.

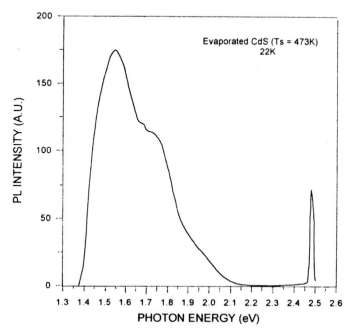

**Figure 4.** Low temperature (22 K) PL spectrum of TE ($T_s = 473$ K) film.

In the PL spectra of as deposited CBD-b films, a broad asymmetric band spreading over 1.40 to 2.20 eV appeared. From the shape of this band it is evident that it consists of a minimum three broad bands in the RE and YE regions. On thermal annealing, the GE band evolved and the RE band intensity increased much faster than the increase of YE band intensity. Though the integrated PL intensity in cubic and near stoichiometric CBD-a films is similar to the hexagonal near stoichiometric CBD-b (TT) films, the GE band in the latter is more intense.

In the CBD-a films, the YE bands are more intense. Generally, the GE band in CdS has been attributed to interstitial sulphur.[10,11] RE has been ascribed to sulphur vacancies, surface damage or the neutral complex $[V_{Cd}^{2-} + V_{S}^{2+}]^0$ (see Ref. 10) and the YE bands to interstitial Cd atoms.[12] In the present study the GE band is designated to the emission near 2.4 eV spectral region. Though this band is believed to be associated to the interstitial S, we could not detect this band in our sulphur rich CBD-b films. Even in the near stoichiometric CBD-a films, the band that appeared with much less intensity. A sharper and intense evolution of this band is observed only in hexagonal films [e.g., in CBD-b (TT) and TE ($T_s = 473$ K) films]. From these observations it seems that this band is closely associated to the hexagonal phase as predicted by Lozada-Morales and Zelaya-Angel.[13] We designate all the bands appeared from 1.60 eV to 2.20 eV as yellow emissions (YE). From the PL spectra of our CBD and TE films, it is clear that there are several transitions in this spectral range, which are more or less resolved in TE films. Though sometimes these bands were associated to interstitial S atoms,[3] weaker evolution of these bands in S rich CBD-b films in comparison with the other near stoichiometric CBD and TE films leads us to support the argument of Agata et al.[12] that they are associated to interstitial Cd atoms. The argument is further supported by the fact that the GE band in TE ($T_s = 473$ K) films was more intense than in the TE ($T_s = RT$) films.

The evolution of so-called RE band (near 1.45 eV) is more prominent in CBD-b (TT) films, indicating its origin as sulphur vacancies. However, annealing induced surface damage or the neutral complex $[V_{Cd}^{2-} + V_{S}^{2+}]^0$ might have some contributions to this. The so-called RE and YE bands were more resolved in TE films than in CBD films which might be due to a more well defined grain structure and hence a reduced grain boundary area in them.

## IV. Conclusions

The GE band in CdS is closely associated with its hexagonal crystalline phase. Yellow emissions are related to interstitial Cd atoms. The association of RE band to the S vacancies is verified further. Discrete states with well characterized energy levels are observed in the PL spectra of evaporated CdS films, where as, a continuum of states is evident for CBD films. Such a difference indicates a lower concentration of defect levels in TE films in comparison with the CBD films which might make TE films more favourable for the applications in devices. Apart from the GE band, all the other bands in CdS films are either associated with the grain boundaries or the trap levels associated with the surface and bulk of the films.[14]

## Acknowledgments

The work is supported by CONACyT (Mexico) grants (Project No. 1351-PA and 5269N). U. Pal thanks CONACyT for the Catedra Patrimonial (No. 481100-2-940460).

# References

1. J. Britt and C. Ferekids, *Appl. Phys. Lett.* **62**, 2851 (1993).
2. Ch. Bouchenaki et al., *J. Cryst. Growth* **101**, 797 (1990).
3. O. Zelaya-Angel et al., *Solid State Commun.* **94**, 81 (1995).
4. 6-0314 ASTM X-Ray Powder Data File.
5. J. Woods and K. H. Nocholas, *Br. J. Appl. Phys.* **15**, 1361 (1964).
6. M. K. Sheinkman, I. B. Ermolovich, and G. L. Belen'kill, *Sov. Phys. Solid State* **10**, 2069 (1969).
7. P. Besomi and B. Wessels, *J. Appl. Phys.* **51**, 4305 (1980).
8. R. H. Bube, *Photoconductivity of Solids* (Wiley, New York, 1960).
9. G. Godrillo, *Sol. Energy Mater. And Solar Cells* **25**, 41 (1992).
10. S. Achour and G. H. Talat, *Thin Solid Films* **144**, 1 (1986).
11. M. Gracia-Jiménez, J. L. Martínez, E. Gómez, and A. Zehe, *J. Electrochem. Soc.* **131**, 2974 (1984).
12. M. Agata, H. Kurase, S. Hayashi, and K. Yamamoto, *Solid State Commun.* **76**, 1061 (1990).
13. R. Lozada-Morales and O. Zelaya-Angel, *Thin Solid Films* **281-282**, 386 (1996).
14. P. Besomi and B. Vesseles, *J. Appl. Phys.* **51**, 4305 (1980).

# Intrinsic Localized Modes in the Bulk and at the Surface of Anharmonic Chains

V. Bortolani,[1] A. Franchini,[1] and R. F. Wallis[2]

[1] INFM and Dipartimento di Fisica
Universitá di Modena
Via Campi 213/A, 41100 Modena
ITALY

[2] Department of Physics
University of California
Irvine, California 92717
USA

## Abstract

In this paper we will review recents results relative to localized modes induced by anharmonicity in one-dimensional lattices. We will show that localized modes exists in monoatomic chains with and without a local inhomogeneity in the anharmonic force field. We will compare the discrete and the quasi-continuum intrinsic even and odd localized solutions. This analysis is carried out by taking into account harmonic and quartic anharmonic interactions. We will also present results for the diatomic chains showing the presence of surface modes and gap modes related to the maximum of the acoustic band. In this analysis will be also studied the effect of the cubic anharmonicity. One of the major effects of the cubic anharmonicity is to produce gap modes that split off from the bottom of the optical branch.

## I. Introduction

In recent years a great interest has grown on the intrinsic effects of the anharmonicity in many different scientific areas. The different problems studied in these fields are all described by nonlinear equations that allow solutions with a strongly localized character, due to the nonlinear nature of the equations. In particular, for systems described by nonlinear equations such as the Korteg de Vries equation, the nonlinear Schrödinger equation, the Sine-Gordon equation are possible soliton solutions. In this paper we review the results of our studies of solitary waves in solids with quartic and cubic anharmonicity, considering monoatomic and diatomic one-dimensional chains. Our interest in this field was induced by the findings of Sievers and Takeno[1] who proved the existence of stationary intrinsic localized modes of odd parity in discrete monoatomic chains and in the continuum limit, and by the Page results[2] on even parity solutions. More recently, monoatomic systems with substitutional impurities were studied by Kivshar[3] in the quasi-continuum limit, proving the existence of odd intrinsic localized modes at the impurity site. The diatomic chains were studied by Kiselev et al.,[4] for Born-Mayer potentials.

## II. Monoatomic Chains

We started studying the discrete monoatomic chain with harmonic and quartic anharmonic interactions, with an inhomogeneous force field, using the rotating wave approximation (RWA) to remove the high frequency harmonics. Our aim was to examine the effects on the intrinsic localized modes produced by a local modification of the anharmonic part of the force field between atoms. The Hamiltonian of the homogeneous chain has the form

$$H = \tfrac{1}{2}m \sum_n \dot{u}_n^2 + \tfrac{1}{2}K_2 \sum_n (u_n - u_{n+1})^2 + \tfrac{1}{4}K_4 \sum_n (u_n - u_{n+1})^4, \qquad (1)$$

where $m$ is the mass of the particles and $u_n$ is the longitudinal displacement of the $n$-th atom. The equations of motion are easily found to be

$$m\ddot{u}_n + K_2(2u_n - u_{n+1} - u_{n-1}) + K_4[(u_n - u_{n+1})^3 + (u_n - u_{n-1})^3] = 0. \qquad (2)$$

We seek for a stationary solution of the nonlinear equation of motion for each lattice site of the form of a vibrational displacement $\xi$ of maximum amplitude $A$ and frequency $\omega$

$$u_n = A\xi_n \cos(\omega t), \qquad (3)$$

which leads to the equations

$$m\omega^2 A\xi_n \cos(\omega t) = K_2 A[2\xi_n - \xi_{n+1} - \xi_{n-1}] \cos(\omega t) \\ + K_4 A^3 [(\xi_n - \xi_{n+1})^3 + (\xi_n - \xi_{n-1})^3] \cos^3(\omega t). \qquad (4)$$

The RWA allows us the linearization in $\cos(\omega t)$ and gives rise to a system of coupled nonlinear equations in the $\xi_n$ site dependent displacements. In this way we are left with the equations of motion

$$m\omega^2 \xi_n = K_2[2\xi_n - \xi_{n+1} - \xi_{n-1}] + \tfrac{3}{4} K_4 A^2 [(\xi_n - \xi_{n+1})^3 + (\xi_n - \xi_{n-1})^3]. \qquad (5)$$

If we integrate back the equations of motion we obtain the expression for the total energy in the RWA approximation.

$$E = \tfrac{1}{2} m A^2 \omega^2 \left(\sum_n \xi_n^2\right) \sin^2(\omega t) + \tfrac{1}{2} K_2 A^2 \left[\sum_n (\xi_n - \xi_{n+1})^2\right] \cos^2(\omega t) \\ + \tfrac{3}{8} K_4 A^4 \left[\sum_n (\xi_n - \xi_{n+1})^4\right] \cos^2(\omega t). \qquad (6)$$

In present calculations we take a finite monoatomic chain, $n = -256, \ldots, 256$, with free end conditions. The intrinsic localized modes arise from the highest frequency harmonic mode when the anharmonicity is switched on. The anharmonicity is measured by the parameter $S = (3K_4/4K_2)A^2$. We start by considering odd parity

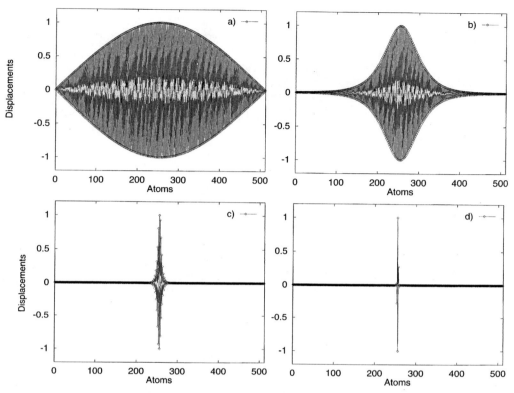

**Figure 1.** Displacement patterns of 512 atoms monoatomic chain: panel (a) refers to the harmonic case, panel (b) is relative to $S = 10^4$, panel (c) refers to $S = 10^2$, and panel (d) is relative to $S = 1$.

solutions for which $\xi_n = \xi_{-n}$. The normalization of the relative displacements pattern is given by choosing $\xi_0 = 1$. The pattern of the harmonic displacements is presented in Fig. 1(a). For clarity we draw the longitudinal displacements vertically. To solve the system of nonlinear equations in Eq. (5) we use a standard routine based on Newton scaled gradient. Given an initial guess of the displacements, in our case the harmonic displacements of the normal mode at frequency $\omega^2_{\max} = 4K_2/m$, through an iterative procedure the routine determines the stable solution for each value of the parameter $S$. The solution obtained for a given $S$ value is used as a guess solution for the anharmonicity parameter increased by $\Delta S$. The same results have been found with an alternative procedure, studying a chain with increasing number of atoms. This method gives also a stability criterion, because the added atoms act as a perturbation on the solution of the previous step.

In figure 1(b) is drawn the solution for a small value of the anharmonicity $S = 10^{-4}$. It is clear that the mode starts to become localized. By increasing the anharmonicity, as shown in Fig. 1(c), corresponding to $S = 10^{-2}$, the mode tends to become narrower and for $S = 1$ [Fig. 1(d)] it is localized on very few atoms at the center of the chain. For a purely anharmonic chain ($S \to \infty$) we get a displacement pattern $(\ldots, 0.023, -0.523, 1, -0.523, 0.023, \ldots)$ which is very close to the analytic result $(\ldots, 0, -\frac{1}{2}, 1, -\frac{1}{2}, 0, \ldots)$ obtained for anharmonicity order $r \to \infty$. We have also studied the even parity localized modes ($\xi_n = -\xi_n$), obtaining very similar results for finite anharmonicity and the fully anharmonic pattern (..., 0, +0.166, −1, +1, −0.166, 0, ...) which is very close to the analytic solution (..., 0, +$\frac{1}{6}$, −1, 1, −$\frac{1}{6}$, 0, ...) obtained for anharmonicity order $r \to \infty$. If we now consider a local modification of the fourth order force constant $K_4$, the equations of motion to be solved in the odd parity case are the following,

$$\Omega^2 \xi_{-256} = (2\xi_{-256} - \xi_{-255}) + S\big[(\xi_{-256} - \xi_{-255})^3\big],$$
$$\vdots$$
$$\Omega^2 \xi_{-n} = (2\xi_{-n} - \xi_{-n+1} - \xi_{-n-1}) + S\big[(\xi_{-n} - \xi_{-n+1})^3 + (\xi_{-n} - \xi_{-n-1})^3\big],$$
$$\vdots$$
$$\Omega^2 \xi_{-1} = (2\xi_{-1} - \xi_0 - \xi_{-2}) + S(\xi_{-1} - \xi_{-2})^3 + S'(\xi_{-1} - \xi_0)^3,$$
$$\Omega^2 \xi_0 = (2\xi_0 - \xi_{-1} - \xi_{+1}) + S'(\xi_0 - \xi_{-1})^3 + S'(\xi_0 - \xi_{+1})^3, \qquad (7)$$
$$\Omega^2 \xi_{+1} = (2\xi_{+1} - \xi_0 - \xi_{+2}) + S(\xi_{+1} - \xi_{+2})^3 + S'(\xi_{+1} - \xi_0)^3,$$
$$\vdots$$
$$\Omega^2 \xi_{+n} = (2\xi_{+n} - \xi_{+n+1} - \xi_{+n-1}) + S\big[(\xi_{+n} - \xi_{+n+1})^3 + (\xi_{+n} - \xi_{+n-1})^3\big],$$
$$\vdots$$
$$\Omega^2 \xi_{+256} = (2\xi_{+256} - \xi_{+255}) + S\big[(\xi_{+256} - \xi_{+255})^3\big],$$

where $\Omega^2 = m\omega^2/K_2$ and the normalization condition chosen is $\xi_0 = 1$. The parameter $S' = (3K'_4/4K_2)A^2$ characterizes the inhomogeneity of the lattice. As $S'/S$ decreases, the intrinsic mode tends to develop a double peaked structure with maximum displacement at $n = \pm 2$, as shown in Fig. 2 for a chain of 512 atoms with global anharmonicity $S = 0.01$ and inhomogeneity $S' = -0.078$. Looking for even solutions, we got the same splitting tendency, but for different values of the parameter $S'$. This phenomenon is similar to the splitting of the soliton mode produced by a mass defect impurity found by Kivshar[3] in the acoustic limit.

We have also studied the existence of localized modes close to the end atom of the finite free ends chain. In this case the initial guess for the displacements is zero except for the last few atoms of the chain. For pure anharmonic interaction, we have found the existence of an even surface mode and of a surface mode similar to the odd symmetry modes, for a frequency $\Omega^2 = 9.6$ in the even case and $\Omega^2 = 6.91$ in the quasi-odd case. We have examined also the case of a chain with a weak harmonic interaction, such as $\frac{1}{10}$ of the quartic anharmonic interaction, and we have found that there is only a slight modification of the displacement patterns that does not destroy the symmetry found for the purely anharmonic case and an obvious shift in frequency.

## II.1 Quasi-Continuum Limit

We made also an analytical investigation in the quasi-continuum approximation, looking for solutions of the equations of motion of the form:

$$u_n(t) = \tfrac{1}{2}(-1)^n \Psi_n(t) e^{i\omega_m t} + \text{c.c.}, \qquad (8)$$

considering the following restrictions on the envelope $\Psi_n(t)$: $\Psi_n(t)$ is slowly varying in time i.e., $\ddot{\Psi}_n \ll \omega_m \dot{\Psi}_n$ and in the Taylor expansion of $\Psi_{n\pm 1}$ we take only terms up to $O(a)^4$.

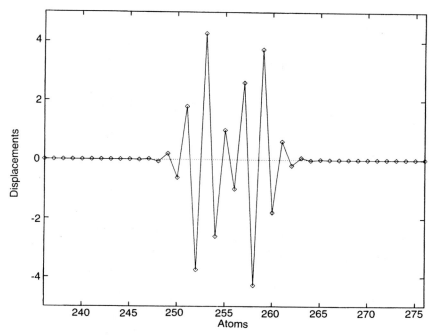

**Figure 2.** Atomic displacements of monoatomic chain with global anharmonicity parameter $S = 0.01$ and puntual anharmonicity defect $S' = -0.078$.

In this way we obtain the perturbed nonlinear Schrödinger equation (NLS):

$$2i\omega_m \Psi_t + \frac{a^2 K_2}{m} \Psi_{xx} + \frac{12 K_4}{m} |\Psi|^2 \Psi = a \frac{12(K_4 - K_{40})}{m} |\Psi|^2 \Psi \delta(x). \tag{9}$$

In the homogeneous case we get

$$\Psi(x,t) = A \frac{e^{i\Omega t}}{\cosh(Bx)}, \tag{10}$$

where

$$\Omega = \frac{3}{4} \omega_m \frac{K_4}{K_2} A^2, \qquad B^2 = 6 \frac{K_4}{K_2} \frac{A^2}{a^2}. \tag{11}$$

To obtain the two different parity we insert a phase in the argument of cosh which is related to the position of the maximum of the displacement $u_n$. Following this procedure is possible to compare the numerical results with those obtained analytically, obtaining a very good agreement between analytical and numerical results in the low anharmonicity region.

In the inhomogeneous case solving the perturbed NLS equation we find the following expression for the envelope function

$$\Psi(x,t) = \frac{A e^{i\Omega t}}{\cosh[B(|x| - x_0)]}, \tag{12}$$

that satisfies the condition

$$\sinh(2Bx_0) = \frac{12(K_4 - K_{40})}{aBK_2} |A|^2, \tag{13}$$

due to the inhomogeneity of force field. In this case the position of the maximum of the envelope function is pinned by the value of $x_0$. Two types of solutions are occurring: if $x_0 > 0$, i.e., $K_4 > K_{40}$, we obtain a double peak envelope, while if $x_0 < 0$, i.e. $K_4 < K_{40}$, the solution is a single peak one. The analytical solutions are very well reproduced by the numerical calculations performed in the low anharmonicity regime.[5]

In particular the position of the maxima of the two peaks displacements envelope obtained numerically is in very good agreement with the analytical prediction of $x_0$: in the case $S = 0.01$ and $S' = -0.078$, analytically we get $x_0 = 3.2a$, while numerically the maximum is on the site $n = 3$.

## III. Diatomic Chains

Our analysis of the diatomic chain was addressed to prove the existence of stationary localized gap and surface modes in presence of cubic and quartic anharmonicity. We seek for a stationary solution of the nonlinear equation of motion for each lattice site composed by a vibrational term $\xi$ of maximum amplitude $A$ and frequency $\omega$ and a static displacement $\varphi$ due to the cubic anharmonicity,

$$u_n = A[\xi_n \cos(\omega t) + \varphi_n]. \tag{14}$$

The total potential energy $\Phi$ in the RWA scheme gives rise to the following effective Hamiltonian

$$H = \sum_{i=-N}^{N} \Big\{ \tfrac{1}{2}\omega^2 A^2 m_i \xi_i^2 \sin^2(\omega t) + \Big[\tfrac{1}{2}K_2 A^2 (\xi_i - \xi_{i-1})^2$$
$$+ K_3 A^3 (\xi_i - \xi_{i-1})^2 (\varphi_i - \varphi_{i-1}) + \tfrac{3}{8}K_4 A^4 (\xi_i - \xi_{i-1})^4$$
$$+ \tfrac{3}{2}K_4 A^4 (\xi_i - \xi_{i-1})^2 (\varphi_i - \varphi_{i-1})^2 \Big] \cos^2(\omega t)$$
$$+ \Big[ K_2 A^2 (\xi_i - \xi_{i-1})(\varphi_i - \varphi_{i-1}) + \tfrac{1}{2}K_3 A^3 (\xi_i - \xi_{i-1})^3$$
$$+ K_3 A^3 (\xi_i - \xi_{i-1})(\varphi_i - \varphi_{i-1})^2 + \tfrac{3}{2}K_4 A^4 (\xi_i - \xi_{i-1})^3 (\varphi_i - \varphi_{i-1})$$
$$+ K_4 A^4 (\xi_i - \xi_{i-1})(\varphi_i - \varphi_{i-1})^3 \Big] \cos(\omega t) \Big\}, \tag{15}$$

where $K_3$ is the cubic anharmonic first nearest interaction force constant. The average energy evaluated using Eq. (15) fulfill the virial theorem

$$2\langle T \rangle = \langle PV \rangle + \Big\langle \sum_{i=-N}^{N} (u_i - u_{i-1}) \frac{\partial \Phi}{\partial (u_i - u_{i-1})} \Big\rangle, \tag{16}$$

where the pressure $P$, induced by the cubic anharmonicity, is evaluated as follows:

$$\langle PV \rangle = -\Big\langle \sum_{i=-N}^{N} \frac{\partial \Phi}{\partial (\varphi_i - \varphi_{i-1})} (\varphi_i - \varphi_{i-1}) \Big\rangle. \tag{17}$$

In the diatomic chain with harmonic and quartic potential, obtained via a Taylor expansion of a Born Mayer potential for KBr (mass ratio 1:2) and LiI (mass ratio 1:18), we found the localized modes shown in Figs. 3 and 4. In particular we found two surface modes: $S_1$ lying in the gap between acoustic and optical harmonic modes, arising from zone border, which is reminiscent of the pure harmonic surface mode, $S_2$ at a

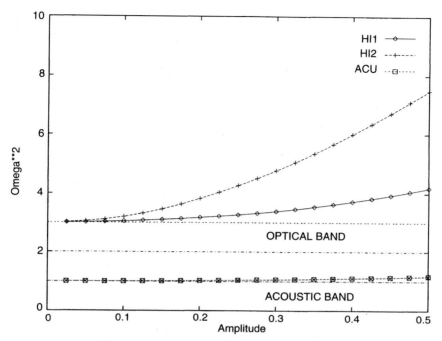

**Figure 3.** Frequency $\omega_0^2$ of the localized modes HI1, HI2, and ACU of a diatomic chain with mass ratio 1:2 *versus* amplitude $A$.

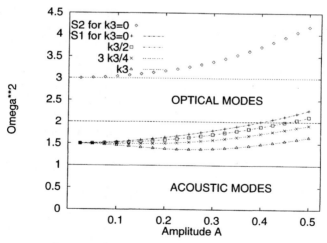

**Figure 4.** Frequency $\omega_0^2$ of the surface modes $S_1$ and $S_2$ of a diatomic chain with mass ratio 1:2 *versus* amplitude $A$. The mode $S_1$ is drawn for increasing values of the cubic anharmonicity force constant $k_3$.

frequency greater than the maximum harmonic, originating from the $\Gamma$ point, entirely due to the anharmonicity, nearly degenerate with the strongly localized mode of odd symmetry. Depending on the mass ratio, the mode $S_1$ at larger amplitudes becomes a

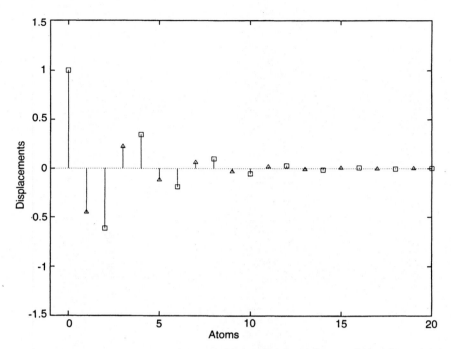

**Figure 5.** Atomic displacements of the surface mode $S_1$ without cubic anharmonicity for a diatomic chain with mass ratio 1:2. The light masses are drawn as triangles, the heavy masses as squares.

resonant mode in the optical continuum, as shown in Fig. 4. These surface modes can be significantly modified by introducing the cubic anharmonicity which produces a lattice parameter deformation because of the pressure induced at the surface and the frequency strongly decreases. The $S_1$ mode at increasing $k_3$ does not enter the optical continuum, while the $S_2$ mode disappear, for any mass ratio. The displacement pattern of the $S_1$ mode is characterized by the motion of two neighboring atoms in the same direction, while in the $S_2$ mode the atoms move out of phase, as shown in Fig. 5 and Fig. 6. Aside these surface modes, there are acoustic gap modes arising from the zone border, in the $K_2 - K_4$ model of even (ACU1) and odd (ACU2) symmetry, localized on light or heavy atoms at the center of the chain, drawn in Figs. 7 and 8.

These modes are characterized by the motion in the same direction of couple of neighboring atoms. They are very sensitive to the cubic anharmonicity: The introduction of the cubic interaction $k_3$ give rise to optical gap modes of even (OT2) and odd (OT1) parity, shown in Figs. 9 and 10, while the acoustic gap modes disappear. The displacement pattern of the optical gap modes is similar to the one of acoustic gap modes, but the odd and even modes are centered on light and heavy atom respectively. The main characteristic of these gap optical modes is that their frequency initially decreases with the increasing amplitude, but for larger amplitudes raises again and enters the optical harmonic continuum, *i.e.*, the mode becomes a broad resonance. This behavior confirms that higher anharmonic terms are required to describe correctly the system for large amplitudes, because the full potential analysis of the same system of Kiselev *et al.*,[4,7] does not show this effect. In the small amplitude region we fully agree with Kiselev results.

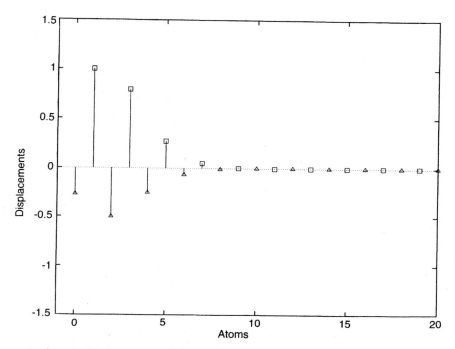

**Figure 6.** Atomic displacements of the surface mode $S_2$, as in Fig. 5.

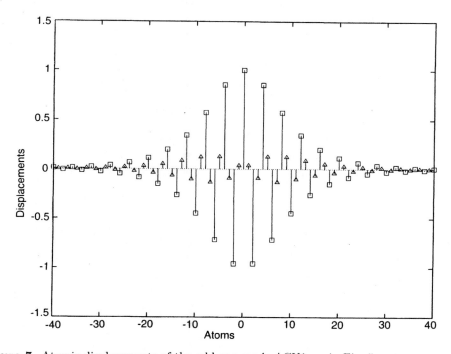

**Figure 7.** Atomic displacements of the odd gap mode ACU1, as in Fig. 5.

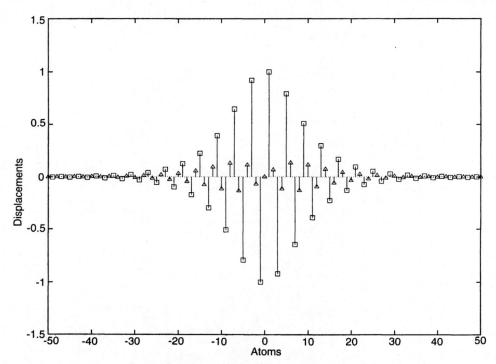

**Figure 8.** Atomic displacements of the even gap mode ACU2, as in Fig. 5.

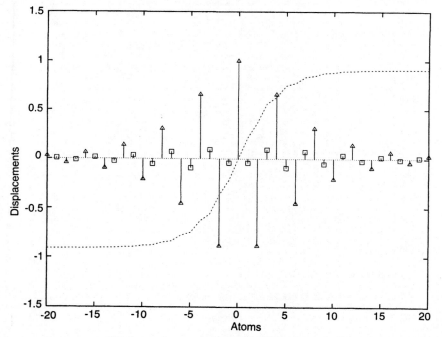

**Figure 9.** Atomic displacements of the odd gap mode OT1 with cubic anharmonicity. Triangles and squares indicate the light and heavy masses, respectively. The dotted line refers to the static displacements $\varphi_n$.

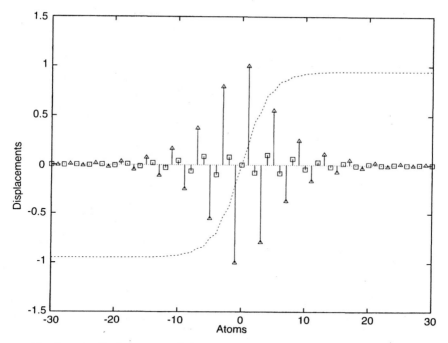

**Figure 10.** Atomic displacements of the even gap mode OT2 with cubic anharmonicity, as in Fig. 9.

Looking to the frequencies greater than $\omega_m$ we have found also odd high frequency ILM modes centered on the light atom (HI1) or on the heavy atom (HI2), which originate both from the $\Gamma$ point. The frequency of this mode HI2 is nearly degenerate with the $S_2$ surface mode. The increasing of the amplitude $A$ produces a narrowing of the envelope shape of these modes and an increase of their frequency, that becomes more evident increasing the mass difference. With increasing cubic anharmonicity the HI1 and HI2 frequencies tend to merge into the optical harmonic continuum.

## IV. Conclusions

In conclusion we have shown a strong evidence of the existence of localized modes at least for the one-dimensional lattices. However, the evidence is so strong that we think that this modes should be observed on real crystals as stepped surfaces with high Miller indices or rows of atoms deposited on surfaces.

## References

1. A. J. Sievers and S. Takeno, *Phys. Rev. B* **39**, 3037 (1989).
2. J. B. Page, *Phys. Rev. B* **41**, 7835 (1990).
3. Y. S. Kivshar, *Phys. Lett. A* **161**, 80 (1991).
4. S. A. Kiselev, S. R. Bickham, and A. J. Sievers, *Phys. Rev. B* **50**, 9135 (1993).
5. R. F. Wallis, A. Franchini, and V. Bortolani, *Phys. Rev. B* **50**, 9851 (1994).

6. A. Franchini, V. Bortolani, F. Corsini, and R. F. Wallis, *Nuovo Cimento*, to be published.
7. S. R. Bickham, S. A. Kiselev, and A. J. Sievers, *Phys. Rev. B* **47**, 14 206 (1993).

# Molecular Orientation Dependence of Dynamical Processes on Metal Surfaces: Dissociative Adsorption and Scattering, and Associative Desorption of Hydrogen

Ayao Okiji,[1,*] Hideaki Kasai,[2] and Wilson Agerico Diño[1]

[1] Department of Applied Physics
Osaka University
Suita, Osaka 565
JAPAN

[2] Department of Material and Life Science
Osaka University
Suita, Osaka 565
JAPAN

## Abstract

The effect of molecular rotational motion on such dynamical processes as dissociative adsorption and associative desorption of hydrogen molecules on metal surfaces is presented in the first half. It is shown that two effects, *viz.*, *steering* effect, and *R-T (rotational-translational) energy transfer* effect, which originate from the molecular rotational degrees of freedom, give rise to a non-monotonous behavior of the adsorption probability of $H_2(D_2)/Cu(111)$ as a function of its initial rotational state when it undergoes dissociative adsorption. It is also shown that a very prominent initial cooling, a mild heating, and then a final cooling of the rotational temperature with respect to the surface temperature, for increasing rotational energies, is observed for the desorption of $H_2(D_2)$ from $Cu(111)$. In the second half, the dissociative scattering of hydrogen molecules from metal surfaces is used as a representative dynamical process occurring on surfaces in the hyperthermal beam energy region to show how molecular orientation dependence manifests itself.

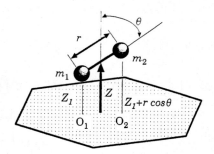

**Figure 1.** The model system showing a diatomic molecule (molecular orientation with respect to the surface normal given by $\theta$, surface to center of the mass (CM) distance $Z$, and bond-length $r$) approaching a flat surface perpendicularly. $m_1$ and $m_2$ correspond to the masses of the constituent atoms of the diatomic molecule, located at distance $Z_1$ and $Z_1 + r\cos\theta$ from the surface, respectively. $O_1$ and $O_2$ are the surface projection of the two constituent atoms.

## I. Dissociative Adsorption and Associative Desorption

Recently, time of flight (TOF) distributions for hydrogen molecules associatively desorbing from Cu(111) could be determined rotationally state resolved.[1,2] The distributions show a non-monotonous behavior with respect to the detected rotational state. Molecules with intermediate rotational states (e.g., $j = 4, 5, 6$) reach the detector faster than those that do not rotate at all ($j = 0$) or those in the higher rotational states ($j > 10$). By invoking detailed balance,[3] the dissociation behavior of hydrogen on Cu can be derived from these distributions. The non-monotonous behavior of the TOF peaks can be related to a non-monotonous sticking (dissociation) probability. At low initial rotational states $j$, rotation inhibits sticking. While at high $j$, rotation promotes sticking. When an impinging hydrogen molecule approaches a Cu surface (Fig. 1), it encounters an orientation dependent potential energy (hyper-)surface (PES).[4-6] In order for the molecule to dissociate and be adsorbed on the Cu surface, it must be able to find the *path of least resistance* (path of the least potential), and have enough energy to reach the surface. This process depends on what the initial rotational- and vibrational-states of the impinging hydrogen molecule are, what its incidence energy is, and how long it stays under the influence of the anisotropic PES.[7-9] In the case of rotation, there are two opposing factors working against and for the dissociation process, viz., steering (dynamical reorientation) and R-T (rotational-translational) energy transfer. The first, steering, which originates from the orientational dependence of the dissociative adsorption process, pertains to a dynamical reorientation of the impinging molecule towards a more favorable orientation (a predominantly parallel orientation). Steering dominates when the impinging molecule does not have sufficient rotational energy to assist in its adsorption. The second, R-T energy transfer, which originates from the strong coupling between the rotational and translational degrees of freedom, pertains to the effective transfer of rotational energy to translational energy. R-T energy transfer dominates when the impinging hydrogen molecule has sufficient rotational energy to assist in its adsorption. The combined effect of these two factors is an initial decrease and then, eventually, an increase in the sticking probability as a function of the initial rotational state of the

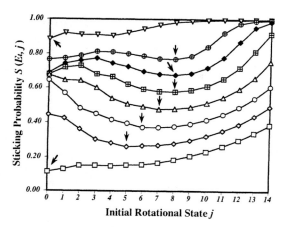

**Figure 2.** Numerical results for the $j$-dependent sticking probability curves for $D_2$ in the vibrational ground state and fixed incidence energies, $E_t$. Arrows point to the corresponding minima for each curve. (squares: $E_t = 0.55$ eV; diamonds: $E_t = 0.575$ eV; circles: $E_t = 0.60$ eV; up triangles: $E_t = 0.625$ eV, filled full squares: $E_t = 0.65$ eV; filled full diamonds: $E_t = 0.675$ eV, crossed circles: $E_t = 0.70$ eV; and down triangles: $E_t = 0.80$ eV).

impinging molecule for a fixed incidence energy (cf., for example, curve corresponding to 0.6 eV in Fig. 2).

To show the significance of the incidence energy $E_t$ in determining which of the two factors (steering or R-T energy transfer) dominates for a fixed initial rotational state $j$ of the impinging molecule, we performed quantum mechanical calculations by the coupled channel method and considered the reaction of a $D_2$ molecule impinging a flat Cu(111) at normal incidence (Fig. 1).[7,9] The $D_2$ molecule is restricted to its vibrational ground state ($\nu = 0$) throughout the adsorption (desorption) process. Our orientationally anisotropic model potential is based on qualitative features of available PES plots for $H_2(D_2)$/Cu-surface systems.[4–9] The energy barrier for a perpendicular-oriented $D_2$ molecule was set at $V_{max} = V_\perp \approx 0.9$ eV, and gradually decreases to a value $V_{min} = V_\parallel \approx 0.5$ eV for parallel-oriented molecules. We can see in Fig. 2 that the location of the minimum for each sticking probability curves shifts as the incidence energy is varied. This is because, for incidence energies comparable to the minimum energy barrier, $V_{min}$, steering will not be effective. Thus, R-T energy transfer will be dominant and we see only an increase in the sticking probability as the initial rotational state is increased for low incidence energies (cf., curve corresponding to 0.55 eV in Fig. 2). As the incidence energy is gradually increased, the efficacy of steering also increases and we see corresponding minima appearing (cf., curves corresponding to $0.575 \sim 0.70$ eV in Fig. 2), that shift towards higher initial rotational states. As the incidence energy is increased to a value that becomes comparable with the energy barrier maximum, $V_{max}$, the efficacy of R-T energy transfer increases and we see a corresponding shift in the curve minimum towards lower initial rotational states.

When the adsorption probability of molecules colliding with a surface is independent of the distribution of molecular internal states, orientations, and velocities, equilibrium statistical mechanics predicts that the molecular quantum state distributions in desorption will be determined solely by the surface temperature $T_s$. However,

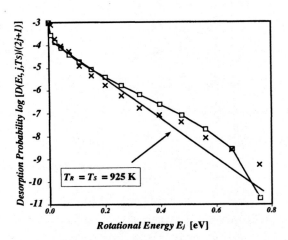

**Figure 3.** Boltzmann plot of desorption probability vs. rotational energy $E_j$ of the state $j$. The line $T_R = T_S = 925$ K is plotted for reference. The crosses correspond to experimental data (Ref. 1).

this is often not the case, as we show here for hydrogen on Cu. Thus it would also be interesting to study how the desorption probabilities behave, as such studies could elucidate the nature of those special forces and configurations experienced by the desorption flux when we relate them to the adsorption probabilities. We performed quantum mechanical calculations for the desorption probability by the coupled channel method. In figure 3, a Boltzmann plot of our numerical results for the desorption probability of $D_2$ molecules as a function of the rotational energy is presented. A Boltzmann distribution would appear as a straight line. We see that the desorption probability is not represented by a single temperature, and the mean rotational energy is less than $T_s$. These qualitative features are also observed experimentally.[1,2,10]

In order to relate the desorption results in Fig. 3 to the adsorption probability results in Fig. 2 we show, in Fig. 4, a Boltzmann plot of our numerical results for the desorption probability of $D_2$ molecules as a function of the rotational energy for a fixed translational energy. Recall from the principle of microscopic reversibility, and energy conservation,[3] that the dynamic behavior of the adsorption probability will be reflected in the distribution of molecular quantum states in desorption in the following manner

$$D(E_t, j) \propto S(E_t, j) \exp\left(-\frac{E_t + E_j}{k_B T_s}\right), \qquad (1)$$

where $E_t$ is the translational energy of the molecule, and $j$ is its rotational state with a corresponding rotational energy $E_j$. It is noted that the numerical results by the coupled channel method for the desorption probability and those for the adsorption probability maintain this relation. The initial decrease in the sticking probability curve corresponding to an incidence energy of $E_t = 0.6$ eV in Fig. 2 is reflected as a decrease in the rotational temperature in desorption (Fig. 4). The final increase in the sticking probability curve is reflected as an increase in the rotational temperature relative to the former decrease. One notices that the increase in rotational temperature is not as dramatic as that expected from the corresponding sticking probability

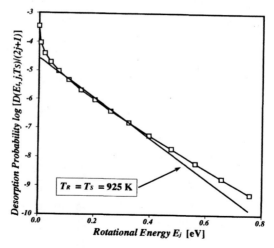

**Figure 4.** Boltzmann plot of the desorption probability *vs.* rotational energy $E_j$ of the state $j(\nu = 0)$, for fixed translational energy $E_t = 0.60$ eV. The line $T_R = T_S = 925$ K is plotted for reference.

curve in Fig. 2. This is because the Boltzmann factor in Eq. (1) decreases much more rapidly as compared to the increase in the sticking probability. If we then consider the distributions from all the sticking probabilities corresponding to all incidence energies, which is equivalent to integrating Eq. (1) with respect to the molecular translational energy $E_t$, we can rationalize the desorption results shown in Fig. 3. The initial decrease in the rotational temperature with respect to the surface temperature for low rotational energies, $E_j \leq 0.05$ eV, is due to an initial decrease in the sticking probability for low initial rotational states, and the mild increase in the rotational temperature for higher rotational energies, $0.05 \text{ eV} \leq E_j \leq 0.5$ eV, is due to an increase in the sticking probability for higher rotational states. The final decrease again of the rotational temperature can be understood by considering again the relation between sticking probability and the desorption probability [Eq. (1)]. As the rotational energy $E_j$ appearing in the Boltzmann factor increases, the only relevant contributions will come from sticking probabilities corresponding to those incidence energies $E_t \ll V_{\min}$. In this energy region, the sticking probabilities are not much different from 0, even for $j = 14$. As a result, we will observe this final decrease in the rotational temperature for the desorption probability of hydrogen molecules in their vibrational ground state $(\nu = 0)$.

## II. Dissociative Scattering

Now we consider the dissociative scattering of a $H_2$ from a metal surface within the framework of the system shown in Fig. 1. In the range of the translational energies involved (a few hundred eV) in the experiments,[11,12] the velocity of the molecular center-of-mass (CM) motion is comparable to the Fermi velocity as an order of magnitude, and much faster than molecular rotation. Then, the coupling between the molecular CM motion and degrees of freedom of metal electron system should be taken into account explicitly. Furthermore, it is assumed in the following calculation that the PES shape loses its effect on the CM motion and thus the surface fills its role

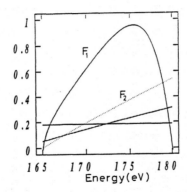

**Figure 5.** Dissociation probability as a function of the final energy of the center of mass motion. The curve $F_1$ shows the orientation ($\theta$) independent part of the transition probability $\omega_{if}$, and, the curves $F_2$ show the corresponding orientation ($\theta$) dependent part (dots: $\theta = 90°$, dash-dot: $\theta = 70°$, solid: $\theta = 10°$). Used parameter values for the molecule are roughly related to those of the molecule at equilibrium interatomic distance, and $\varepsilon_{\rm CM} = 200$ eV, the energy for CM motion of the impinging molecule, and $\varepsilon_F = -5$ eV (Fermi energy), $D = 20$ eV (occupied band width) for the metal substrate, and $T_s = 500$ K (surface temperature). The vacuum level is chosen as the origin of energy.

as a hard wall to the CM motion, and that the orientation of the molecule is fixed throughout the scattering process.

The simple model described above is then used to investigate the dissociation of a $H_2$ 'scattering from a metal surface into two neutral H atoms. It is assumed that dissociation is triggered/induced by a two-step process involving the excitation of an electron from the single bonding state, $^1\Sigma_g$, of the impinging $H_2$ to the triplet antibonding state, $^3\Sigma_u$, during the scattering process.[13] Initially, (1) one of the electrons in the filled bonding orbital of the impinging molecule, $^1\Sigma_g$, is excited into a state above the surface Fermi level, after which (2) an electron from the energy band of the metal substrate is excited into the antibonding state of the molecule [or (2) and then (1)], with the impinging molecule finally ending up in the triplet state, $^3\Sigma_u$.

In figure 5, the corresponding orientation-independent and orientation-dependent parts of the dissociation probability are shown as functions of the final translational energy of the impinging molecules. The orientation independent part of the dissociation probability peaks can be seen at around 175 eV of the final translational energy (cf., curve $F_1$). And from the orientation dependent part, a parallel (90°) orientation preference for dissociation can be seen at around 175 eV (cf., curve $F_2$). Shown in the upper region of Fig. 6 are the corresponding total dissociation probabilities as a function of the final energy of the CM motion after scattering from the metal surface, for the case where the final product are neutral atoms. In the calculations, the impinging neutral molecules are considered to have an initial orientation of 90°, 70°, and 10° with respect to the surface normal. One can see here a strong orientation dependence of the total dissociation probability (peak at around 175 eV), with the parallel orientation (90°) preferred. It can be also seen here that the orientation independent part of the dissociation probability determines where, along the final energy of the CM motion of the impinging molecules, a dissociation peak can be observed. On the other hand, the orientation dependent part of the dissociation probability determines which initial orientation yields the highest dissociation peak.

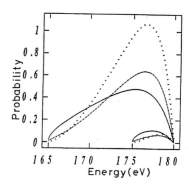

**Figure 6.** Comparison between the dissociation probability curves as a function of the final energy of the CM motion of the molecule scattered from (upper region) a metal surface with $D = 20$ eV, and (lower right-hand region) another metal surface with $D = 10$ eV. The other parameters used in the calculation are the same as those in Fig. 5, (dots: $\theta = 90°$, dash-dot: $\theta = 70°$, solid: $\theta = 10°$).

Upon changing the characteristics of the substrate, by changing the position of the bottom of the substrate energy band $-D$ from $-20$ eV to $-10$ eV relative to the vacuum level (*i.e.*, the value of the filled-band width, $\varepsilon_F + D$, is changed from 15 eV to 5 eV), the behavior of the orientation independent and the orientation dependent part of the dissociation probability is affected considerably. We observed a remarkable decrease in yield, a shift of the dissociation peak towards the higher final translational energies in the orientation independent part, and a near perpendicular (10°) orientation preference for dissociation in the orientation dependent part. As a result, besides a marked decrease in yield, two additional things happen to the total dissociation probability. First, there is a general shift of the dissociation peaks towards the higher final translational energy region. And second, there is a change in orientation preference for dissociation, from parallel orientation (90°, when $D = 20$ eV) to near perpendicular (10°, when $D = 10$ eV).

The same calculation was done for the case when the final products are ions, $H^+$ and $H^-$. In this case, it is assumed that the process is triggered/induced by a two-step process involving the excitation of an electron from the singlet bonding state, $^1\Sigma_g$, of the impinging $H_2$ to a singlet antibonding state, $^1\Sigma_u$, during the scattering process.[13] Except for a decrease in the magnitude of the dissociation yield, the same strong orientation dependence of the dissociation probability is observed as that shown in the upper region of Fig. 6. When the characteristics of the substrate are changed (in the manner mentioned above), the same change is observed in the orientation preference as that shown in Fig. 6.

# References

* Present address: Wakayama National College of Technology, Gobo, Wakayama 644, Japan.
1. H. A. Michelsen, C. T. Rettner, D. J. Auerbach, and R. N. Zare, *J. Chem. Phys.* **98**, 8294 (1993).
2. C. T. Rettner, H. A. Michelsen, and D. J. Auerbach, *J. Chem. Phys.* **102**,

4625 (1995).
3. H. Kasai and A. Okiji, *Prog. Surf. Sci.* **44**, 101 (1993).
4. K. Tanada, M. E. Thesis, Osaka University, 1993.
5. B. Hammer, M. Scheffler, K. W. Jacobsen, and J. K. Nørskov, *Phys. Rev. Lett.* **73**, 1400 (1994).
6. J. A. White, D. M. Bird, M. C. Payne, and I. Stich, *Phys. Rev. Lett.* **73**, 1404 (1994).
7. W. A. Diño, H. Kasai, and A. Okiji, *J. Phys. Soc. Jpn.* **64**, 2478 (1995).
8. W. A. Diño, H. Kasai, and A. Okiji, *Surf. Sci.* **363**, 52 (1996).
9. W. A. Diño, H. Kasai, and A. Okiji, *Phys. Rev. Lett.* **78**, 286 (1997).
10. G. D. Kubiak, G. O. Sitz, and R. N. Zare, *J. Chem. Phys.* **81**, 6397 (1984); **83**, 2538 (1985).
11. U. van Slooten, D. R. Anderson, A. W. Kleyn, E. A. Gislason, *Surf. Sci.* **274**, 1 (1992).
12. A. Nesbitt et al., *Chem. Phys.* **179**, 215 (1994).
13. H. Kasai, A. Okiji, and W. A. Diño, *Springer Series in Solid-State Sciences* **121**, 99 (1996).

# Gaps in the Spectra of Nonperiodic Systems

R. A. Barrio,[1] Gerardo G. Naumis,[1] and Chumin Wang[2]

[1] Instituto de Física
Universidad Nacional Autónoma de México
Apartado Postal 20-364, 01000 México, D. F.
MEXICO

[2] Insituto de Invegtigaciones en Materiales
Universidad Nacional Autónoma de México
Apartado Postal 70-360, 04510 México, D. F.
MEXICO

## Abstract

There are bipartite networks, in which the tight-binding spectrum for electrons presents a gap and localized states at the center of the band. This anomalous situation is analyzed by renormalizing one of the sublattices. Then, the states near the center of the band map into the lower band edge, and the localized states of interest become low-energy excitations. The existence of a gap between the ground state and the rest of the excitation spectra is then revealed to be due to frustration of perfectly coherent antibonding states. A detailed analysis of electrons in the random binary alloy in two dimensions and in the Penrose lattice is performed and compared with numerical calculations. Predictions from this theory agree with former numerical calculations and recent experimental findings in quasi-crystals. The same theoretical considerations might be applied to other cases, such as electron, phonon and magnetic excitation spectra in disordered solids.

When studying a physical system, one is frequently perturbing it by a small amount. Therefore, if the system is set to a ground state, the few first excited states, or low-energy excitations, are the ones that dictate the response of the system to the perturbation. There are cases in which the symmetries of the system predict the existence of excitations arbitrarily close to the ground state. In particular, if the ground state is extended and if the Hamiltonian presents a continuous symmetry that is broken by the ground state, then Goldstone's theorem[1] asserts that there must exist a continuum of low-energy excitations. However, the theorem does not predict when these excitations must be absent.

In this paper we shall be interested in studying nonperiodic systems using a simple tight-binding nearest-neighbor Hamiltonian for $s$-electrons. In some cases, namely when the ordered lattice that defines disorder is bipartite, gaps in the center of the energy spectra appear that are not expected, and are generally attributed to failures due to finite size effects in the computations.

We suggest that these gaps should be present also in the infinite nonperiodic system by introducing a notion of non-uniformity of the system when considering Hamiltonian interactions. We demonstrate this by using a mapping that takes the states near the gap to the lowest eigenvalues of the mapped spectra, or low-energy excitations. Then, it becomes clear that if a system is non-uniform, then one can assure that *there are no* low-energy excitations. Then, there must be a gap separating the ground state from the rest of the spectrum of excitations.

The key ingredient is the non-uniformity of the system and we shall start by defining this concept. In general, the first excited states correspond to totally coherent wave functions throughout the system, with very little spatial variations of their amplitude.[2] Therefore, these states are truly extended. Suppose that for some reason, an excited state lowers its energy by varying its amplitude in different regions of the system. This results in a kind of localization of the wave function because the amplitude is substantial only in certain preferred regions of the space. In this situation one could say that the system itself is non-uniform, because one can identify the regions where the wave function concentrates its amplitude. Notice that the system could be homogeneous and yet be non-uniform in the sense just explained. The important phenomenon that defines the non-uniformity is that the excitation reduces its eigen-energy by localizing its wave function.

Non-uniform systems can be found in many totally different situations, ranging from the very obvious ones, as in the case of an admixture of two very different materials, that tend to segregate, to totally unexpected ones. As a good example, we will treat one of the latter systems in detail, that illustrates the subtlety and importance of this concept. This is the case of electronic excitations described by a tight-binding nearest-neighbor Hamiltonian in the vertex problem of a Penrose lattice, (PL).[3]

The electronic spectrum of the PL,[4] shows a $\delta$-function of degenerate states at the Fermi level (at the center of the spectrum), separated from the rest of the states by a sizeable gap of $\sim 0.107$, in units of the hopping parameter ($t$). The wave functions of the states near the center of the spectrum are localized, as opposed to the case of Anderson localization,[5] where the localization, due to disorder, is stronger near the band edges.

The reason for this unexpected features is better analyzed if one notes that the PL is bipartite,[6] so it can be divided in two alternating sublattices. Any given site on one sublattice has all of its neighbors on the other sublattice. Therefore, one of the sublattices can be renormalized[7] and a new lattice ($H2$) is defined [see Figs. 1(a) and 1(b)]. If the Hamiltonian in the original lattice is

$$H = \sum_i \epsilon_i |i\rangle\langle i| + t \sum_{\langle i,j \rangle} |i\rangle\langle j|,$$

with eigenvalues $E$, the corresponding Hamiltonian in $H2$ ($H^2$) has an electronic band which is a function of $E^2$. Then the center of the spectrum in $H$ is found to be the

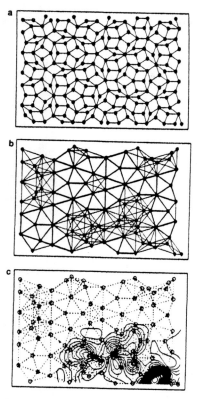

**Figure 1.** (a) Selected section of a large Penrose lattice. The atoms are at the vertices of rhombuses. Notice that the system is homogeneous. (b) The same portion shown in (a), where one of the alternating sublattices has been renormalized. The interactions defined by the renormalized Hamiltonian are represented by lines. The system ($H2$) becomes non-uniform, since there are regions with triangular cells (thick lines) and other regions with second-neighbor interactions. (c) Contour map of the electronic charge for a localized state near the band gap. The amplitude of the wave function is only appreciable in one region without triangles.

minimum eigenvalue of $H^2$ ($E^2 = 0$ if the self-energy $\epsilon_i = 0$) and both ends of the spectrum of original PL coincide to be the highest eigenvalue of $H^2$.

Furthermore, in $H2$ the degenerate states corresponding to $E = 0$ present an antibonding (i.e., the phase difference on nearest-neighbors is $\pi$) in confined regions of the lattice. In a uniform lattice these states should be perfectly coherent and extended. However, $H2$ contains odd rings [see Fig. 1(b)], and the, antibonding wave function is frustated. Thus, it is expected that by varying its amplitude to favor regions of lower frustration, the system can reach a state of lower energy (nearer to $E = 0$).

This is indeed the case, through numerical calculations[8] have shown that the states near the gap are confined around sites with five-fold symmetry. We claim that this modulation of the amplitude arises because the regions of frustration are not uniform throughout the lattice. In figure 1(b) one notices two mutually exclusive regions in $H2$, one with triangles, and the other with cells presenting first and second neighbor interactions. The wave function facing frustration tends to be localized to avoid the region with triangles. This is shown in Fig. 1(c), where the contours of constant charge for the state at $E = 0.18t$ are concentrated around the region without triangles. Note that this particular state is truly localized staying in only one region, since its energy is very close to the gap edge. If it were critical, it should present amplitude in all regions without triangles. The limit for having extended state predicted by a calculation[4] assuming that the amplitude is constant throughout the lattice gives $E = 0.91t$, where one expects a mobility edge. Optical conductivity experiments in real icosahedral quasi-crystals[9] reveals that the conductivity is thermally activated and the measurements can be quantitatively explained by assuming the existence of a pseudo gap and of a mobility edge.

One realizes immediately that all the conditions defining a non-uniform system are met in this case, and it remains to investigate the functional dependence of the eigenvalues on the localization length. The reason for this dependence being the frustration in $H2$.

In the case of constant amplitude, the change in energy due to frustation $(E_f)$ is[10] $E_f = 2tN_f/N$, where $N_f$ is the total number of frustrated bonds, and $N$ is the total number of bonds. If the wave function is localized in an area of radius $\lambda$, $E_f$ changes to

$$E_f = 2t(\pi\lambda^2)\rho_f. \qquad (1)$$

Here $\rho_f$ is the mean number of frustrated bonds per unit area. In fact, the wave function localizes itself in regions where $\rho_f$ is a minimum, since $E_f$ acts as a repulsive potential in zones of high frustration. Then, there must be an eigenstate $E \leq E_f$, with wave function localized in zones of lower frustration.

This behavior can be confirmed by looking at the inverse participation ratio (IPR),[11] that might be interpreted as the inverse of the area occupied by the wave function $(\pi\lambda^2)$, if one defines the dimensionless localization length $\lambda$ properly. Using Eq. (1) one gets, IPR $\sim 2t\rho_f/E$. Indeed, numerical calculations in large PL's show[4] an IPR that could be adjusted by a $1/E$ law with an error of 6.55%. Adding a term of the form $1/E^2$ does not give a much better approximation (the error is 6.3%). Therefore, the gain in energy by localizing the wave function of the excitations is $E = A\lambda^2$ where $A = 0.0021$, in units of $t$.

These localized states form exponential tails in the density of states,[12] as expected. Indeed, this can be demonstrated calculating

$$\frac{\partial G_{ii}(E)}{\partial E} = -G_{ii}^2(E),$$

where $G$ is the Green's function for electrons. Assuming that the wave function is normalizable, and that the energy is far from a rapidly-varying potential region, one obtains

$$\frac{d\omega(E)}{dE} = -\frac{2\Omega(d)\lambda(E)^d F}{E - E_0}\omega(E), \qquad (2)$$

where $\omega(E)$ is the density of states, $\Omega$ is the solid angle in $d$ dimensions and $F/(E-E_0)$ is an approximate expression for the real part of the Green's function away from a pole $E_0$. For simplicity we have assumed that the off-diagonal elements of the Green's function $(G_{ii})$ decay exponentially with exponent $\lambda$, with respect to the distance between sites $(|r_i - r_j|)$. It is seen that Eq. (2) gives the correct form of the density of states. Numerical calculations in the PL agree well with the prediction of Eq. (2) for the integrated density as well.[12]

This argument alone implies that $E$ could be arbitrarily near zero, since the wave function can be localized in a sufficiently small area. However, it is known that the minimum eigenvalue of a localized electron ($E_c$) is inversely proportional to the square of the length of confinement, then $E_c = C/\lambda^2$, where $C$ depends on the geometry of the confinement well and it is of the order of unity. Therefore, there must be an additional term in the dependence of $E$ on $\lambda$ due to confinement, and one should write $E = A\lambda^2 + C/\lambda^2$.

Note that there is a competition between frustration, which favors localization, and quantum confinement that puts a price when localizing the wave function. This competition gives a minimum eigenvalue ($\Delta$) obtained by taking the derivative of the energy with respect to $\lambda$ and equating it to zero,

$$\Delta = 2\sqrt{A} = 0.092t.$$

Numerical calculations[4,8] give a gap of 0.107, very close to the above value. It is worth pointing out that real quasi-crystals are alloys that present a pseudogap at the Fermi level.[13] The same calculation gives $\lambda_{min} = 4.67$, if the bond length is one. Notice that this value is very close to the distance between five-fold sites in the PL, which is the cube of the golden section. This further corroborates the fact that the wave function localizes around five-fold sites.[8]

It is worth remarking that the above explanation of the appearance of a gap in the center of the spectrum of a PL is not exclusive to this system. There are other physical systems in which similar gaps appear and one can demonstrate that the concept of non-uniformity also allows a sound explanation. For instance, consider a binary alloy of two very different materials where segregation is expected. This system is inhomogeneous and also non-uniform, since one could envisage the situation in which a given low-energy excitation (phonon, magnon, or electron) finds energetically favorable to be localized in one of the materials.

More specifically, consider a random binary alloy (RBA) with diagonal disorder in the split band regime. In this system there is a large difference between the self-energy of the two kinds of atoms (A and B). If the concentration of the component with smaller self-energy (A say) is below the percolation threshold, it is observed[14] that the electronic spectrum of subband A, in square and cubic lattices, presents a $\delta$-function in the middle and gaps separating the rest of the excitations from the center, analogous to the situation in the PL.

The explanation of this anomalous situation has been elusive, or not really convincing for a long time. Although the situation in the RBA is different from the PL, they share the same property of being non-uniform with respect to the excitations. The very large self-energy of B atoms acts as a barrier potential which provides a mechanism to localize electrons in regions with A atoms. Since the alloy is random, below the percolation limit one can assure the existence of clusters of A atoms of all sizes in which the wave function has larger amplitude.

Moreover, the analogy with the PL is even more profound that just the aspect of the excitation spectrum, one could translate the analysis of the Penrose spectrum by saying that the role of the frustrated bonds is played by the B atoms, with large self-energy. Therefore, there exists a decreasing function $E(l/\lambda)$, which ultimately competes with the increasing function due to quantum confinement, giving rise to the existence of the gap. One should point out that the RBA are defined in cubic or square lattices, so the $H2$ transformation is still valid. The $\delta$-function in the center of the spectrum arises from local configurations of A atoms surrounded, either by B atoms, or by sites in which the coherent wave function $(+0-0+0-\ldots)$ has nodes (0) in A atoms. These local arrangements that fulfill the requirement $E = 0 = \sum \psi_{\text{neigh}}$ are far apart, so the state at $E = 0$ is highly degenerate. Above the percolation limit, there is certainly an infinite cluster of A atoms and then the gaps should disappear if only this geometrical effect is present (Lifshitz tails). However, we have performed numerical calculations in very large square lattices that show that the gap persists well above the percolation limit. We regard this phenomenon as an expression the non-uniformity of the system.

Hamiltonians similar to the tight-binding used here can successfully model other very different physical systems and excitations, such as phonons, magnons, etc. Therefore, the ideas exposed here can be applied to other general situations, as the absence of spin wave in dilute magnets, or the disordered $xy$ model. Some of these problems are of great current interest and importance, as systems with localized phonons or layered systems with complicated antiferromagnetic interactions.

Summarizing, we have explained the appearance of localized states and gaps by using very general physical arguments, that are applicable to a wide range of different systems. The important concept is the non-uniformity of the system that introduces a peculiar dependence of the energy of the excitations when localizing their wave functions. We have presented two cases, the PL in which a detailed study of the spectrum reveals that the cause of the non-uniformity is the frustration of bonds, and the RBA, in which the large self-energy of some sites favors localization.

Obviously, the renormalization process is not essential to the physics explained here. This is the reason why we speak about low-energy excitations in general and we think that the ideas exposed here contribute as a complement to Goldstone theorem, that is, it is possible to assure when a system cannot have low-energy excitations.

## Acknowledgments

We benefited from enlightening discussions with R. J. Elliott, R. B. Stinchcombe, T. Fujiwara, and M. F. Thorpe. Financial support by UNAM, through projects DGAPA IN-104595 and IN-104296 is greatly acknowledged.

## References

1. R. M. White, and T. H. Geballe, *Long Range Order in Solids* (Academic Press, New York, 1979), p.37.
2. M. F. Thorpe, in *Excitations in Disordered Systems*, edited by M. F. Thorpe, NATO Advanced Summer Institute Series, Vol. B78, (Plenum, New York, 1982), p. 85.
3. M. Arai, T. Tokihiro, T. Fujiwara, and M. Kohmoto, *Phys. Rev. B* **38**, 1621 (1988).
4. G. C. Naumis, R. A. Barrio, and C. Wang, *Phys. Rev. B* **50**, 9834 (1994).
5. J. M. Ziman, *Models of Disorder* (Cambridge University Press, 1979).
6. M. Kohmoto and B. Sutherland, *Phys. Rev. Lett.* **56**, 2740 (1986).
7. R. A. Barrio and C. Wang, *J. Non-Cryst. Solids*, **153** & **154**, 375 (1993).
8. T. Rieth and M. Schreiber, *Phys. Rev. B* **51**, 15 827 (1995).
9. D. N. Basov et al., *Phys. Rev. Lett.* **73**, 1865 (1994).
10. M. H. Cohen, in *Topological Disorder in Condensed Matter*, edited by F. Yonezawa and T. Ninomiya, Springer Series in *Solid State Science*. **46**, (Springer-Verlag, New York, 1983), p. 122.

11. P. Ma and Y. Liu, *Phys. Rev. B* **39**, 9904 (1989).
12. G. G. Naumis, R. A. Barrio and C. Wang, *Proc. V Int. Conf. on Quasicrystals*, edited by C. Janot and R. Mosseri, (World Scientific, Singapore, 1995), p. 431.
13. Z. M. Stadnik and G. Stroink, *Phys. Rev. B* **47**, 100 (1993).
14 S. Kirkpatrick, and T. P. Eggarter, *Phys. Rev. B* **6**, 3598 (1972).

# Electronic Theory of Colossal Magnetoresistance Materials

R. Allub* and B. Alascio*

Centro Atómico Bariloche
8400 San Carlos de Bariloche
ARGENTINA

## Abstract

We study a model based on the double exchange mechanism and diagonal disorder to calculate magnetization and conductivity for $La_{1-x}Sr_xMnO_3$ type crystals as a function of temperature. The model represents each $Mn^{4+}$ ion by a spin $S = \frac{1}{2}$, on which an electron can be added to produce $Mn^{3+}$. We include a hopping energy $t$, and a strong intra-atomic exchange interaction $J$. To represent in a simple way the effects of disorder we assume a Lorentzian distribution of diagonal energies of width $\Gamma$ at the Mn sites. We calculate the mobility edge and the Fermi level as functions of magnetization. We add the spin entropy to build up the free energy of the system. In the strong coupling limit, $J \gg t, \Gamma$, the model results can be expressed in terms of $t$ and $\Gamma$ only. We use the results of the model to draw "phase diagrams" that separate ferromagnetic from paramagnetic states and also "insulating" states where the Fermi level falls in a region of localized states from "metallic" where the Fermi level falls in a region of extended states. We then add the contributions to the conductivity of extended states to those of localized states to calculate the resistivity for different concentrations and the magnetoresistance. We conclude that the model can be used successfully to represent the transport properties of the systems under consideration.

## I. Introduction

The recent discovery of "colossal" magnetoresistance (CMR) in $La_{1-x}Sr_xMnO_3$ type compounds[1] and its relation to possible applications to magnetoresistance (MR) devices has generated strong interest in these materials.

Before the discovery of CMR, Jonker and Van Santen[2] established a temperature-doping phase diagram separating metallic ferromagnetic from insulating antiferromagnetic phases. Zener[3] proposed a "double exchange" (DE) mechanism to understand the properties of these compounds and the connection between their magnetic and transport properties. This DE mechanism was used by Anderson and Hasegawa[4] to calculate the ferromagnetic interaction between two magnetic ions, and by de Gennes[5] to propose canting states for the weakly doped compounds. Kubo and Ohata[6] used a spin wave approach to study the temperature dependence of the resistivity at temperatures well below the critical temperature and a mean field approximation at $T$ near $T_c$. Mazzaferro, Balseiro, and Alascio[7] used a mixed valence approach similar to that devised for TmSe combining DE with the effect of doping to propose the possibility of a metal insulator transition in these compounds.

Recently, a wealth of experimental results have been obtained on the transport, optical, spectroscopic and thermal properties of these materials under the effects of external magnetic fields and pressures.[8]

Theoretically, Furukawa[9] has shown that DE is essential to the theory of these phenomena, while Millis et al.[10] have argued that DE alone is not sufficient to describe the properties of some of the alloys under consideration and have proposed that polaronic effects play an important role. In a previous work[11] we have explored a semi-phenomenological model that includes the effect of disorder in the transport properties. Müller-Hartmann and Dagotto[12] have pointed out that a new phase appears in the proper derivation of the effective hopping, but have not studied its effect in the physical properties of the systems under consideration.

Although there has been considerable theoretical effort, there has been no comparison with experiment except in the work of Furukawa[9] where the connection between magnetization and conductivity brought about by double exchange is clearly shown, but the comparison is reduced to the paramagnetic phase of a single sample. Critical temperatures and other thermodynamical quantities have been compared to experiment also mainly by Furukawa, but there is no way in which one can parametrize the difference between different concentration samples for instance. Because DE connects intimately magnetization and conductivity and the later depends strongly on the defect structure of the materials (grain boundaries, impurities, Bloch walls, etc.) it is important to find out whether such comparison is possible.

In our previous paper[11] we treat the Hamiltonian proposed for these systems using an alloy analogy approximation to the exchange terms and including the effects of disorder by introducing a continuous distribution of the diagonal site energies.

Here we continue that treatment by proposing a free energy that allows to determine the magnetization as a function of temperature. We then proceed to find the Fermi energy and the mobility edge (ME) as functions of temperature. Finally, we calculate the contribution to the conductivity of extended and localized electrons under different external conditions. We compare our results with experiments in single crystals of $La_{1-x}Sr_xMnO_3$ reported by Moritomo et al.,[14] for three different values of external pressure. We draw the magnetoresistance for different values of the magnetic field.

In Section II we review, for completeness the main results of our previous paper and we propose a free energy that allows us to obtain magnetization as a function of temperature and from this the Fermi energy and mobility edge also as a function of temperature. Finally, assuming that for the $La_{1-x}Sr_xMnO_3$ samples in consideration the extended and localized states dominate the conductivity we obtain an expression for the conductivity as a function of temperature.

Section III is devoted to drawing phase diagrams in terms of the parameters of the model, to the comparison of the conductivity obtained in Section II with experiment and to the discussion of results.

## II. Model

In our previous paper[11] we considered a simplified model Hamiltonian given by

$$H_m = \sum_{i,\mu} \varepsilon_i c^\dagger_{i\mu} c_{i\mu} - t \sum_{\langle i,j\rangle,\mu} c^\dagger_{i\mu} c_{j\mu} + U \sum_i c^\dagger_{i\uparrow} c_{i\uparrow} c^\dagger_{i\downarrow} c_{i\downarrow} - J \sum_i \vec{S}_i \cdot \vec{\sigma}_i, \qquad (1)$$

where we use the same notation.

We simplify the last term to the $z$-component of the local and itinerant spin at each site. Since then the $S_{zi}$ are good quantum numbers the states of the system are characterized by itinerant electrons moving on a frozen distribution of localized up or down spins. To obtain site Green functions and thus local density of states for this problem, we ignore at the start the site dependence of the diagonal energies: i.e., we set $\varepsilon_i = \varepsilon$ and we use an alloy analogy approximation to obtain the effect of $J$ (assumed larger than $t$) in the electronic band structure of the model. Using the renormalized perturbation expansion[15] in the manner described in Ref. 11, we obtain the corresponding local Green functions and the average density of states for spin up and down. The densities of states for each spin split into two bands centered at $E_\pm = (\varepsilon \pm J)$ with weights and widths that depend on the number of sites with each spin $S_z = \frac{1}{2}$ or $-\frac{1}{2}$ i.e., they depend on the magnetization of the system. The electronic structure of the compounds consists of essentially four bands, two for spin up and two for spin down, The splitting between the up and down bands is given by the intra-atomic exchange energy $J$, their weight and width by the normalized magnetization $m = 2\langle S\rangle$. The Fermi level falls always in the lower bands so that the transport properties are determined by these bands. Consequently, for $J \gg \sqrt{K}t$, where $K$ is the connectivity, using the site density of states [Eq. (11) in Ref. (11)] the averaged density of states per site reduces to

$$\rho_{0\mu}(\omega) = \frac{\nu_\mu(K+1)\sqrt{4Kt^2\nu_\mu - (\omega - E)^2}}{2\pi|(K+1)^2 t^2 \nu_\mu - (\omega - E)^2|}. \qquad (2)$$

where $E = (\varepsilon - J)$ and $\nu_\mu = \frac{1}{2}(1 + \mu m)$ ($\mu = \pm$ for up and down spin respectively).

The densities of states depend on the magnetization though the $\nu_\mu$ factors: the width and weight of the up (down) band increases (decreases) with magnetization as shown in Fig. 1.

At this point, we introduce the effect of the disorder originated by the substitution of some of the rare earth ions by Sr, Ba, or Ca. We assume that this can be described within the model by making the diagonal energies site dependent. As is well known, since Anderson's original paper[16] a distribution of diagonal energies produces localization of the electronic states from the edges of the bands to an energy within them which is called "mobility edge" (ME). The precise position of the ME is difficult to calculate and different localization criteria result in different values for it.[17] However, we do not aim here to an absolute value for the ME but rather to its change with respect to the Fermi level when the magnetization changes from saturation to zero. For this reason we assume that there is no localization before disorder and for simplicity, we use a Lorentzian distribution of energies[18] (width $\Gamma$) and the Ziman criterion of localization.[19]

From the ensemble-averaged Green function we obtain the densities of states

$$\rho_\mu(\varepsilon) = \int_{-\infty}^{+\infty} \rho_{0\mu}(\varepsilon') L(\varepsilon - \varepsilon')\, d\varepsilon', \qquad (3)$$

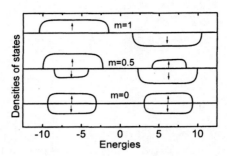

**Figure 1.** Partial densities of states for spin up and down for $K = 5$, $E = \mp 6$, $t = 1$, and different values of the magnetization.

where $L(x)$ is a Lorentz distribution given by

$$L(x) = \frac{\Gamma}{[\pi(x^2 + \Gamma^2)]}. \tag{4}$$

Within this *comparative* approach one can make the further approximation of replacing in Eq. (3) $\rho_{0\mu}$ by a square density of states with the same width $W_\mu = 2t\sqrt{K\nu_\mu}$ and the same weight $\nu_\mu$ to obtain,

$$\rho_\mu(\varepsilon) = \frac{\nu_\mu}{2\pi W_\mu}\left[\arctan\frac{(W_\mu - \varepsilon)}{\Gamma} + \arctan\frac{(W_\mu + \varepsilon)}{\Gamma}\right], \tag{5}$$

which allows for analytical expressions for the number of particles $n$, and the internal energy $E$ as functions of the magnetization $m$, and the Fermi energy $\varepsilon_F$. In some instances, when the Fermi level falls too near the band edge, this approximation can differ from the more realistic case where the density of states increases as $\sqrt{\varepsilon}$.

To proceed further, we need an expression for the entropy of these system. Again for *comparative* purposes, we resort to the simplest possible form compatible with our earlier approximations, that of a spin one half array of sites:

$$S = \ln(2) - \nu_+ \ln(2\nu_+) - \nu_- \ln(2\nu_-). \tag{6}$$

More accurate forms of the entropy valid in the mixed valence regime can be used, see for example Ref. 20.

In the presence of a magnetic field $H$, the free energy per site is then,

$$G = E - TS - \mu_B m H, \tag{7}$$

where $T$ is the temperature and $\mu_B$ is the magnetic moment per site.

We proceed as follows: for each $n$, we use (assuming $k_B T \ll W_\sigma$)

$$n = \sum_\mu \int_{-\infty}^{\varepsilon_F} \rho_\mu(\varepsilon)\, d\varepsilon, \tag{8}$$

to obtain a relation between $n, m$ and $\varepsilon_F$ from which $\varepsilon_F$ can be determined numerically.

The free energy is then a function of $m$ and $T$ only and allows, by minimization, to determine $m(T)$. The resulting $m(T)$ does not differ essentially from the law of corresponding states for spin $\frac{1}{2}$. Having obtained $m(T)$ for each value of the parameters we can determine the up and down mobility edges ($B_+$ and $B_-$) and the Fermi energy.

Following Mott and Davies[21] we calculated the transport properties assuming that two forms of d.c. conduction are possible: thermally activated hopping and excitation to the mobility edge. When the difference ($\Delta$) between the mobility edge and the Fermi level is not too large as compared to $k_BT$, the conductivity is dominated by particles in the extended states, and is given by the usual relaxation time form,

$$\sigma_d = \frac{e^2}{3a^3} \sum_\mu \int_{-\infty}^{\infty} v_\mu^2(\varepsilon)\tau_\mu(\varepsilon)\rho_\mu(\varepsilon)\left[-\frac{\partial f(\varepsilon)}{\partial \varepsilon}\right] d\varepsilon, \qquad (9)$$

in which $a$ is the Mn-Mn distance in the simple cubic lattice, $f(\varepsilon)$ is the Fermi function. We assume that the relaxation time $\tau_\mu$ is a step function equal to zero for $\varepsilon < B_\mu$ and takes a value $\tau_0$ related to the minimum metallic conductivity for $\varepsilon > B_\mu$, where according to Ref. 18 $B_\mu = -\sqrt{t^2K^2\nu_\mu - \Gamma^2}$. Further, replace $v_\mu(\varepsilon)$ by its average $v_\mu^2 = (W_\mu^2 a^2/2\hbar^2)$ to obtain:

$$\sigma_d = \frac{e^2\tau_0}{6\hbar^2 a} \sum_\mu W_\mu^2 \int_{B_\mu}^{\infty} \rho_\mu(\varepsilon)\left[-\frac{\partial f(\varepsilon)}{\partial \varepsilon}\right] d\varepsilon. \qquad (10)$$

An Anderson transition takes place when $B_\mu$ vanishes. For $(t^2K^2\nu_\mu - \Gamma^2) < 0$ all eigenstates became localized.

To estimate the contribution of the localized states to the conductivity, according to Mott and Davies[21] we write the following expression for the variable range hopping conductivity:

$$\sigma_h = e^2 \sum_\mu N_\mu(\varepsilon_F)\overline{R}_\mu^2 \nu_{ph} \exp(-D_\mu/T^{1/4}), \qquad (11)$$

where $\nu_{ph}$ is a phonon related constant, $N_\mu(\varepsilon_F)$ is the density of states per cm$^3$, $\overline{R}_\mu$ is the hopping distance for each spin and is given in terms of the inverse localization length $\alpha_\mu(\varepsilon)$ by

$$\overline{R}_\mu = \frac{3^{5/4}}{4(2\pi)^{1/4}}[N_\mu(\varepsilon_F)\alpha_\mu(\varepsilon_F)k_BT]^{-1/4}. \qquad (12)$$

In our case the localization length at the Fermi level depends on the distance to the mobility edge. Here again we follow Mott and Davies[21] to take

$$\alpha_\mu(\varepsilon_F) = \alpha_0 \left(\frac{B_\mu - \varepsilon_F}{B_\mu}\right)^{2/3}, \qquad (13)$$

and $D_\mu$ is given by

$$D_\mu = D_0 \left[\frac{\alpha_\mu^3}{k_B N_\mu(\varepsilon_F)}\right]^{1/4}, \qquad (14)$$

where $\alpha_0$ and $D_0$ are constants which we take to be $\alpha_0 = 0.358\ A^{-1}$, $D_0 = 1.66$.

Since $B_\mu - \varepsilon_F$ is a sensitive function of the magnetization,[11] the localization length and the conductivity are also strongly dependent on the magnetization. Figure 2 shows $(B_+ - \varepsilon_F)$ and the localization length $(\alpha_+)^{-1}$ as functions of $m$.

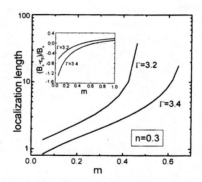

**Figure 2.** Localization length ($\alpha_+$) vs. magnetization for $n = 0.3$, $K = 5$, $t = 1$, and two different values of $\Gamma$. Inset: Difference ($\Delta$) between the mobility edge and the Fermi energy (in units of $B_+$) as a function of $m$ for the same values of $\Gamma$.

## III. Results and Discussion

For $n = 0.5$, $\varepsilon_F = 0$ independently of $m$ and we find an analytical expression for the free energy, from which we obtain the ferromagnetic transition temperature $T_c$:

$$T_c = \left[(\Gamma^2 + 30t^2)\arctan(\sqrt{10}t/\Gamma) - \Gamma\sqrt{10}t\right] / \left(8\pi\sqrt{10}t\right), \tag{15}$$

where we take $K = 5$ to describe the simple cubic lattice of the Mn sites.

We define a characteristic temperature $T_M$ at which the mobility edge crosses the Fermi level. However, this crossing does not imply any discontinuous change in the resistivity, the only non-analyticity occurs at $T_c$. For $n = 0.5$, it is easy to get an explicit expression for $T_M$:

$$T_M = \left[4\pi \ln\left(\frac{1 + m_c}{1 - m_c}\right)\right]^{-1} \sum_\mu \mu\left[(3A_\mu^2 - \Gamma^2)\arctan(A_\mu/\Gamma) - 2A_\mu \Gamma \ln\left(A_\mu^2 + \Gamma^2\right)\right], \tag{16}$$

where $A_\mu = 2t\sqrt{Km_\mu}$, $m_\mu = \frac{1}{2}(1 + \mu m_c)$, and $m_c = (2\Gamma^2/t^2K^2) - 1$.

In what follows we consider $n < 0.5$ and identify $n$ with the number of holes, which we take to be equal to the concentration of divalent component of the alloy. We define as insulator the state where the Fermi level falls below the ME [$\Delta = (B_+ - \varepsilon_F) > 0$]. So that, for small $\Gamma$ the Fermi level falls above the ME ($\Delta < 0$) and only the metallic state appears. When $\Gamma$ increases, $\Delta$ reduces and, finally $\Delta = 0$ for a critical value $\Gamma_- = \sqrt{0.5K^2t^2 - \varepsilon_F^2}$ (where $m_c = 0$ and $T_c = T_M$). When $\Gamma$ is increased from $\Gamma_-$, $T_M$ reduces and finally $T_M = 0$ at a critical value $\Gamma_+ = \sqrt{K^2t^2 - \varepsilon_F^2}$. Above $\Gamma_+$ the system remains insulating at all temperatures. Consequently, only for $\Gamma_- < \Gamma < \Gamma_+$ the transition between metallic and insulating regimes appears.

Figure 3 shows $T_c$ and $T_M$ as functions of $n$ for $\Gamma = 3t$. As a consequence of the density of states being modified by disorder, the Curie temperatures decrease with $\Gamma$, while the increase with $n$ is just a consequence of the energetics of the bands.

In figure 4 we show the resistivity as a function of $T/T_c$ for $\Gamma = 2t$ and different values of the concentration. The upper panel ($n = 0.1$) is such that the contribution from extended states is negligible. In fact, the temperature dependence enters the resistivity through $D_\mu/T^{1/4}$ so that in the temperature range where $m$ does not vary with temperature ($T \ll T_c$, and $T \geq T_c$), $D_\mu$ remains also constant and the typical

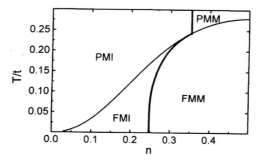

**Figure 3.** $T/t$ vs. $n$ phase diagram. Ferromagnetic $T_c$ (thin lines) and metal-to-insulator $T_M$ (thick lines) transition temperatures vs. doping $n$, for $K = 5$, $t = 1$, and $\Gamma = 3$. Regions labeled as FMM (ferromagnetic metal: $m \neq 0$ and $\Delta < 0$), FMI (ferromagnetic insulator: $m \neq 0$ and $\Delta > 0$), PMM (paramagnetic metal: $m = 0$ and $\Delta < 0$), and PMI (paramagnetic insulator: $m = 0$ and $\Delta > 0$).

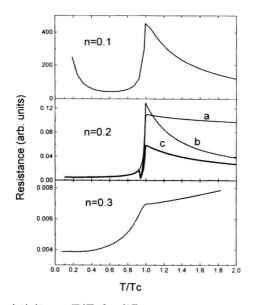

**Figure 4.** Zero field resistivity vs. $T/T_c$ for different concentrations and $\Gamma = 2t$.

variable range dependence sets in. In figure 5 we illustrate this behavior along with experimental results by Tokura et al.[13]

The intermediate panel in Fig. 4 corresponds to a situation where both contributions to the resistivity become comparable at some temperatures near the critical temperature. Curve (a) corresponds to the variable range hopping resistivity, curve (b) corresponds to the resistivity as calculated from the extended states only and curve (c) is the total resistivity. At temperatures higher than $T_c$ the resistivity is dominated by the smaller resistivity contribution. The discontinuity at $T$ near $T_c$ is an artifact of the approximation we use to the hopping conductivity: We take the variable range hopping expression evaluated at the Fermi level, instead of an average

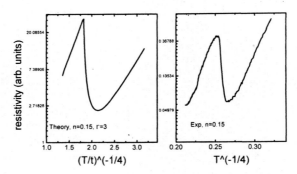

**Figure 5.** Resistivity vs. $T^{-1/4}$ in $La_{1-n}Sr_nMnO_3$ ($n = 0.15$). Left panel: theory as obtained from Eq. (11). Right panel: experiment as obtained from Ref. 13.

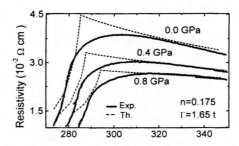

**Figure 6.** Pressure dependence of resistivity (solid lines) for the $La_{1-n}Sr_nMnO_3$ ($n = 0.175$) taken from Ref. 14. The dashed lines represent the fits with Eq. (10) for $\Gamma = 1.65t$, $K = 5$, and different values of $t$: $t = 1572$ K to fit the curve corresponding to 0 GPa, $t = 1586$ K to 0.4 GPa, and $t = 1622$ K to 0.8 GPa.

over a range $k_BT$ of energies near $\varepsilon_F$ and take infinite resistivity when $\varepsilon_F$ crosses the mobility edge. Finally, the lower panel shows the resistivity due to the extended states alone as correspond to higher doping.

Figure 6 shows the pressure dependence of the resistivity and we compare with the measurements of Moritomo et al.[14] We fix $\Gamma = 1.65t$ and $n = 0.175$, and we take $t = 1572$ K to fit the curve corresponding to 0 GPa, $t = 1586$ K to 0.4 GPa, and $t = 1622$ K to 0.8 GPa. We multiply the values of each calculated resistivity by a constant to fit the experimental values at $T_c$. This amounts to an election of band parameters and relaxation time quite consistent with bandwidths and lower limit metallic conductivity.

Figure 7 shows the temperature dependence of the magnetoresistance for various fields. We take $\Gamma = 1.8t$, $t = 1789$ K and select $\tau_0 = 0.96 \times 10^{-14}$ and $\mu_B = 0.964 \times 10^{-20}$ erg/Gauss. These values are again consistent with data known to be valid for $La_{1-n}Sr_nMnO_3$.

In summary, we conclude that the model allows for a complete description of the phenomena in consideration. The theory presented here allows to characterize the resistivity behavior of different samples by two parameters, one associated to the degree of disorder ($\Gamma$), and the other to the hopping energy $t$. The values of the

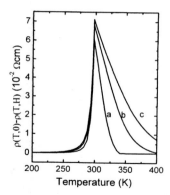

**Figure 7.** Magnetoresistance $\rho(T,0) - \rho(T,H)$ vs. temperature for several values of $H$. (a) $H = 3$ T, (b) $H = 8$ T, and (c) $H = 15$ T. We take here $t = 1789$ K, $\Gamma = 1.8$, $t$, $K = 5$, and $n = 0.175$ as an example appropriate to compare with $La_{1-n}Sr_nMnO_3$.

hopping energy $t$ can be affected by displacement of the oxygen atoms, or by polaronic or other many body effects. The most natural source of disorder is the substitution of rare earth by Sr, Ca or Ba, but polaronic or other many body effects may act in a similar way. Despite this simple approach we have obtained satisfying results, since they reproduce well most of the experimental data. Nevertheless, we must point out that other interactions should be incorporated in a more complete description of these materials. Static and dynamic lattice effects can modify not only the values of both parameters $t$ and $\Gamma$, but also the thermodynamics of the transition, leading to first-order transitions as those found in many of the compounds.[23] The connection to the dynamics of the lattice has been recently very elegantly demonstrated by Zhao et al.[24] Coulomb interactions between ions, in combination with lattice effects could also produce charge ordering and lead to the reentrant behavior found in Ref. 23. Last but not least, super-exchange interactions between the localized spins, give rise to new phases and consequently to quite different and new behaviors. The electron doped materials like $Ca_{1-x}Y_xMnO_3$ (Ref. 25) are excellent candidates to study these phenomena.

## References

[*] Member of the Carrera del Investigador Científico del Consejo Nacional de Investigaciones Científicas y Técnicas (CONICET).
1. R. von Helmholt, et al., *Phys. Rev. Lett.* **71**, 2331 (1993).
2. G. H. Jonker and J. H. van Santen, *Physica* **16**, 337 (1950); J. H. van Santen and G. H. Jonker, *Physica* **16**, 599 (1950).
3. C. Zener, *Phys. Rev.* **82**, 403 (1951).
4. P. W. Anderson and H. Hasegawa, *Phys. Rev.* **100**, 675 (1955).
5. P. G. de Gennes, *Phys. Rev.* **118**, 141 (1960).
6. K. Kubo and N. Ohata, *J. Phys. Soc. Jpn.* **33**, 21 (1972).
7. J. Mazzaferro, C. A. Balseiro, and B. Alascio, *J. Phys. Chem. Solids* **46**, 1339 (1985).
8. Y. Okimoto et al., *Phys. Rev. Lett.* **75**, 109 (1995); S. W. Cheong et al., Proceed-

ings of the *Physical Phenomena at High Magnetic Fields-II* Conference, Tallahassee, Florida. World Scientific, to be published; M. C. Martin *et al.*, to be published; R. Mahendiran, R. Mahesh, A. K. Raichaudhuri, and C. N. R. Rao, *Solid State Commun.* **94**, 515 (1995); H. L. Ju *et al.*, *Phys. Rev. B* **51**, 6143 (1995); M. K. Gubkin *et al.*, *JETP Lett.* **60**, 57 (1994).
9. N. Furukawa, *J. Phys. Soc. Jpn.* **63**, 3214 (1994).
10. A. J. Millis, P. B. Littlewood, and B. I. Shrainman, *Phys. Rev. Lett.* **74**, 5144 (1995).
11. R. Allub and B. Alascio, *Solid State Commun.* **99**, 613 (1996).
12. E. Müller-Hartmann and E. Dagotto, to appear in *Phys. Rev. B*, preprint cond-mat/9605041.
13. Y. Tokura *et al.*, *J. Phys. Soc. Jpn.* **63**, 3931 (1994).
14. Y. Moritomo, A. Asamitsu, and Y. Tokura, *Phys. Rev. B* **51**, 16491 (1995).
15. See, *e.g.*, E. N. Economou, *Green's Functions in Quantum Physics*, edited by P. Fulde, Springer Series in Solid State Sciences (Springer-Verlag, Berlin), Vol. 7.
16. P. W. Anderson, *Phys. Rev.* **109**, 1492 (1958).
17. D. C. Licciardello and E. N. Economou, *Phys. Rev.* **11**, 3697 (1975).
18. P. Lloyd, *J. Phys. C* **2**, 1717 (1969).
19. J. M. Ziman, *J. Phys. C* **2**, 1230 (1969).
20. A. A. Aligia Ph.D. Thesis, Instituto Balseiro, 1984.
21. N. F. Mott and E. A. Davies, *Electronic Processes in Non-Crystalline Materials* (Oxford University Press, 1971).
22. H. Y. Hwang, *et al.*, *Phys. Rev. Lett.* **75**, 914 (1995).
23. H. Kuwahara, *Science* **270**, 961 (1995).
24. Guo-meng Zhao, K. Conder, H. Keller, and K. A. Muller, *Nature* **381**, 676, (1996).
25. J. Briatico *et al.*, *Czechoslovak J. Phys.* **46**, S4 2013 (1996).

# First-Principles Study of Phase Equilibria in the Ni-Cr System

J. M. Sanchez and P. J. Craievich

*Center for Materials Science and Engineering*
*The University of Texas at Austin*
*Austin, TX 78712*
*USA*

## Abstract

The relative importance of various free energy contributions to phase stability in the Ni-Cr system is investigated using total energy calculations in the local density approximation combined with the cluster expansion and the cluster variation method. In particular, we develop a formalism to calculate the vibrational free energy in disordered alloys exhibiting short-range order. The methodology is well adapted to be used in conjunction with first-principles total energy calculations. It is found that the coupling between configurational and displacive degrees of freedom is crucial for the accurate description of the equilibrium phase diagram of the Ni-Cr system, with electronic excitations playing a less important role and the effects of short-range order being essentially negligible.

## I. Introduction

The cluster expansion (CE)[1-3] method combined with the local density approximation (LDA)[4] for the total energy calculations of ordered compounds has played a key role in the computation of equilibrium phase diagrams from first principles. In the CE method, the energy is expanded in a rapidly convergent bilinear form that involves the products of effective cluster interactions (ECI) times multisite cluster functions that describe the configuration of the alloy. These multisite cluster functions can be shown to form a complete and orthogonal basis in configurational space.[1] The ECIs, in turn, can be calculated from the energy of a few (typically 10 to 20) ordered structures, which can be determined from self-consistent band structure calculations in the LDA. Once the ECIs for a given lattice are known, it is possible to determine the energy of ordered and disordered alloys with any degree of short-range order.

The procedure described above has been extensively used over the last decade to compute the phase diagrams of a variety of binary and ternary alloys. Whereas the approach produces a reasonable qualitative description of the equilibrium phase diagram, in most cases the agreement with experimental data is quantitatively poor. Among the reasons that have been quoted for the discrepancies between theory and experiments are that, despite the dominant role of the configurational free energy, other contributions such as electronic excitations, static relaxations and lattice vibrations can significantly affect the computation of temperature-composition phase diagrams. At present, however, these additional contributions to the free energy of formations have not been systematically included in order to calculate quantitatively correct equilibrium phase diagrams. In particular, no general formalism exist for the treatment of the vibrational free energy that properly accounts for short-range order effects in alloys.

Nevertheless, the effect of electronic excitations, static relaxations and vibrational modes in alloy phase stability has been addressed by several authors. For example, Wolverton and Zunger[5] have recently carried out a detailed study of electronic excitations in the Ni-V and Pd-V systems. For pure systems, the work of Moroni et al.[6] suggests a dominant effect of electronic excitations in the bcc $\leftrightarrow$ hcp allotropic transition, although, as pointed out by Craievich et al.[7] for the case of Ti, this contribution is in general insufficient to account for the observed transition temperature. Static atomic relaxations have also been shown to be important in the phase stability of semiconductor alloys, such as Si-Ge,[8] and in fcc transition metals alloys such as Cu-Au and Cu-Pd.[9]

With regard to vibrational modes, the semiempirical approach developed by Moruzzi et al.[10] for pure metals has been widely used. This approach was implemented for phase diagram calculations by Sanchez et al.[11] and subsequently used extensively by several authors.[12–17] Although the approach is undoubtedly a crude approximation to describe vibrational modes in alloys, the different implementations of the method exemplifies the application of the CE to the vibrational free energy and/or the Debye temperature.[11] More recently Garbulsky et al.[18] (see also Ref. 19) used the CE to describe the vibrational free energy of formation in the harmonic approximation of a model system with central pair potentials. However, a direct cluster expansion of the vibrational free energy is not applicable to most real alloys since, for a given lattice, low symmetry configurations are generally mechanically unstable.[7,20]

Here we present the results of first-principles calculations of the configurational free energy of formation (with and without short-range order) and of the contribution due to electronic excitations and vibrational modes in the Ni-Cr system. A methodology is also developed for calculating the vibrational contribution to the free energy based on the characterization of the ECIs as a function of atomic displacements. Application of the theory to the Ni-Cr system yields a temperature-composition phase diagram in excellent agreement with experiment.

## II. Configurational Free Energy

The configuration of a binary alloy with $N$ sites is described, as usual, by a vector $\vec{\sigma} = \{\sigma_1, \sigma_2, \sigma_3, \ldots, \sigma_N\}$ of $N$ spin-like occupation numbers $\sigma_i$ that take values $\pm 1$ according to the atomic species occupying lattice site $i$. The energy of these configurations is expressed in terms of ECIs using the standard cluster expansion method.[1–3]

In this representation, the configurational energy is given by:

$$E_c(\vec{\sigma}) = \sum_\alpha V_\alpha \, \Phi_\alpha(\vec{\sigma}), \tag{1}$$

where the expansion coefficients $V_\alpha$ are the ECIs for the cluster of lattice sites labeled by $\alpha$, and where the cluster functions $\Phi_\alpha(\vec{\sigma})$ form a complete and orthonormal basis in configurational space.[1-3] In general, the expansion in Eq. (1) converges fast and, thus, the sum includes only a few clusters of relatively small sizes. The values of the ECIs for these clusters can be determined by structure inversion, whereby the ECIs in the expansion are fitted to the energies of a set of ordered structures. The energies of these structures are usually obtained from first-principles band structure calculations.

The clusters used in the expansion for the present study were the same as described in Ref. 17 and include pair and many body interactions up to third and fourth neighbors in the *fcc* and *bcc* phases, respectively (see also Ref. 20). The ECIs were fitted to the energies of a set of ordered structures calculated using the linear muffin-tin orbital method in the atomic sphere approximation[22] (LMTO-ASA), with the Barth-Hedin exchange-correlation potential.

## II.1 Short-Range Order and Electronic Excitations

The effects of SRO and electronic excitations on the equilibrium phase diagram were investigated by carrying out finite temperature calculations using the first-principles ECIs determined from the cluster expansion in three different cases: (i) in the Bragg-Williams (BW) approximation (*i.e.*, neglecting SRO) without electronic excitations; (ii) in the cluster variation method (CVM)[25] without electronic excitations; and (iii) in the BW approximation with electronic excitations.

The CVM calculations for the *fcc* phase were carried out using a 13-points cluster, comprised of a point and its twelve nearest-neighbors, and the 14-points *fcc* unit cell.[26] For the *bcc* lattice the maximum clusters consisted of the 6-points octahedron, the 9-points *bcc* unit cell and the 8-points rhombohedron.[27]

Electronic excitations were included by adding to the configurational energy at $T = 0$ K the contribution arising from the one-electron energy at finite temperatures. We note that, in these calculations, we used the self-consistent electronic density of states obtained at $T = 0$ K. Neglecting the effect of temperature on the density of states has been previously established to be a good approximation.[5] The cluster expansion method was applied to both the energy and entropy contributions arising from electronic excitations over a range of temperatures. The clusters used in these expansions were the same as those used in the expansion of the total energy for the determination of the ECIs.

The results obtained in the BW approximation with and without electronic excitations are shown in Figs. 1 and 2 for, respectively, the energy and free energy of formation. The symbols in these figures are experimental data at 1550 K (Ref. 13). Also shown in Figs. 1 and 2 (dotted lines) are the energy and free energy of formation obtained by Tso *et al.*[24] using a phenomenological model. In the work of Tso *et al.*[24] the parameters of the model were determined by a least square fit to all available experimental measurements of thermodynamic potentials and phase diagram in Ni-Cr.

From Fig. 1 we see good agreement between the calculated and experimental energies of formation indicating the accuracy of the cluster expansion and of the calculated ECIs. In contrast, the results of Fig. 2 show a large discrepancy between

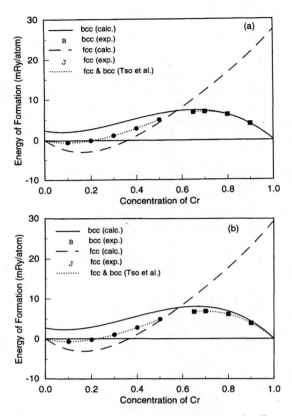

**Figure 1.** Calculated configurational energy of formation in the Bragg-Williams approximation at 1550 K: (a) neglecting and (b) including electronic excitations. The dotted lines are from the phenomenological model of Tso et al.[24] and the symbols are experimental data at the same temperature.[23]

the calculated and the experimental free energies of formation for both the *fcc* and *bcc* phases. Also apparent from Fig. 1 is the fact that electronic excitations have a small effect on the energy of formation of the Ni-rich *fcc* phase, producing only a slight increase in the energy of formation of the Cr-rich *bcc* phase. The effects on the free energy of formation are also small although slightly more pronounced due to electronic entropy contributions (see Fig. 2).

The phase diagrams calculated for the different cases are compared in Fig. 3. Figure 3(a) compares the BW and CVM approximations ignoring electronic excitations while Fig. 3(b) shows the results of the BW approximation with electronic excitations. The dotted line in Fig. 3 corresponds to a metastable miscibility gap in the *bcc* phase calculated in the BW approximation. The experimental Ni-Cr phase diagram from Ref. 28 is reproduced in Fig. 4 for comparison.

The similarities between the BW and CVM phase diagrams [(Fig. 3(a)] indicate that SRO effects are negligible in Ni-Cr, at least at high temperatures. Electronic excitations, on the other hand, have a noticeable effect on the phase diagram [see Fig. 3(b)] resulting, for example, in a drop of ∼ 1000 K in the critical temperature of the metastable miscibility gap.

# First-Principles Study of Phase Equilibria in the Ni-Cr System

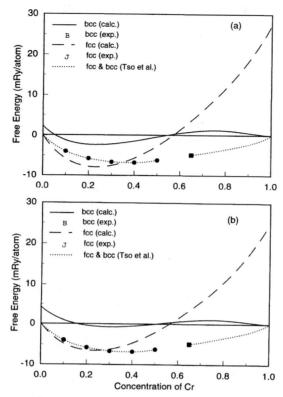

**Figure 2.** Calculated configurational free energy of formation in the Bragg-Williams approximation at 1550 K: (a) neglecting and (b) including electronic excitations. The dotted lines are from the phenomenological model of Tso et al.[24] and the symbols are experimental data at the same temperature.[23]

Also apparent from Figs. 3 and 4 is the significant difference in the temperature scale of the calculated and experimental phase diagrams indicating that, at this level of approximation, the theory is only in qualitative agreement with experiment. We note that the discrepancies between calculated and experimental phase diagrams are in line with those noted in Fig. 2 for the free energy. In the light of the agreement found for the energy of formation at $T = 0$ K, we conclude that the errors in the calculated free energies are primarily due to the fact that vibrational contributions have been neglected.

## III. Vibrational Free Energy

In order to calculate the vibrational contribution to the free energy we consider the total partition function $Z_T$ for a binary system in the classical limit:

$$Z_T = \left(\frac{k_B T}{\hbar}\right)^{3N/2} (m_A)^{3N_A/2} (m_B)^{3N_B/2} Z, \quad (2)$$

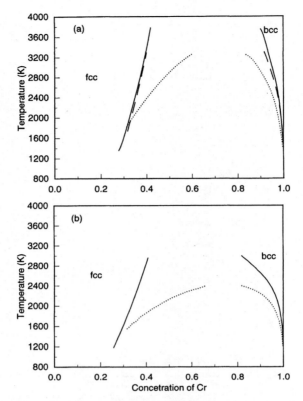

**Figure 3.** Calculated Ni-Cr phase diagram: (a) in the Bragg-Williams (solid lines) and CVM (dashed lines) approximations without electronic excitations and (b) in the Bragg-Williams approximation including electronic excitations. The dotted line corresponds the a metastable miscibility gap in the *bcc* phase calculated using the Bragg-Williams approximation.

where $N_I$ and $m_I$ are, respectively, the number and mass of atoms of type $I$ ($I = A, B$), $N = N_A + N_B$, and where the factor $Z$ is given by:

$$Z = \frac{1}{N_A! N_B!} \int d\vec{R}^N \, e^{-E(\vec{R}_N)/k_B T}, \tag{3}$$

with $E(\vec{R}_N)$ the energy of the system with the $N$ particles located at the positions $\vec{R}_N = \{\vec{R}_1, \vec{R}_2, \ldots, \vec{R}_N\}$. Thus, in Eq. (2), the integral over the momentum degrees of freedom has been explicitly carried out, and the remaining integral in $Z$ is over the positions of all $N$ particles, $\vec{R}_N$. We note that the integral over $\vec{R}_N$ includes the configurational degrees of freedom. Furthermore, $Z$ is the relevant partition function needed to calculate the free energy of formation.

In general, there is no simple representation for the energy $E(\vec{R}_N)$. However, for a given temperature and concentration range, the dominant excitations contributing to the partition function Eq. (3) correspond to displaced atomic configurations on a lattice; *i.e.*, the relevant excitations are characterized by a set of occupation numbers $\vec{\sigma}$ and a set of small displacements $\vec{u}_N$ from some reference lattice. In that case,

# First-Principles Study of Phase Equilibria in the Ni-Cr System

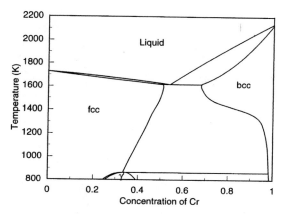

**Figure 4.** Experimental phase diagram for the Ni-Cr system.[28]

Eq. (3) can be approximated by:

$$Z = \sum_{\vec{\sigma}} \int' d\vec{u}^N e^{-[E_c(\vec{\sigma}) + \Delta E(\vec{\sigma}, \vec{u}_N)]/k_B T}, \quad (4)$$

where the integration over each displacement $\vec{u}_i$ is limited to the Wigner-Seitz cell of the $i$-th lattice point. In Eq. (4), $E_c(\vec{\sigma})$ is the configurational energy of the system in the absence of displacements and the term $\Delta E(\vec{\sigma}, \vec{u}_N)$ provides the coupling between the configurational and displacive degrees of freedom.

A fundamental difficulty in evaluating the partition function in Eq. (4), that has not been properly recognized in previous treatments of vibrational free energies, is that the harmonic approximation in the atomic displacements *is not* a good approximation to $\Delta E(\vec{\sigma}, \vec{u}_N)$. In fact, for Ni-rich (Cr-rich) compositions in the Ni-Cr system, the harmonic approximation for $\Delta E(\vec{\sigma}, \vec{u}_N)$ yields non-positive definite force constant matrices for most configurations $\vec{\sigma}$ in the *bcc* (*fcc*) lattices. This behavior of $\Delta E(\vec{\sigma}, \vec{u}_N)$ along a tetragonal distortion path connecting the *bcc* and *fcc* lattices at constant volume is illustrated in Fig. 5 for a NiCr compound. Thus, in general, we expect $\Delta E(\vec{\sigma}, \vec{u}_N)$ to be given by a highly anharmonic form in the atomic displacements:

$$\Delta E(\vec{\sigma}, \vec{u}_N) = \sum_i \left(\frac{\partial \Delta E}{\partial u_i^x}\right)_{\vec{u}_N = 0} u_i^x + \frac{1}{2} \sum_{i,j} \left(\frac{\partial^2 \Delta E}{\partial u_i^x \partial u_j^{x'}}\right)_{\vec{u}_N = 0} u_i^x u_j^{x'}$$
$$+ \frac{1}{3!} \sum_{i,j,k} \left(\frac{\partial^3 \Delta E}{\partial u_i^x \partial u_j^{x'} \partial u_k^{x''}}\right)_{\vec{u}_N = 0} u_i^x u_j^{x'} u_k^{x''} + \cdots, \quad (5)$$

where a summation is implied over repeated indices $x$, $x'$ and $x''$ ($= 1, 2, 3$). We also note that the derivatives of $\Delta E(\vec{\sigma}, \vec{u}_N)$ in the expansion of Eq. (5) are configuration dependent and, therefore, do not have neither the translational nor the point group symmetry of the undecorated lattice. We note that although the harmonic approximation is in general not applicable to $\Delta E$, this approximation can be used to describe the *configurational average* of the energy, $\langle \Delta E \rangle_c$ over a concentration range where the reference phase (*bcc* or *fcc* in Ni-Cr) is stable.

An upper bound for the free energy can be established by using standard perturbation theory:[29]

$$F \leq F_0 + \langle E - E_0 \rangle_0, \quad (6)$$

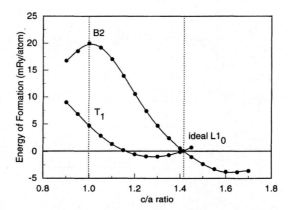

**Figure 5.** Energy of a NiCr compound along tetragonal distortion path. At $c/a = \sqrt{2}$ the structure corresponds to the ideal $L1_0$ fcc ordered phase. The path connecting to the B2 structure corresponds to a deformation along the unique $L1_0$ axis. The deformation for the other path is perpendicular to the unique axis

where $E = E_c(\vec{\sigma}) + \Delta E(\vec{\sigma}, \vec{u}_N)$ is the full Hamiltonian of the system and $E_0$ is a suitably chosen trial Hamiltonian. We introduce a trial Hamiltonian of the form:

$$E_o = E_c^0 + \tfrac{1}{2}\sum_{i,j} K_{i,j}^{x,x'} u_i^x u_j^{x'}, \tag{7}$$

where $E_c^0$ is a configurational energy on the perfect lattice given by ECIs $V_\alpha^0$ [different from the ECIs in Eq. (1)], and where the $K_{i,j}^{x,x'}$ are harmonic force constants with the symmetry of the undecorated lattice. As usual, both $V_\alpha^0$ and $K_{i,j}^{x,x'}$ play the role of variational parameters which must be determined self-consistently by minimizing the upper bound to the free energy. The reference free energy $F_0$ is given by:

$$F_0 = F_c^0(\{V_\alpha^0\}) + \frac{k_B T}{2}\sum_{\vec{q},\nu}\ln\left(\frac{\lambda_{\vec{q}}^\nu}{\pi k_B T}\right), \tag{8}$$

where $\lambda_{\vec{q}}^\nu$ are the eigenvalues of the force constant matrix $(K_{i,j}^{x,x'})$ for reciprocal lattice vector $\vec{q}$ and mode $\nu$ $(=1,2,3)$, and where the configurational free energy $F_c^0$ is:

$$F_c^0(\{V_\alpha^0\}) = -k_B T \ln\left(\sum_{\vec{\sigma}} e^{-E_c^0/k_B T}\right). \tag{9}$$

Keeping only quadratic terms in the displacements for *average* of the Hamiltonian $\langle \Delta E \rangle_0$, the minimization of the upper bound to the free energy with respect to the variational parameters yields:

$$V_\alpha^0 = V_\alpha + \frac{k_B T}{2}\frac{\partial}{\partial \langle \Phi_\alpha \rangle_0}\left[\sum_{\vec{q},\nu}\ln\left(\lambda_{\vec{q}}^\nu\right)\right], \tag{10}$$

$$K_{i,j}^{x,x'} = \left\langle \frac{\partial^2 \Delta E}{\partial u_i^x \partial u_j^{x'}}\right\rangle_0, \tag{11}$$

and

$$F \approx F_c^0(\{V_\alpha^0\}), \tag{12}$$

where in Eq. (10), $\langle \Phi_\alpha \rangle_0$ is the correlation function associated to cluster $\alpha$, and where the derivatives in Eq. (11) evaluated at $\vec{u}_N = 0$.

We note that $K_{i,j}^{x,x'}$ depends on concentration and SRO and, therefore, it must be calculated self-consistently. A convenient way of approaching this self-consistency is to cluster expand the total spectral density of states of the force constant matrix:[20]

$$n(\lambda) = \sum_\alpha n_\alpha(\lambda) \langle \Phi_\alpha \rangle_0, \tag{13}$$

in which case the ECIs renormalized by lattice vibrations become:

$$V_\alpha^0 = V_\alpha + \frac{k_B T}{2} \int_0^\infty n_\alpha(\lambda) \ln(\lambda) \, d\lambda. \tag{14}$$

Given the ECIs and the force constants $K_{i,j}^{x,x'}$, the evaluation of $F_c^0$ can be carried out using standard techniques such as the cluster variation method or Monte Carlo simulations. The procedure requires the diagonalization of $K_{i,j}^{x,x'}$ which can be conveniently carried out in reciprocal space since the matrix has the symmetry of the undecorated lattice.

In order to assess the approximation involved in using Eq. (6) we note that taking the average of the squared displacements ($\langle u^2 \rangle_0$) as the small parameter in the theory, an expansion of the free energy correct to order of $\langle u^2 \rangle_0$ yields:

$$F = F_0 + \frac{1}{2} \sum_{i,j} \left[ \left\langle \frac{\partial^2 \Delta E}{\partial u_i^x \partial u_j^{x'}} \right\rangle_0 - \frac{1}{k_B T} \left\langle \frac{\partial \Delta E}{\partial u_i^x} \frac{\partial \Delta E}{\partial u_j^{x'}} \right\rangle_0 - K_{i,j}^{x,x'} \right] \langle u_i^x u_j^{x'} \rangle_0$$
$$+ O(\langle u^4 \rangle_0), \tag{15}$$

where as before the derivatives are evaluated at $\vec{u}_N = 0$. Thus, a slightly improved approximation over that of Eq. (6) can be obtained with the force constant matrix given by:

$$K_{i,j}^{x,x'} = \left\langle \frac{\partial^2 \Delta E}{\partial u_i^x \partial u_j^{x'}} \right\rangle_0 - \frac{1}{k_B T} \left\langle \frac{\partial \Delta E}{\partial u_i^x} \frac{\partial \Delta E}{\partial u_j^{x'}} \right\rangle_0. \tag{16}$$

These second order force constants can be obtained by cluster expanding the configurational average of the square of $\Delta E$ which, to second order in the atomic displacements, is given by:

$$\langle \Delta E^2 \rangle_0 = \sum_{i,j} \left\langle \frac{\partial \Delta E}{\partial u_i^x} \frac{\partial \Delta E}{\partial u_j^{x'}} \right\rangle_0 \langle u_i^x u_j^{x'} \rangle_0. \tag{17}$$

It is perhaps worthwhile to point out that the second order correction to the force constants does not require additional total energy calculations beyond those needed for an accurate cluster expansion representation of $\langle \Delta E \rangle$.

## III.1 Force Constants in Ni-Cr

For pure $fcc$ Ni it is possible to obtain a reasonably accurate description of the phonon spectrum using a general nearest-neighbor force constant matrix. In the case of $bcc$ Cr the same level of accuracy in the description of the phonon spectrum is obtained including a general force constant between nearest-neighbor and a central force between next-nearest-neighbor. In both cases, the force constants are fully determined by the three cubic elastic constants $c_{11}$, $c_{12}$, and $c_{44}$.[30] In Figs. 6 and 7 we compare

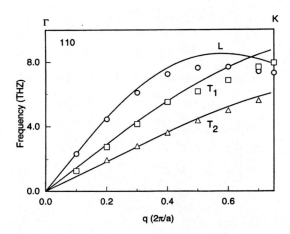

**Figure 6.** Phonon dispersion for pure Ni along the (110) direction calculated from the experimental elastic constants. Symbols are experimental data.[31]

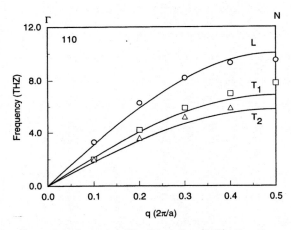

**Figure 7.** Phonon dispersion for pure Cr along the (110) direction calculated from the experimental elastic constants. Symbols are experimental data.[32]

calculated and experimental[31,32] phonon dispersions along the (110) direction in reciprocal space for, respectively, pure Ni and Cr. The calculated dispersion relations were obtained using the experimental elastic constants of Ni and Cr. Similar agreement between experiment and theory is found for other high symmetry directions in reciprocal space.

Assuming the same range of interactions as in the pure elements, the first order average force constants required for the calculation of the vibrational free energy of *fcc* and *bcc* Ni-Cr alloys [see Eq. (11)] can be be obtained from the average of the three cubic elastic constants $\langle c_{11} \rangle_c$, $\langle c_{12} \rangle_c$, and $\langle c_{44} \rangle_c$. These average elastic constants, in turn, follow from a cluster expansion of the *configurational average* of the energy ($\langle \Delta E \rangle_c$) as a function of homogeneous deformations. Such an expansion has recently been carried out by Craievich et al.[7] for a Bain tetragonal distortion in Ni-Cr. The procedure consists of calculating the energy of several ordered compounds for a set

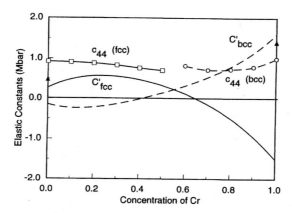

**Figure 8.** Elastic constants for random *fcc* and *bcc* Ni-Cr alloys. Open squares correspond to the shear elastic constant $c_{44}$ obtained by fitting the vibrational free energy at 1500 K. Full triangle are experimental values of $C'$.

of homogeneous deformations and carrying out the cluster expansion with ECIs that are function of the homogeneous strains. Figure 8 shows the results for the average elastic constant $\langle C' \rangle_c = (\langle c_{11} \rangle_c - \langle c_{12} \rangle_c)/2$ for random *fcc* and *bcc* Ni-Cr alloys as a function of concentration.[20] We note that for pure Ni and Cr the calculated values of $C'$ are in reasonably good agreement with experiment (full symbols in Fig. 8).

The two remaining elastic constants (bulk modulus and $\langle c_{44} \rangle$) were estimated from experimental data. The bulk modulus and its concentration dependence is relatively simple to obtain by the cluster expansion method, which essentially predicts a linear dependence with concentration. However, the LMTO-ASA approximation used here gives values for the bulk moduli of the pure elements that are significantly larger than the experimental ones. In view of this, we assumed that the bulk modulus for the alloy is given by the composition average of the experimental values for pure Ni and Cr.

The LMTO-ASA approximation is also inadequate for the calculation of $c_{44}$. Therefore, the concentration dependence of $c_{44}$ was adjusted to fit an experimental estimate of the vibrational free energy at one temperature (1550 K). This experimental estimate was obtained by subtracting from the measured total free energy of formation at 1550 K the calculated configurational free energy (including electronic excitations) at the same temperature. The resulting $\langle c_{44} \rangle$ are shown by the open symbols in Fig. 8. We note that the composition dependence of $\langle c_{44} \rangle$ needed to fit the estimated vibrational free energy is relatively small.

The calculated ECIs, elastic constants and electronic free energy, provide all the elements needed for the calculation of the total free energy of formation. From this free energy, it is straightforward to calculate the equilibrium-temperature composition phase diagram. The calculated phase diagram in the Bragg-Williams approximation is shown in Fig. 9. We see that including vibrational modes brings the calculated phase diagram in excellent agreement with experiment.

## IV. Conclusions

The methodology developed for the treatment of configurational and displacive degrees of freedom is particularly well adapted for utilization with first-principles electronic

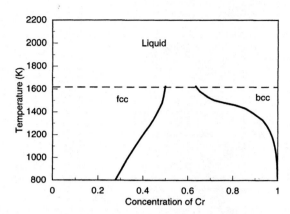

**Figure 9.** Calculated Ni-Cr phase diagram in the Bragg-Williams approximation including electronic excitations and vibrational modes

structure methods and brings, for the first time, first-principles phase diagram calculations into a solid quantitative basis. In particular the theory satisfactorily resolves the long-standing problem of treating vibrational modes in alloys that are known to be mechanically unstable for many (if not most) atomic configurations with lower symmetry than the stable phase.

All parameters entering the theory, with the exception of the bulk modulus and shear elastic constant $c_{44}$, were obtained from first-principles electronic structure calculations. For the bulk modulus of the alloy we used a composition average of the bulk moduli of pure Ni and Cr, and the shear elastic constant $c_{44}$ was fitted to an estimate of the vibrational free energy at one temperature. The reliance on experimental data for the calculations presented here are due primarily to limitations of the LMTO-ASA method. It is expected that a full potential electronic structure method should provide reliable values for these elastic constants.

In Ni-Cr, however, the concentration dependence of the bulk modulus and $c_{44}$ play a secondary role. Most of the concentration dependence of the vibrational free energy follows from that of $C' = (c_{11} - c_{12})/2$, which was obtained from first-principles ECIs as a function of a tetragonal Bain distortion.

Finally, he treatment of configurational, electronic and vibrational free energies in the Ni-Cr system was shown to provide a quantitative description of thermodynamic potentials and phase diagram that is an excellent agreement with experiments.

# References

1. J. M. Sanchez, F. Ducastelle, and D. Gratias, *Physica A* **128**, 334 (1984).
2. J. M. Sanchez, *Phys. Rev. B* **46**, 14 013 (1993).
3. J.M. Sanchez, in *Theory and Applications of the Cluster Variation and Path Probability Methods*, edited by J. L. Morán-López and J. M. Sanchez, (Plenum, New York, 1996).
4. W. Kohn and L. J. Sham, *Phys. Rev. A* **140**, 1133 (1965).
5. C. Wolverton and A. Zunger, *Phys. Rev. B* **52**, 8813 (1995).
6. E. G. Moroni, G. Grimvall, and T. Jarborg, *Phys. Rev. Lett.* **76**, 2758 (1996).

7. P. J. Craievich, J. M. Sanchez, M. Weinert, and R. E. Watson, *Phys. Rev. B* **55**, 787 (1997).
8. S. de Gironcoli, P. Giannozzi, and S. Baroni, *Phys. Rev. Lett.* **66** 2116 (1991)
9. Z. W. Lu et al., *Phys. Rev. B* **44**, 512 (1991).
10. V. L. Moruzzi, J. F. Janak, and K. Schwartz, *Phys. Rev. B* **37**, 790 (1988).
11. J. M. Sanchez, J. P. Stark, and V. L. Moruzzi, *Phys. Rev. B* **44**, 5411 (1991).
12. M. Asta, R. McCormack, and D. de Fontaine, *Phys. Rev. B* **48**, 748 (1993).
13. J. D. Becker and J. M. Sanchez, *Mater. Sci. Eng. A* **170**, 161 (1993).
14. T. Mohri, S. Takizawa, and K. Terakura, *J. Phys. Condens. Matter* **35**, 1473 (1993).
15. C. Amador, W. R. L. Lambrecht, and B. Segall, *Mat. Res. Symp. Proc.* **253**, 297 (1992).
16. J. M. Sanchez, in *Structural and Phase Stability of Alloys*, edited by J. L. Morán-López, F. Mejía-Lira, and J. M. Sanchez, (Plenum Press, New York, 1992).
17. C. Colinet et al., *J. Phys. Condens. Matter* **6**, L47 (1994).
18. G. D. Garbulsky and G. Ceder, *Phys. Rev. B* **53**, 8993 (1996).
19. A. Chiolero and D. Baeriswyl, *J. Stat. Phys.* **76**, 347 (1994)
20. P. J. Craievich, Ph.D. Thesis, The University of Texas at Austin, 1997.
21. P. J. Craievich, M. Weinert, J. M. Sanchez, and R. E. Watson, *Phys. Rev. Lett.* **72**, 3076 (1994).
22. H. L. Skriver, *The LMTO Method* (Springer-Verlag, Berlin, 1984).
23. R. Hultgren et al., *Selected Values of the Thermodynamic Properties of Binary Alloys* (Am. Soc. Metals, Metals Park, OH, 1973).
24. N. C. Tso, M. Kosugi, and J. M. Sanchez, *Acta Metall.* **37**, 121 (1989).
25. R. Kikuchi, *Phys. Rev.* **81**, 988 (1951).
26. J. M. Sanchez and D. de Fontaine, *Phys. Rev. B* **17**, 2926 (1978).
27. J. M. Sanchez, M. C. Cadeville, V. Pierron-Bohnes, and G. Inden, *Phys. Rev. B* **54**, 8958 (1996).
28. T. B. Massalski, J. L. Murray, L. H. Bennett, and Hugh Baker, *Binary Alloy Phase Diagrams; Selected Values of the Thermodynamic Properties of Binary Alloys* (Am. Soc. Metals, Metals Park, OH, 1973).
29. R. P. Feynman, *Statistical Mechanics, A set of Lectures* (W. A. Benjamin Inc., Reading, Massachusetts, 1972).
30. B. T. M. Willis and A. W. Pryor, *Thermal Vibrations in Crystallography* (Cambridge University Press, Cambridge, 1975).
31. R. J. Birgeneau, J. Cordes, G. Dolling, and A. D. B. Woods, *Phys. Rev.* **136**, 1359 (1964).
32. W. M. Shaw and L. D. Muhlestein, *Phys. Rev. B* **4**, 969 (1971).

# Theoretical Aspects of Porous Silicon

M. R. Beltrán,[1] C. Wang,[1] M. Cruz,[2] and J. Tagüeña-Martínez[3]

[1]*Instituto de Investigaciones en Materiales*
*Universidad Nacional Autónoma de México*
*Apartado Postal 70-360, 04510 México D. F.*
*MEXICO*

[2]*Escuela Superior de Ingeniería Mecánica y Eléctrica-UC*
*Instituto Politécnico Nacional*
*MEXICO*

[3]*Centro de Investigación en Energía*
*Universidad Nacional Autónoma de México*
*Apartado Postal 34, 62580 Temixco, Morelos*
*MEXICO*

## Abstract

In this work, we study the electronic and optical properties of porous silicon (PS) using a supercell model, where an $sp^3s^*$ tight-binding Hamiltonian is used and empty columns of atoms are produced in an otherwise perfect silicon structure, passivated with hydrogen atoms. As it is considered that quantum confinement is one of the causes of the optoelectronic properties of PS, we perform a detailed analysis of the consequences of confinement on its band structure. Our results show that the band gap broadens and the minimum of the conduction band shifts towards the gamma point as the porosity is increased. The polarized light absorption study shows that the optically active zone in the reciprocal space broadens significantly due to disorder, relaxing the k-wavevector selection rule. We found that introducing non-vertical interband transitions to take into account the PS disordered nature, we get a very good agreement with experimental data.

## I. Introduction

The nanostructures have opened a new area in materials research, since they present interesting phenomena such as efficient luminescence and localization of carriers. In particular, an important example of these structures is porous silicon (PS), due to its possible applications to the optoelectronic industry discovered at the beginning of the nineties.[1] PS emerges as a new material built upon crystalline silicon wafers through a corrosive treatment with HF. This treatment leads to a coral or sponge-like structure with branches whose diameters are nanometric, which presents very different and interesting properties with respect its crystalline counterpart, such as visible efficient luminescence at room temperature. This luminescence is particularly surprising for an indirect gap type material where vertical dipole interband transitions are forbidden.

It is considered that quantum confinement is an essential ingredient for the optoelectronic properties of PS,[1] although the surface contribution is also very important and cannot be neglected:[2] the surface to volume ratio in porous silicon samples is approximately 200 m$^2$/cm$^3$ (Ref. 3). Furthermore, in PS the branches intermingle and are randomly distributed; this disorder partially breaks the **k**-wavevector selection rule and allows new radiative transitions even without significant phonon assistance.[4] In spite of intensive theoretical[5] and experimental[6] research done during the last five years, it is still not possible to clearly elucidate the origin of this luminescence, thus further microscopic analysis should be performed to clearly identify its source.

The calculation of the electronic properties of PS requires to choose on one hand a geometrical structure and on the other a Hamiltonian. PS is a very complex material but the idealized view adopted in most theories is to consider it as a periodic network of parallel crystalline silicon columns or crystallites. The main disadvantage of this approach is that it ignores disorder and interconnections between branches, which are fundamental, for example, to describe the transport properties. Besides, the interconnections compete with the quantum confinement, modifying electronic behaviors, such as the effective mass and the localization properties.[9]

From the Hamiltonian's choice point of view, the models can be classified in two mayor categories: first-principles and semiempirical frameworks. The first-principle calculations have been performed mainly for quantum dots (clusters)[7] or quantum wires.[8] Semiempirical or tight-binding calculations are simple enough to be applied to larger systems with complex morphologies. On the other hand, the use of phenomenological parameters, which are very well known for silicon, includes some many-body effects that are neglected in a first-principles Hamiltonian.

Recently, we have introduced a semi-empirical model to study the porosity and pore morphology effects on the electronic band structure of PS.[9] Using this model, we have also calculated the oscillator strength and the dielectric function for different porosities. We have compared the calculated dielectric function with experimental data for bulk c-Si, ultrathin c-Si films and PS.

In the next Section we describe the microscopic supercell model. Following this, we present and discuss the results. Finally, some conclusions are given.

## II. Theory

We use a tight-binding Hamiltonian with an $sp^3s^*$ basis, which is the minimum basis capable of describing an indirect 1.1 eV band gap along the X-direction for bulk crystalline silicon, where Vogl, Hjalmarson, and Dow's parameters[10] are taken. We use cubic 8-atom supercells as a unit basis to build up larger ones joining them in the X-Y plane. Empty columns are produced removing columns of atoms within the supercell in the [001] (Z) direction.

As PS exhibits a very large surface mainly hydrogen passivated,[1] we saturate the pore surface with hydrogen atoms. The Si-H bond length is taken as 1.48 Å. The on-site energy of the H atom is considered to be $-4.2$ eV, since the free H atom energy level, $-13.6$ eV, is so close to the s-state energy level of a free Si atom, $-13.55$ eV (Ref. 11), therefore the on-site energy of H is taken to be the same as that of silicon, as in Ref. 12. The H-Si orbital interaction parameters are taken as $ss\sigma_{\text{H-Si}} = -4.075$ eV, $sp\sigma_{\text{H-Si}} = 4.00$ eV, which are obtained by fitting the energy levels of silane.[7]

Apart from providing valuable information of the electronic behavior, the electronic structure calculations are a key factor for the study of optical properties through the oscillator strength analysis.[13,14] We believe that this analysis could give insights into the controversy of having an indirect gap in PS as it is suggested by induced absorption experiments,[6] even when high luminescence efficiency is observed. We start by defining the dimensionless interband oscillator strength following Koiller, Osorio, and Falicov:[13]

$$f_{v,c} = 2m|\langle v|\mathbf{p}|c\rangle|^2/(E_c - E_v), \tag{1}$$

where $|v\rangle$ and $|c\rangle$ are valence- and conduction-band eigenstates, respectively. In the tight-binding scheme eigenstates $|e\rangle = \sum_{i,\mu} a^e_{i,\mu}|i,\mu\rangle$, where $i$ is the site index and $\mu$ identifies the orbital, then the dipole matrix in Eq. (1) can be expressed as

$$\langle v|\mathbf{p}|c\rangle = \sum_{i,j,\mu,\nu} (a^v_{i,\mu})^* a^c_{j,\nu} \langle i\mu|\mathbf{p}|j\nu\rangle. \tag{2}$$

The dipole matrix elements in Eq. (2) may be rewritten in terms of the Hamiltonian ($H$) and the position ($\mathbf{r}$) operators, using the commutation relation $\mathbf{p} = im/\hbar[H,\mathbf{r}]$,

$$\langle i\mu|\mathbf{p}|j\nu\rangle = im\hbar \sum_{l,\lambda} \left( \langle i\mu|H|l\lambda\rangle\langle l\lambda|\mathbf{r}|j\nu\rangle - \langle i\mu|\mathbf{r}|l\lambda\rangle\langle l\lambda|H|j\nu\rangle \right). \tag{3}$$

Since the polarizability of a free atom is much smaller than that of the corresponding semiconductor,[15] Eq. (3) can be simplified as:[13]

$$\langle i\mu|\mathbf{p}|j\nu\rangle = im\hbar\langle i\mu|H|j\nu\rangle \mathbf{d}_{i,j}, \tag{4}$$

where $\mathbf{d}_{i,j} = \langle j\nu|\mathbf{r}|j\nu\rangle - \langle i\mu|\mathbf{r}|i\mu\rangle$ is the distance between the gravity centers of the orbitals $\mu$ and $\nu$ placed at atoms $i$ and $j$, respectively, and it is independent on orbitals if the crystal field is symmetric. Notice that the contribution to the dipole matrix coming from two orbitals at the same atom is neglected.

The final stage of the calculation is to obtain the dielectric function ($\epsilon = \epsilon_1 + i\epsilon_2$). When pores are introduced, we take also into account the oscillator strength from non-vertical transitions, considering that there is disorder in PS and therefore a relaxation of the electron k-wavevector conservation should be included, i.e., the existence of a random perturbative potential, which can always be expanded as a Fourier series, makes a broadening of the perturbed wavefunction in the **k**-space. The imaginary part of the dielectric function is proportional to[11]

$$\epsilon_2 \propto \sum_{k,k'} f_{k,k'}\, \delta\left(\epsilon_{k'} - \epsilon_k - \hbar\omega\right), \tag{5}$$

where $f_{k,k'}$ are the oscillator strengths defined in Eq. (1), $k$ and $k'$ correspond to the valence and the conduction band states, respectively.

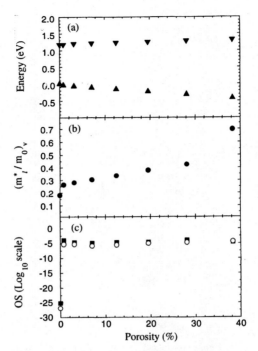

**Figure 1.** (a) Shifts of the conduction (solid down triangles) and valence (solid up triangles) band edges, (b) trend of the light-hole effective mass, (c) logarithm of the oscillator strength calculated following Ref. 13 (solid squares) and Ref. 14 (open circles). All these results are plotted *versus* the porosity and they are calculated for a fixed pore shape and 128-atom supercells.

## III. Results

In figure 1 we present a summary of the effects of the porosity on the electronic and optical properties of PS. All these calculations were obtained from 128-atom supercells removing square columns of 1, 4, 9, 16, 25, 36, and 49 atoms. Fig. 1(a) shows the variation of the conduction and the valence band edges as a function of the porosity. Notice that the band gap broadens with the porosity; this fact is not a conventional quantum confinement effect because there are Bloch wave functions on the confinement plane. However, the electron wave functions have nodes at the pore surfaces and these extra nodes cause a sort of quantum confinement and consequently a band gap broadening, since wave functions with wavelengths longer that the distance between nodes will not be accessible for the system. Furthermore, the band edges shift asymmetrically; the fact that the valence band edge shifts more rapidly is consistent with the difference in the band curvatures and it has been already observed.[12]

In figure 1(b) the light-hole effective mass as a function of the porosity is presented. In spite of an expected increase of the effective mass with the porosity, one can clearly observe two jumps. The first occurs when the pores are introduced, which is due to the appearance of new scattering centres. The other is found at the high porosity regime within the model and it is caused by the fact that we reached a percolation limit, and the pore edges almost touch each other.

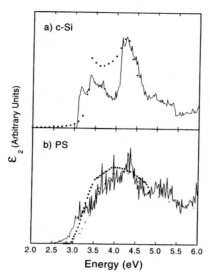

**Figure 2.** Imaginary part of the dielectric function calculated from interband transitions in an 8-atom supercell without [solid line in Fig. 2(a) and with solid line in Fig. 2(b)] a 1-atom columnar pore, for light excitations polarized in the X-direction. The theoretical results are compared with experimental c-Si data [solid circles in Fig. 2(a)],[16] thin 6 Å films results [open circles in Fig. 2(b)][17] and measured PS data [solid squares in Fig. 2(b)].[18]

Finally, Fig. 1(c) shows the oscillator strength *versus* the porosity, where transitions between wave functions at the valence band maximum and those at the conduction band minimum are considered. One can see that there is a jump of 20 orders of magnitude predicted by both approaches[13,14] when a pore is introduced. This jump is due to the fact that in supercell schemes, not all the states at the $\Gamma$ point have wave functions with $\mathbf{k} = 0$ symmetry. Rather, many are the product of the folding of the Brillouin zone, in particular, the states at the minimum of the conduction band for a 128-atom supercell. When a pore is introduced, the supercell becomes the unit cell of the system, and all $\Gamma$ states are real ones, in spite of the existence of extra nodes in the wavefunction. After the jump, an almost constant slope is found, instead of the monotonic increase previously observed in Fig. 1(b). This fact could be explained by a competing mechanism between the significant enlargement of the optically active zone in the k-space caused by the localization, and the sum rule $\sum_k f_{k',k} = 1$, where the sum is over all occupied as well as unoccupied states.

To produce Fig. 2 we have considered transitions between states of the valence band and those of the conduction band, for X-direction polarized light. They are calculated in 8-atom supercells[9] with and without columnar pores, all of which have been saturated by hydrogen atoms. Figure 2(a) shows the dielectric function for crystalline silicon calculated with vertical interband transitions (solid line), *i.e.*, $\mathbf{k}_i^c = \mathbf{k}_f^v$, and it is compared with the experimental data reported in Ref. 16 (solid circles). The calculation has been performed by considering 205 379 k-points in the first Brillouin zone. It can be seen that the theory gives reasonably well the energy range and the shape, in spite that no d-orbital is considered. It is important to notice that the absorption onset corresponds to the optical c-Si gap, which is much larger than the indirect band gap (1.1 eV).

In figure 2(b) the calculated dielectric function for a supercell with a 1-atom columnar pore is shown, including non-vertical transitions, *i.e.* $\mathbf{k}_i^c \neq \mathbf{k}_f^v$, to consider the disorder effects in PS which have been excluded by the supercell model. The calculation has been performed only for 729 k-points in the first Brillouin zone, since adding non-vertical transitions is a very lengthy calculation. Notice that the dielectric function shows a smaller band gap associated to PS [Fig. 2(b)] than that of c-Si [Fig. 2(a)], although a band gap broadening is observed in PS [Fig. 1(a)]. This happens because for c-Si only the optical band gap is being revealed by the optical absorption spectra, while for PS the dielectric function shows the real band gap, since the relaxation of the electron wavevector conservation, introduced by the disorder present in PS, allows interband transitions between all the states. Furthermore, a tail is observed in Fig. 2(b) in contrast with the c-Si case, and this is due again to the disorder, as occurred in *a*-Si.[17] Finally, it is worth mentioning that although the porosity simulated by this supercell is only 12.5%, while the experimental data are obtained from 70% porosity samples,[18] the comparison of the dielectric function between theoretical and experimental data is in good agreement, since the characteristic confinement lengths are similar in both cases

## IV. Conclusions

We have shown that a simple microscopic quantum mechanical treatment, such as a phenomenological tight-binding technique, is capable of reproducing the essential electronic properties of PS, and that the quantum confinement concept is adequate to explain all the electronic and optical properties calculated in this work. However, as we have mentioned, the quantum confinement concept should be extended to include intermediate situations, where nanometric silicon columns intermingle. We call this quasi-confinement[9] in PS where electrons can find ways out through the necks between the pores. The quantum confinement is partial, and becomes complete only at the limit where the pores touch each other. This behavior is not only a peculiarity of our model, but we think that it occurs in the real PS where columns intermingle, producing alternative connections, where the carriers eventually find paths from a quantum wire to another. This assertion is supported by transmission electron microscopy view-graphs that show wire interconnections.[19] This quasi-confinement concept is important to explain the jump observed in the effective mass at high porosity regime [Fig. 1(b)].

This model is also capable to reproduce essential features of the dielectric function of PS. Actually, from the comparison with the experimental data we can conclude that the tight-binding supercell model, when non-vertical transitions are included, gives the correct energy range and the shape of the dielectric response of PS. The need of introducing non-vertical interband transitions to take into account the PS disordered nature, reveals a significant enlargement of the optically active **k** zone. This enlargement can be explained by the Heisenberg's uncertainty principle, due to the localization of the wavefunctions caused by the disorder of the shape and distribution of pores. Furthermore, this enlargement together with the trend observed towards a direct band gap,[9] could reconcile the apparent controversy between the efficient luminescence of PS and its indirect gap observed experimentally.

It is interesting to notice the fact that a very thin c-Si film has similar experimental dielectric function behavior to PS, since the quantum confinement is essential in both cases. As a first approximation we are describing the pores as columns and saturating the surface with hydrogen atoms. Clearly, both approximations can be extended to include other saturators, surface relaxation and amorphization, these studies are currently in progress.

## Acknowledgments

We would like to thank the technical support of Sara Jiménez Cortés. This work has been partially supported by projects from DGAPA-IN104595, CONACyT-0205P-E9506, DGAPA-IN101797, CONACyT-4229-E and CRAY-UNAM-SC008697.

## References

1. L. T. Canham, *Appl. Phys. Lett.* **57**, 1046 (1990); A. G. Cullis and L. T. Canham, *Nature* **353**, 335 (1991); L.T. Canham *et al.*, *J. Appl. Phys.* **70**, 422 (1991).
2. D. I. Kovalev *et al.*, *Appl. Phys. Lett.* **64**, 214 (1994).
3. G. Bomchil, R. Herino, K. Barla, and J. C. Pfister, *J. Electrochemi. Soc.* **130**, 1611 (1983).
4. C. Delerue *et al.*, *Phys. Rev. Lett.* **75**, 2228 (1995).
5. George C. John and Vijay A. Singh, *Physics Reports* **263**, 93 (1995).
6. P. Fauchet, "Porous Silicon: Photoluminescent Devices" in *Light Emission in Silicon*, edited by D. Lockwood, to appear in the Semiconductors and Semimetals Series (Academic Press, New York, 1996).
7. Fu Huaxiang, Ye Ling, and Xide Xie, *Phys. Rev. B* **48**, 10978 (1993).
8. F. Buda, J. Kohanoff, and M. Parinello, *Phys. Rev. Lett.* **69**, 1272 (1992).
9. M. Cruz, C. Wang, M. R. Beltrán, and J. Tagueña-Martínez, *Phys. Rev. B* **53**, 3827 (1996).
10. P. Vogl, H. P. Hjalmarson, and J. D. Dow, *J. Phys. Chem. Solids* **44**, 365 (1983).
11. Walter A. Harrison, in *Electronic Structure and the Properties of Solids* (Dover Pub., New York, 1989), p. 50 and p. 100.
12. Shang Yuan Ren and John D. Dow, *Phys. Rev. B* **45**, 6492 (1992).
13. B. Koiller, R. Osório, and L. M. Falicov, *Phys. Rev. B* **43**, 4170 (1991).
14. A. Selloni, P. Marsella, and R. Del Sole, *Phys. Rev. B* **33**, 8885 (1986).
15. L. Brey and C. Tejedor, *Solid State Commun.* **48**, 403 (1983).
16. G. E. Jellison, *Optical Materials* **1**, 41 (1992).
17. H. V. Nguyen, Y. Lu, S. Kim, M. Wakagi, and R. W. Collins, *Phys. Rev. Lett.* **74**, 3880 (1995).
18. N. Koshida *et al.*, *Appl. Phys. Lett.* **63**, 2774 (1993).
19. O. Teschke, F. Alvarez, L. Tessler, and R. L. Smith, *Appl. Phys. Lett.* **63**, 1927 (1993).

# The Principle of Self-Similarity and Its Applications to the Description of Noncrystalline Matter

Richard Kerner

*Laboratoire de Gravitation et Cosmologie Relativistes*
*Université Pierre et Marie Curie - CNRS URA 769*
*Tour 22, 4-ème étage, Boite 142*
*4, Place Jussieu, 75005 Paris*
*FRANCE*

## Abstract

We discuss he concept of self-similarity and illustrate it on several examples from mathematics and physics. Next we show how various growth and agglomeration processes can be described by means of systems of algebraic or differential equations which rule the evolution and behavior of the probabilities of elementary agglomeration processes. These methods are then used for modeling of several growth and agglomeration processes, usually leading to noncrystalline states of condensed matter, such as fullerenes, quasi-crystals or network glasses

## I. The Concept of Self-Similarity: A Few Examples

The self-similarity is one of those principles that elude precise definition, although it seems to be very natural. In spite of this fact, we believe that after considering a few examples, it will become quite clear what we mean by this concept. The most general (and incidentally the vaguest) definition of what we have in mind when we evoke self-similarity while looking at any space-filling structure can be summarized by the words "more and more of the same," which express rather a general feeling than an exact definition. This is the idea we grasp immediately when we observe a growing crystal, or when we try to analyze closely the structure of a crystalline lattice or a snowflake already formed.

We have a similar feeling when we observe certain more complicated structures, like fractals displaying at any magnification scale similar, although never strictly identical images. Another good example is provided by quasi-crystals, in which any local configuration can be found again at a finite distance from the original one, although they lack both translational and rotational invariance characteristic for crystalline matter.

Even in the apparently totally disordered systems such as network glasses, one can easily observe the repetition of characteristic local structures such as rings or star-like configurations, as well as certain typical correlations between them. In this case, although it is hard to find exactly the same configuration, the repetition and self-similarity become evident if we use the probabilistic description of the network. This is akin to what we see in a forest, where starting from certain size of samples considered we get the entire statistical information, so that there is no need to consider any larger neighborhoods.

Perhaps this is the most characteristic feature of the self-similarity we have in mind: the fundamental information about the structure of a complicated system is contained already in quite small samples, and we can reproduce all the essential features by adding up and repeating similar subsets *ad infinitum* even if they are not strictly identical like in crystalline lattice, but just very much alike, as in fractals, quasi-crystals or network glasses.

In what follows, we shall give examples of self-similar growth processes and present simple mathematical techniques describing them. Then we apply these techniques to the models of growth of Fibonacci chains, fullerenes and covalent network glasses.

## II. Mathematical Description of Self-Similar Growth

There is perhaps nothing as self-similar as a crystal, in which the same and only building block (called an elementary cell) is repeated and added over and over again, tiling the entire space. It is also well known that in spite of the infinite variety of the internal structures of the elementary cells, the overall crystal symmetries resulting from their translation-invariant arrangements can all be classified in thirty-two crystal groups.[1]

However, the self-similarity has been introduced long before by ancient Greeks as an important notion in geometry; curiously enough, it describes the features shared by crystals and quasi-crystals alike.

Its essence can be illustrated on the example of the golden ratio. If we consider a rectangle with the sides $a$ and $b$, and if we cut out a square of the side $a$ on one of its sides (*e.g.*, the square $a^2$ if $a < b$), then the remaining figure is also a rectangle with the sides $(b-a)$ and $a$. Among all rectangles [*i.e.*, with arbitrary ratios $(b/a)$] there is only one that displays the self-similarity property:

$$a : (b-a) = b : a. \tag{1}$$

The ancient Greeks gave this particular shape the name of golden rectangle, or the golden ratio. The requirement expressed by Eq. (1) defines the shape of the golden rectangle in a unique way, because Eq. (1) has only one solution with $a, b$ both positive, and $a < b$. Let us call $(b/a) = \tau$. Then we can write:

$$(\tau - 1)^{-1} = \tau \Rightarrow 1 = \tau^2 - \tau \Rightarrow \tau^2 - \tau - 1 = 0.$$

The above equation has two real solutions, but only one of them is positive:

$$\tau = \tfrac{1}{2}(1 + \sqrt{5}) = 1.618\,034\ldots,$$

# The Principle of Self-Similarity

the other root being equal to $\frac{1}{2}(1-\sqrt{5}) = -0.618\,034\ldots < 0$. This value of $\tau$ defines the famous golden number, whose inverse is equal to itself less one.

Starting from a small golden rectangle (the seed), one can grow on it a bigger one by adding an appropriate square, thus getting golden rectangles of an arbitrarily large size. Inscribing quarters of circles into these squares, one can create a shape quite faithfully rendering the form of certain spiral shells found in Nature. The scale factor between the consecutive (growing) quarters of circles is equal to $\tau$. This golden number can be defined as the limit of an infinite Diophantine fraction:

$$\frac{1}{1},\quad \frac{1}{1+1},\quad \frac{1}{1+\dfrac{1}{1+1}},\quad \frac{1}{1+\dfrac{1}{1+\dfrac{1}{1+1}}},\quad \text{etc.} \tag{2}$$

The infinite limit can be approximated by the following series of fractions:

$$\tfrac{1}{2}=0.5;\quad \tfrac{2}{3}=0.667;\quad \tfrac{3}{5}=0.6;\quad \tfrac{5}{8}=0.625;\quad \tfrac{8}{13}=0.6154;\quad \tfrac{13}{21}=0.619\,05\ldots.$$

The consecutive values of this series of fractions approach the inverse of the golden number, $\tau^{-1} = 0.618\,034\ldots$. It is also important to note the oscillating character of this approximation: once it is greater than the limit, next time it is smaller, then greater again, although the amplitude of variations decreases very rapidly. The numbers appearing in these fractions form the so called Fibonacci series: 1, 2, 3, 5, 8, 13 ..., and are formed with a simple algorithm: each term (except the first two) is the sum of the two former ones: $3 = 2+1$, $5 = 3+2$, $8 = 5+3$, etc. Fibonacci claimed that the growth of new branches follows this scheme in many species of trees.

Another example of growth process obeying Fibonacci's rule is the creation of one-dimensional quasi-crystalline lattice, realized by means of geometric representation as follows. Suppose that we have at our disposal two types of building blocks, one short and another long, symbolized by $S$ and $L$, respectively. The above algorithm can be applied to the following transformation of these entities: after each next step of growth from a "seed," which may be either of two elementary blocks, each short item $S$ transforms into a long one $L$, while the long one $L$ transforms itself into a pair $SL$, like in the following diagram:

$$L \to SL \to LSL \to SLLSL \to LSLSLLSL \to SLLSLLSLSLLSL \ldots.$$

The ratio of the total number of $L$ blocks to the $S$ blocks tends again to the golden limit $1,618\,034\ldots$.

Now, what is the exact relation between the Fibonacci series and the golden number? It becomes clear if we translate the problem into modern terms.

The algorithm defining the Fibonacci series can be viewed upon as a mapping of the space of two-dimensional vectors into itself, $\mathbf{x} \mapsto \mathcal{A}\mathbf{x}$, given in a chosen basis by the following $2 \times 2$ matrix (often called the inflation matrix):

$$\begin{pmatrix} x_1 \\ x_2 \end{pmatrix} \mapsto \begin{pmatrix} 0 & 1 \\ 1 & 1 \end{pmatrix} \begin{pmatrix} x_1 \\ x_2 \end{pmatrix} = \begin{pmatrix} x_2 \\ x_1 + x_2 \end{pmatrix}. \tag{3}$$

Then, starting with the vector $\begin{pmatrix} 0 \\ 1 \end{pmatrix}$, and applying to it the matrix $\mathcal{A}$ many times, we get the series of vectors whose coordinates are the Fibonacci numbers:

$$[1,1];\quad [1,2];\quad [2,3];\quad [3,5];\quad [5,8];\quad [8,13],\quad \text{etc.}$$

Each consecutive second coordinate is the sum of the coordinates of previous vector. This can be written as the simple rule:

$$a_{n+2} = a_{n+1} + a_n$$

with $a_0 = 0$ and $a_1 = 1$. Note that we need to fix two first terms in order to define the subsequent series. The limit of the ratio $a_{n+1}/a_n$ for $n \to \infty$ is equal to the golden number $\tau = 1.618\,034\ldots$. This can be easily shown with the matrix notation introduced above. First, let us find the eigenvalues and the corresponding eigenvectors of the matrix $\begin{pmatrix} 0 & 1 \\ 1 & 1 \end{pmatrix}$. If we have $\begin{pmatrix} 0 & 1 \\ 1 & 1 \end{pmatrix} \begin{pmatrix} v_1 \\ v_2 \end{pmatrix} = \lambda \begin{pmatrix} v_1 \\ v_2 \end{pmatrix}$, then the eigenvalues $\lambda$ must satisfy the characteristic equation $\lambda^2 - \lambda - 1 = 0$, identical with Eq. (2) defining the golden ratio $\lambda_1 = \tau = \frac{1}{2}(1 + \sqrt{5})$, as well as the second solution which is negative, $\lambda_2 = -\frac{1}{2}(1 - \sqrt{5}) = 1 - \tau = -0.618\,03\ldots$.

Let us denote the eigenvectors corresponding to these eigenvalues by $\mathbf{v}_1$ and $\mathbf{v}_2$. The initial vector giving rise to the Fibonacci series is decomposed as

$$\begin{pmatrix} 0 \\ 1 \end{pmatrix} = \alpha \mathbf{v}_1 + \beta \mathbf{v}_2.$$

Now, the action of the matrix $\mathcal{A}$ on its eigenvectors amounts to simple multiplication by the corresponding eigenvalue. Therefore, each time when the matrix (operator) $\mathcal{A}$ acts on any linear combination of the two eigenvectors $\mathbf{v}_1$ and $\mathbf{v}_2$, the result is the multiplication of the first eigenvector $\mathbf{v}_1$ by $\lambda_1 = \tau$ and the second one, $\mathbf{v}_2$, by $\lambda_2 = (1 - \tau)$. After $n$ consecutive steps we get:

$$\mathcal{A}^n \begin{pmatrix} 0 \\ 1 \end{pmatrix} = \alpha \tau^n \mathbf{v}_1 + \beta \left( -\frac{1}{\tau} \right)^{n+1} \mathbf{v}_2. \qquad (4)$$

We observe that $\tau > 1$, whereas $|-\frac{1}{\tau}| < 1$. Because of this, the term proportional to $\mathbf{v}_1$ is growing after each application of the operator $\mathcal{A}$, while the second contribution (along the vector $\mathbf{v}_2$) is steadily decreasing, changing its sign after each action of $\mathcal{A}$. This is why we observed the oscillating approach to the limit of ratios in the Fibonacci series. With growing $n$ the result is approaching closely the direction of the first eigenvector $\mathbf{v}_1$, which explains why the golden ratio between the consecutive entries is attained as a limit:

$$\lim_{n \to \infty} \frac{a_{n+1}}{a_n} = \tau.$$

It is often useful to describe the great ensemble of discrete processes of this kind by the continuous limit, which usually leads to a set of ordinary nonlinear differential equations. Let us illustrate it on the example we are already familiar with, *i.e.*, the analysis of the statistical behavior of the Fibonacci series.

The linear transformation defined by the inflation matrix $\mathcal{A}$ can be also used as a generator of infinitesimal transformations in a plane $(x, y)$ if we extend its action onto $\mathbf{R}^2$, which corresponds to a continuous limit of the discrete growth problem considered above. Then we may write down the following system of differential equations:

$$\frac{d}{dt} \begin{pmatrix} x \\ y \end{pmatrix} = \begin{pmatrix} 0 & 1 \\ 1 & 1 \end{pmatrix} \begin{pmatrix} x \\ y \end{pmatrix} = \begin{pmatrix} y \\ x+y \end{pmatrix} \quad i.e., \quad \frac{dx}{dt} = y, \quad \frac{dy}{dt} = x + y. \qquad (5)$$

The trajectories of this system in the $(x, y)$-plane can be found from the single differential equation involving $x$ and $y$ only, $(dy/dx) = (x + y)/y$

The singular point $(x, y) = (0, 0)$ is called a "saddle point," because the incoming and outgoing trajectories meet there.

The asymptotes are the straight lines whose slope coincides with the slope of the trajectories $dy/dx$ which approach them. The equation defining them reads therefore $(y/x) = (dy/dx) = (x+y)/y$, so that the asymptotes are: $y = \tau x$ and $y = -\tau^{-1} x$. Starting from any point of the plane $(x, y)$ (*i.e.*, with arbitrary initial conditions), one always ends up on the straight line $y = \tau x$, whose slope is defined by the golden ratio, because this is the asymptote to which the outgoing trajectories converge.

In the next Section we shall apply these simple mathematical ideas to the description and analysis of several growth processes observed in nature, leading to various noncrystalline states of matter.

## III. A Model of Growth of Fullerene Molecules

The above ideas have been quite successfully applied to explain the growth of fullerene molecules[2-4] in a model proposed recently in the series of papers co-authored by Karl-Heinz Bennemann.[5-7] The model of nucleation and growth of the fullerene $C_{60}$ molecules follows strictly the ideas of self-similarity and convergence algorithms exposed above.

We propose the following model of agglomeration processes leading to the formation of fullerene molecules. These molecules are found in great abundance in the soot falling from the electric arc discharge between two graphite electrodes, at the temperature about 2800°C, in helium atmosphere (acting as a moderator) under the pressure of 0.4 atm. Each fullerene $C_{60}$-molecule contains 60 carbon atoms disposed in rings, with 20 hexagons and 12 pentagons arranged like in a soccer ball. Each pentagon is surrounded by hexagons only, while each hexagon is surrounded alternatively by three pentagons and by three hexagons.

In the hot flame surrounding the electric arc one finds many carbon clusters, the acetylene groups $C_2$, the molecules $C_3$, $C_4$ etc., up to $C_6$ benzene rings, $C_{10}$ naphtalene double rings, and even the $C_{12}$ molecules built up from three rings, two hexagons and one pentagon (see Fig. 1).

In an achieved fullerene molecule all the vertices are of the same type, in which two hexagons and one pentagon meet together. Two pentagons can never share an edge, neither three hexagons can share a common vertex. That is why, when a new ring is formed in one of the cavities of a $C_{12}$ molecule by addition of a $C_2$ or a $C_3$ molecule abundant in the hot gas, out of four possible stable configurations (excluding the formation of unstable seven-sided rings), only two are appropriate for the consecutive fullerene formation, the other two containing wrong combinations of polygons as shown in Fig. 1.

The same is true at each consecutive step of agglomeration, consisting in the creation of a new polygon. Every time we get only half of configurations that are proper for the fullerene building, the other half being lost because it contains the wrong configurations, such as two neighboring pentagons, or three hexagons sharing a common vortex.

Starting from the stage of 3-ring molecule $C_{12}$, we need still 29 more rings in order to complete a fullerene molecule with the total of 32 rings; but in order to build all the remaining 29 rings, it is sufficient to complete only about 22 or 23 polygons. It is obvious that when in such a closed structure we have already 31 or 30 correct polygons in place, the remaining one or two rings are also there; one can safely assume that this reasoning is still true at the level of 26 or 27 completed rings, when only no more than 5 or 6 polygons are missing.

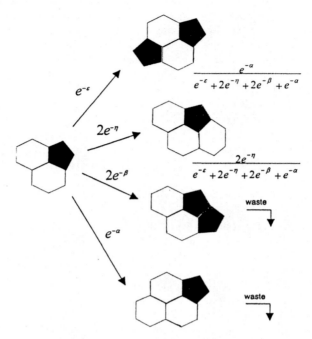

**Figure 1.** Formation of new four-polygon clusters from a $C_{12}$ molecule.

This means that if at the beginning about 25% of carbon available in the hot gas is contained in the $C_{12}$, $C_2$, and $C_3$ molecules, the final yield of fullerenes is given by the geometric progression with the ratio $\frac{1}{2}$, which gives the estimate

$$25\% \times (0.5)^{23} \sim 10^{-8}.$$

This is more than seven orders of magnitude below the observed 10% yield. Maintaining the idea that on the average the yield of the fullerene-like molecules at the subsequent agglomeration stages behaves as a geometric progression, we can easily evaluate the ratio $q$ that leads to the observed final yield:

$$35\% \times (q)^{23} \simeq 10\% = 0.1 \text{ then } q \simeq 0.961. \tag{6}$$

In order to explain the experimental facts, we must assume that at each agglomeration step the "proper," *i.e.*, fullerene-like configurations, are highly preferred to the "wrong" ones. Their pure combinatorial factors being the same, the only reasonable explanation could be given by the difference of energies related to the respective polygon construction; these energies should be contained in the corresponding Boltzmann factors as follows:

$e^{-\epsilon}$ : for a pentagon created between two hexagons;
$e^{-\beta}$ : for a pentagon created between a pentagon and a hexagon;
$e^{-\alpha}$ : for a hexagon created between two hexagons;
$e^{-\eta}$ : for a hexagon created between a pentagon and a hexagon.

It is easy to compute the normalized probabilities of producing the "good" clusters that are proper for the subsequent construction of $C_{60}$ after each agglomeration step.

# The Principle of Self-Similarity

For example, at the stage of construction when the four-polygon molecules are being formed, out of which only two are proper for fullerene formation, as shown in Fig. 1, their total probability is easily found to be

$$\frac{e^{-\epsilon} + 2e^{-\eta}}{e^{-\epsilon} + 2e^{-\eta} + e^{-\alpha} + 2e^{-\beta}}.$$

This rational expression depends only on the ratios of the Boltzmann factors involved; let us denote $y = e^{-(\eta-\epsilon)}$, $z = e^{-(\alpha-\epsilon)}$, and $t = e^{-(\beta-\epsilon)}$; then the above expression can be written as

$$F_4 = \frac{1 + 2y}{1 + 2y + 2t + z}.$$

Another interesting feature related to th self-similarity is the average ratio of pentagons among all rings in the fullerene-like clusters. At the stage of four-polygon clusters it is computed as follows:

$$G_4 = \frac{1}{2}\frac{e^{-\epsilon}}{e^{-\epsilon} + 2e^{-\eta}} + \frac{1}{4}\frac{2e^{-\eta}}{e^{-\epsilon} + 2e^{-\eta}} = \frac{e^{-\epsilon} + e^{-\eta}}{e^{-\epsilon} + 2e^{-\eta}} = \frac{1+y}{1+2y}.$$

We were able to compute these characteristic expressions up to the 10th step of agglomeration, when the clusters made up from 11 polygons contain from 28 to 30 carbon atoms, obtaining two series of functions of three variables, $y$, $z$, and $t$, denoted by $F_n$ and $G_n$.

We need three independent equations in order to solve for $(y, z, t)$. These equations can be easily produced if we suppose that on the average the yield at each step is to verify, and close to the evaluation we made shortly before, *i.e.*, if we set

$$F_{n+1}/F_n \simeq 0.961.$$

But with the help of the second set of functions we can be more ambitious and try to find the numerical value from the first principles. The functions $F_n$ should behave as a geometrical sequence; so we shall require that

$$F_{n+2}/F_{n+1} = F_{n+1}/F_n.$$

The functions $G_n$ give the average ratio of pentagons in all fullerene-like clusters containing $n$ polygons. It is reasonable to suppose that at the very early stages of agglomeration this ratio is very close to the ultimate limit, and that the convergence is also very rapid, *i.e.*, exponential.

In the final product—a $C_{60}$ molecule—this ratio is equal to $\frac{3}{8} = 0.375$. It seems therefore reasonable to suppose that functions $G_n$ obey the exponential law:

$$G_n = G_{\text{end}}(1 - e^{\lambda n}).$$

Here too, we can eliminate the unknown characteristic exponent $\lambda$ and the final value $G_{\text{end}} = 0.375$ by comparing several expressions with different values of $n$, arriving at the following law of self-similarity:

$$\frac{G_{n+2} - G_{n+1}}{G_{n+1} - G_n} = \frac{G_{n+1} - G_n}{G_n - G_{n-1}}.$$

With two equations for $F_n$ and one for $G_n$ constructed with $n = 8, 9, 10$, and 11 we have solved for $y$, $z$, and $t$:

$$y = 0.691, \quad z = 0.122, \quad \text{and} \quad t = 0.032.$$

This leads to the constant average ratio $F_{n+1}/F_n = 0.957$, which is very close to what has been anticipated, and gives the final yield of about 9.5%, which is also satisfactory.

For the obtained values of the parameters $y, z$, and $t$ the characteristic exponent $\lambda$ is equal to 0.635, and the rates defined by $G_n$'s are:

$$G_7 = 0.370; \quad G_8 = 0.372; \quad G_9 = 0.373; \quad G_{10} = 0.3736.$$

Which is fairly rapid, taking into account that without discriminating Boltzmann factors the same ratios would behave as:

$$G_6 = 0.3417; \quad G_7 = 0.3444; \quad G_8 = 0.3467; \quad G_9 = 0.3492; \quad G_{10} = 0.3512, \quad \text{etc.}$$

It is interesting to note that all the numerical values (the final yield of $C_{60}$ molecules, the ratio of pentagons in the final product and the values of the Boltzmann factors) are obtained here by applying exclusively the self-similarity principle - quite a remarkable result.

Finally, knowing that the temperature around the arc at which the process takes place is about 3000 K, we find the energy differences:

$$E_2 - E_1 = 0.104 \text{ eV}; \quad E_3 - E_1 = 0.588 \text{ eV}; \quad E_4 - E_1 = 0.965 \text{eV},$$

which suits reasonably well our ideas about the forces needed to bend the graphite network creating local curvature around one of the carbon atoms.

## IV. Stochastic Matrix Description of Growth Processes

The growth of clusters composed of certain number (usually quite low, two, three, or four) of elementary blocks can be described with the stochastic matrix method, which contains and generalizes the approaches described above. The main idea can be illustrated on the one-dimensional Fibonacci lattice, growing in an irreversible manner via addition of one of the two possible elementary blocks, $L$ or $S$, to the extremity of the already constituted chain. We shall suppose that like it often happens during the crystalline or amorphous growth from a solution, the atoms or molecules that once stick to the rim of a cluster, remain there, constituting the growing bulk.

Suppose that the growth of the chain of $L$ and $S$ blocks occurs in the infinite reservoir of free "molecules" of both types. Let us denote by $n$ the concentration of $L$'s, and by $(1-n)$ the concentration of the $S$ blocks. Inside a typical chain the probability of finding the corresponding block will be denoted by $\langle p_L \rangle$ and $\langle p_S \rangle$; there is no reason to suppose that $\langle p_L \rangle$ must be always equal to $n$.

The growth of a Fibonacci chain can be described by successive approximations that progressively take into account the influence of the closest neighbors, then the second neighbors, and so on.

Let us start with the first-neighbor interaction only. We know that the $S$-$S$ pairs must be excluded; therefore, supposing an infinite energy barrier for the $S$-$S$ couple and finite energy barriers $E_1$ and $E_2$ for the couples $L$-$L$ and $L$-$S$ respectively, at some finite temperature $T$ the sticking probabilities are proportional to:

$$\begin{aligned} L \to L + L &: \langle p_L \rangle n e^{-E_1/kT}, \\ L \to L + S &: \langle p_L \rangle (1-n) e^{-E_2/kT}, \\ S \to S + L &: \langle p_S \rangle n e^{-E_2/kT}, \\ S \to S + S &: 0. \end{aligned} \quad (7)$$

# The Principle of Self-Similarity

These relations can be encoded in the following matrix:

$$\begin{pmatrix} ne^{-\alpha} & (1-n)e^{-\beta} \\ ne^{-\beta} & 0 \end{pmatrix},$$

where we have set $e^{-\alpha} = e^{-E_1/kT}$ and $e^{-\beta} = e^{-E_2/kT}$. The above matrix should be normalized: if we want it to transform the initial probabilities $\langle p_L \rangle$ and $\langle p_S \rangle$ into new ones resulting from the elementary agglomeration step, the columns of the matrix have to be normalized to one. Then we can write:

$$\begin{pmatrix} \dfrac{ne^{-\alpha}}{ne^{-\alpha}+(1-n)e^{-\beta}} & 1 \\ \dfrac{(1-n)e^{-\beta}}{ne^{-\alpha}+(1-n)e^{-\beta}} & 0 \end{pmatrix} \begin{pmatrix} p_L \\ p_S \end{pmatrix} = \begin{pmatrix} p'_L \\ p'_S \end{pmatrix}. \qquad (8)$$

The matrix whose all columns are normalized to 1 is called stochastic matrix. It is easy to see that it always has at least one eigenvalue equal to 1, while all other (in general, complex) eigenvalues have their real part always less than 1. This means that only the eigenvectors corresponding to the eigenvalue 1 survive after many consecutive actions of stochastic matrix, which corresponds to the creation of many consecutive layers.

Introducing the notation $\xi = e^{\beta-\alpha}$, we can rewrite our matrix as

$$\begin{pmatrix} \dfrac{n\xi}{n\xi+(1-n)} & 1 \\ \dfrac{(1-n)}{n\xi+(1-n)} & 0 \end{pmatrix},$$

the characteristic equation is

$$\lambda^2 - \dfrac{n\xi}{n\xi+(1-n)}\lambda - \dfrac{(1-n)}{n\xi+(1-n)} = 0,$$

and the eigenvalues are

$$\lambda_1 = 1, \qquad \lambda_2 = -\dfrac{(1-n)}{n\xi+(1-n)}.$$

As expected, $\lambda_2$ has the absolute value less than 1. The eigenvector corresponding to the eigenvalue 1 represents the common limit to which tends the statistics of chains grown at random:

$$\vec{v}_1 = \begin{pmatrix} \dfrac{n\xi+(1-n)}{n\xi+(2-n)} \\ \dfrac{(1-n)}{n\xi+(2-n)} \end{pmatrix},$$

which means that the density of the $L$-blocks in the resulting chains is

$$\langle p_L \rangle = \dfrac{n\xi+(1-n)}{n\xi+2(1-n)},$$

which is not necessarily equal to the concentration $n$ of $L$'s in the surrounding medium.

At this stage, the model can answer several interesting questions. For example, we can ask at which values of the parameter $\xi = e^{\beta-\alpha}$ the resulting density of $L$'s in the chains is the same as the concentration of $L$-blocks outside, i.e., when $\langle p_L \rangle = n$, or vice versa, for a given value of $\xi$ at what concentration $n$ one gets the best approximation of a Fibonacci chain, i.e., $\langle p_L \rangle = 1/\tau$, etc.

Although we have excluded the formation of $S$-$S$ doublets, there still remains a finite probability of creating the triplets $L$-$L$-$L$, and even quadruplets. In order to exclude this possibility, we have to take into account the second neighbor's influence, i.e., consider the doublets $L$-$L$ $L$-$S$ and $S$-$L$ to which one of the two items, $L$ or $S$ is attached, excluding now the possibility of the transitions $L$-$L \to L$-$L$-$L$ as well as $L$-$S \to L$-$S$-$S$. With this in mind we construct the following $3 \times 3$ stochastic matrix:

$$\begin{pmatrix} 0 & 0 & ne^{-\alpha} \\ (1-n)e^{-\beta} & 0 & (1-n)e^{-\beta} \\ 0 & ne^{-\beta} & 0 \end{pmatrix} \begin{pmatrix} p_{LL} \\ p_{LS} \\ p_{SL} \end{pmatrix} = \begin{pmatrix} p'_{LL} \\ p'_{LS} \\ p'_{SL} \end{pmatrix}. \quad (9)$$

The characteristic equation is now of third order:

$$\lambda^3 - \frac{1-n}{n\xi + (1-n)}\lambda - \frac{n\xi}{n\xi + (1-n)} = 0,$$

of which $\lambda_1 = 1$ is always a solution. The remaining eigenvalues are:

$$\lambda_{2,3} = -\frac{1}{2} \pm \frac{1}{2}\sqrt{\frac{(1-n) - 3n\xi}{n\xi + (1-n)}},$$

which can become complex for certain range of the parameters $n$ and $\xi$. When the complex solutions appear, the growth displays an oscillatory behavior; such oscillations can be often observed in clusters of various sizes, namely via their magnetic properties.[8]

The asymptotic behavior of surviving stable configurations is given by the normalized eigenvector corresponding to the eigenvalue 1; it reads:

$$p_{LS} = p_{SL} = \frac{n\xi + (1-n)}{3n\xi + 2(1-n)} \quad \text{and} \quad p_{LL} = \frac{n\xi}{3n\xi + 2(1-n)}.$$

The density of $L$-blocks is then equal to

$$\langle p_L \rangle = \tfrac{1}{2}(2p_{LL} + p_{SL} + p_{LS}) = \frac{2n\xi + (1-n)}{3n\xi + 2(1-n)},$$

which gives a better approximation then the case of the closest-neighbor interaction. Nevertheless, the so obtained random chains are not exactly the Fibonacci sequences, but just very good approximations.

Our last example concerns covalent network glasses, which are well described with the ball-and-stick models. Nevertheless, it will also capture the most important features of any growth process in two or three dimensions. This technique stems from the approach that has been developed previously and exposed in our recent papers.[9-12]

We shall illustrate the stochastic matrix technique on the simplest example, in which there are only two different elementary blocks present in the surrounding medium (which can represent either a hot liquid about to undergo the glass transition, or a solution in which the precipitation of growing clusters can be observed).

Consider two star-like elementary blocks, symbolizing two types of atoms, with valences $m$ and $m'$. Many network glasses are known that correspond fairly well to this definition, e.g., $GeS_2$-glass, or $GeSe_2$-glass, etc. In our example we shall choose the lowest possible values $m = 2$ and $m' = 3$, the corresponding atoms will be denoted symbolically by $A$ and $B$. We shall consider here the simplest case of purely dendritic

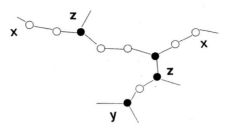

**Figure 2.** A typical cluster with three varieties of sites on the rim

growth, when only one bond can be created between two closest neighbors, excluding the possibility of two- or three-membered rings.

Let us assume that the concentration of $B$ atoms in the surrounding hot medium is $c$. During a slow cooling process, clusters of various sizes appear. A new atom (an $A$ or a $B$ type) coming close to one of these clusters and attaching itself to one of free valences available on the rim of the cluster, may encounter one of the following three situations, as shown in Fig. 2. We shall denote these configurations by $x, y,$ and $z$, with $x$ meaning an $A$-atom on the rim, with one free valence still available, $y$ corresponding to a $B$-atom on the rim with two free valences available, and finally $z$ denoting a $B$ atom whose two valences are already engaged in the bulk of the cluster, and only one is still available for further growth (Fig. 2).

If we suppose that the concentration of the $B$-atoms in the surrounding medium is $c$, then that of the $A$-atoms is $(1 - c)$. Let us suppose that after the average characteristic time $\Delta t$ all the free valences available on the rim of the cluster are saturated by a single $A$ or $B$ atom. It is easy to find the contributions to the probability factors of the six possible transitions:

$$
\begin{aligned}
x + A &\to x & &: P(x,x) \sim 2(1-c)e^{-\epsilon}, \\
x + B &\to y & &: P(x,y) \sim 3ce^{-\eta}, \\
y + A &\to x + z & &: P(y,x) + P(y,z) \sim 4(1-c)e^{-\eta}, \\
x + B &\to y + z & &: P(y,y) + P(y,z) \sim 6ce^{-\alpha}, \\
z + A &\to x & &: P(z,x) \sim 2(1-c)e^{-\eta}, \\
z + B &\to y & &: P(z,y) \sim 3ce^{-\alpha}.
\end{aligned}
$$

Here we took into account the purely statistical factors (2, 3, 4, and 6; according to the number of free bonds available for each transition), and the Boltzmann factors taking into account the corresponding energy barriers, which are

$$
\begin{aligned}
e^{-\epsilon} &= e^{-E_1/kT} & &\text{for } A + A, \\
e^{-\eta} &= e^{-E_2/kT} & &\text{for } A + B, \\
e^{-\alpha} &= e^{-E_3/kT} & &\text{for } B + B.
\end{aligned}
$$

After normalization, we get the following stochastic matrix that transforms the probabilities of finding one of the three configurations on the rim of a cluster, $(p_x, p_y, p_z)$ into a new set of probabilities $(p'_x, p'_y, p'_z)$ after the characteristic time

$\Delta t$ during which an entire new layer of atoms has been grown:

$$\begin{pmatrix} \dfrac{2(1-c)e^{-\epsilon}}{2(1-c)e^{-\epsilon}+3ce^{-\eta}} & \dfrac{(1-c)e^{-\eta}}{2(1-c)e^{-\eta}+3ce^{-\alpha}} & \dfrac{2(1-c)e^{-\eta}}{2(1-c)e^{-\eta}+3ce^{-\alpha}} \\ \dfrac{3ce^{-\eta}}{2(1-c)e^{-\epsilon}+3ce^{-\eta}} & \dfrac{3ce^{-\alpha}}{4(1-c)e^{-\eta}+6ce^{-\alpha}} & \dfrac{3ce^{-\alpha}}{2(1-c)e^{-\eta}+3ce^{-\alpha}} \\ 0 & 1/2 & 0 \end{pmatrix}$$

$$\times \begin{pmatrix} p_x \\ p_y \\ p_z \end{pmatrix} = \begin{pmatrix} p'_x \\ p'_y \\ p'_z \end{pmatrix}. \tag{11}$$

The characteristic equation for the above linear system is of third order; it is easy to prove that this matrix has one eigenvalue equal to 1 and two other eigenvalues which can be real, complex or imaginary depending on the values of the parameters $c$, $\epsilon$, $\eta$, and $\alpha$ involved.

The complex eigenvalues indicate the presence of an oscillatory regime of growth, usually damped by the eigenvalue's real part which is always less than 1. The imaginary part can be interpreted as a circular frequency measured in radians per unit of time, i.e., the number of layers, or the real time divided by the characteristic time $\Delta t$.

The eigenvector corresponding to the eigenvalue 1 determines the distribution $(p_x, p_y, p_z)_\infty$ to which the average statistic tends asymptotically. This is also the statistics of the bulk if they are very big; for clusters of intermediate size, one should rather average over the sum of many layers. Here too, like in differential geometry, the surface is in some sense the differential of the volume, and the circumference is a differential of an area.

It is easy to compute the eigenvector $(p_x, p_y, p_z)_\infty$ in our case:

$$\begin{pmatrix} p_x \\ p_y \\ p_z \end{pmatrix}_\infty = \begin{pmatrix} \dfrac{4(1-c)[2(1-c)+3c\xi]}{4(1-c)[2(1-c)+3c\xi]+9c\xi[2(1-c)+3c\mu]} \\ \dfrac{6c\xi[2(1-c)\xi+3c\mu]}{4(1-c)[2(1-c)+3c\xi]+9c\xi[2(1-c)+3c\mu]} \\ \dfrac{3c\xi[2(1-c)\xi+3c\mu]}{4(1-c)[2(1-c)+3c\xi]+9c\xi[2(1-c)+3c\mu]} \end{pmatrix}.$$

The two- or three-dimensional character of growth enables us to address more pertinent questions to which this model can give a qualitative answer at least. For example, it is interesting to follow the time development of the average density of free bonds on the surface of growing cluster. In our model only the adjunction of the $B$ atoms leads to one new bond creation, whereas when an $A$ atom joins the cluster, the number of free bonds remains strictly the same.

This is why, if at a given moment $t$ the number of free bonds on the surface layer of a cluster was $N(t)$, on the next layer produced after the time interval $\Delta t$, the number of free bonds $N(t + \Delta t)$ is computed as follows:

$$N(t + \Delta t) = N(t)$$
$$\times \left[ 1 + \left( \dfrac{3ce^{-\eta} p_x}{2(1-c)e^{-\epsilon}+3ce^{-\eta}} + \dfrac{3ce^{-\alpha} p_y}{2(1-c)e^{-\eta}+3ce^{-\alpha}} + \dfrac{3ce^{-\alpha} p_z}{2(1-c)e^{-\eta}+3ce^{-\alpha}} \right) \right],$$

which after developing the difference $N(t + \Delta t) - N(t)$ in Taylor series and keeping just the first order term, yields

$$\frac{1}{N}\frac{dN}{dt} = \frac{d\ln N}{dt} = \frac{c}{\Delta t}\left[\frac{3e^{-\eta}p_x}{2(1-c)e^{-\epsilon}+3ce^{-\eta}} + \frac{9e^{-\alpha}p_y}{4(1-c)e^{-\eta}+6ce^{-\alpha}}\right]. \quad (12)$$

After replacing in the above formula the variables $(x, y, z)$ by their asymptotic values $(p_x, p_y, p_z)_\infty$ we get the following expression:

$$\frac{d\ln N}{dt} = \frac{6c\xi}{\Delta t}\frac{2(1-c)+3c\mu}{4(1-c)[(21-c)+3c\mu]+9c\xi[2(1-c)+3c\mu]}. \quad (13)$$

Note the dependence on the concentration $c$: when $c \to 0$, the time derivative of $N$ goes to zero, too, because there is no multiplication of bonds due to the agglomeration of valence 3 $B$-type atoms.

If the growth is supposed to follow a steady rate (one new layer after time $\Delta t$), then in 2 dimensions the circumference should grow proportionally to $t$, while in 3 dimensions the surface grows roughly as $t^2$. This gives the clue as to what values of parameters (i.e., the concentration $c$, the temperature $T$ with given characteristic energies $E_1$, $E_2$, and $E_3$ are necessary to ensure the same rate of growth for $N$.

If the right-hand side of Eq. (12) is greater than the characteristic exponent 1 (in 2 dimensions) or 2 (in 3 dimensions), then the density of free bonds on the surface will grow until they entangle so that farther growth will become impossible; on the contrary, if it is smaller, then the density of free bonds will decrease so that at the end only thin filaments will composed only of chains of $A$-atoms will survive.

The statistical distribution of various sites on the rim of average cluster enables us to evaluate the final concentration of the corresponding elementary building blocks in the resulting bulk matter, be it crystalline or amorphous in nature. This concentration need not be in principle the same as in the surrounding medium, especially if the agglomeration process takes place in a vapor or from a solution. But in the case of glass transition one should suppose that the concentration of any chemical modifier remains exactly the same as in the hot melt above the glass transition temperature, because if the contrary was true, one would observe visible local departures from the homogeneity (fluctuations of density and chemical composition), which is clearly not the case.

This verification enables us to introduce an important constraint on the process, which yields a very useful equation. Let us denote by $\tilde{c}$ the asymptotic value of the concentration of the $B$-type atoms on the rim of typical cluster $\tilde{c} = y_\infty + z_\infty$, which yields

$$\tilde{c} = \frac{9c[2(1-c)\xi + 3c\mu]}{4(1-c)[2(1-c)+3c\xi]+9c[2(1-c)\xi + 3c\mu]}.$$

(We can identify $\tilde{c}$ with $y_\infty + z_\infty$ because each of the sites of the type $y$ or $z$ contains exactly one atom of the $B$-type). The requirement that the asymptotic value of the concentration $\tilde{c}$ in the resulting bulk network be the same as in the surrounding hot liquid before the glass transition takes place amounts to the equation $\tilde{c} = c$, which means that during the glass transition a stationary regime is established, i.e., that the mapping $c \to \tilde{c}(c, T)$ attains one of its fixed points.

When explicited, this condition amounts to the following equation:

$$c(1-c)[18(1-c)\xi + 27c\mu - 8(1-c) + 12c\xi] = 0.$$

Besides the two obvious solutions, $c = 0$ and $c = 1$, which must represent the fixed points of the agglomeration process (if $c = 0$, it will remain so whatever the rules of agglomeration, similarly for $c = 1$), a third fixed point is found when the square bracket vanishes:

$$9[2(1-c)\xi + 3c\mu] - 4[2(1-c) + 3c\xi] = 0,$$

which yields the explicit equation

$$c = \frac{2(9\xi - 4)}{(30\xi - 27\mu - 8)}. \tag{14}$$

This intermediate ($0 < c < 1$) solution represents a mixture of both ingredients ($A$ and $B$ type atoms) at the microscopic level, therefore it can be identified with the amorphous glassy state. Implicitly, it gives the dependence of the glass transition temperature on the concentration $c$.

In a good glass former the amorphous homogeneous configuration is easily obtained with arbitrarily low values of the modifier concentration; the corresponding glass transition temperature is denoted by $T_{0g} = T_g(c = 0)$. The comparison with the experimental value of $T_{0g}$ enables us to fix the difference of the energies $E_1 - E_2$:

$$c = 0 \to 9\xi - 4 = 0, \quad \text{or} \quad e^{(E_1 - E_2)/(kT_{0g})} = \tfrac{4}{9} < 1. \tag{15}$$

Therefore,

$$E_1 - E_2 = kT_{0g} \ln \tfrac{4}{9} < 0, \quad i.e., \quad E_2 > E_1.$$

This is what should be intuitively expected from the glass-forming tendency: at the local level the system behaves undecidedly, in a "frustrated" way, because while the purely statistical factor (9 versus 4) increases the probability of agglomeration of the modifier's atoms (of the $B$-type, with valence 3) versus the probability of pairing of the pure glass-former atoms (of the $A$-type, here with lower valence 2), the Boltzmann factors act in an opposite direction. Recalling the definition of the parameters $\xi$ and $\mu$, we can easily find the derivative $dc/dT$:

$$\frac{dc}{dT} = \frac{18(E_2 - E_1)(30\xi - 27\mu - 8) - (18\xi - 8)[30(E_2 - E_1)\xi - 27(E_3 - E_1)\mu]}{kT^2 [30\xi - 27\mu - 8]^2}. \tag{16}$$

Its inverse, $dT/dc$, gives the slope of the curve describing the dependence of the glass transition temperature on the modifier concentration in binary network glass composed of a 2-valenced glass former (e.g., selenium) and a 3-valenced modifier (e.g., arsenic).

This formula can be simplified in the low concentration limit when $c \to 0$. We know already that in this limit $T \to T_{0g}$, and $\xi \to \tfrac{4}{9}$. If the energy barrier $E_3 - E_1$ is large compared to $E_2 - E_1$, then the corresponding Boltzmann factor can be neglected, especially in the limit of $c \to 0$, when the probability of encounter of two $B$-type atoms is extremely low. In such case our formula can be simplified even more, giving

$$c \simeq \frac{2(9\xi - 4)}{(30\xi - 8)} \quad \text{and} \quad \left.\frac{dT}{dc}\right|_{c \to 0} = \frac{2T_{0g}}{27 \ln(3/2)}, \tag{17}$$

which does not depend on the energy differences.

Another characteristic behavior of the glass transition that can be predicted by our model is the presence of the inflexion points occurring in several thermodynamical functions, in particular, in the curve describing the specific heat $C_p$, defined as $(\partial U/\partial T)_{p=\text{const}}$.

In order to evaluate $U(T)$ as a function of the parameters $E_k$, we shall assume that each covalent bridge can be considered as a harmonic oscillator whose instant energy contains essentially three contributions: its kinetic energy, the potential energy, roughly equal to the kinetic part if the system is not too far from thermal equilibrium, and the third contribution, which is the energy needed to create the bonds, which can also be interpreted as the latent heat.

Our model enables one to evaluate this contribution quite easily. All we need is to count the relative density of three types of bridges, $A$-$A$-type with the energy $E_1$, $A$-$B$-type with the energy $E_2$ and the $B$-$B$-type with the energy $E_3$. But this is very easy to compute if we recall how the creation of those bonds is described by the stochastic matrix: for example, the process leading to the creation of a new $A$-$A$ bond corresponds to the matrix element $\{M_{xx}\}$ acting on the component $p_x$ of the probability vector, giving the contribution $E_1 M_{xx} p_x$, and so on. New bonds of the $A$-$B$-type are created each time a $B$ atom joins the site $x$, or when an $A$ atom joins one of the sites $y$ or $z$, and the $B$-$B$ bonds are created when a $B$-type atom joins a site $y$ or $z$. Summing up yields the following expression for $U_l(T)$:

$$U_l = E_1 M_{xx} p_x + E_2 [M_{yx} p_x + M_{xy} p_y + M_{xz} p_z] + E_3 [M_{yy} p_y + M_{zy} p_y + M_{yz} p_z], \quad (18)$$

using the elements of the matrix $M_{ik}$ defined above. Assuming that the clusters are large enough, we may use the values obtained for the asymptotic regime, $(p_x, p_y, p_z)_\infty$; this enables us to simplify the above expression even more.

The curves representing $C_p(T)$ and the quantity $d(\ln N)/dt(T)$ are very similar and display the characteristic inflexion point in the vicinity of the glass transition temperature. It is noteworthy that the inflexion points of both curves almost coincide. The inflexion point of $C_p(T)$ should correspond to the glass transition temperature; the coincidence suggests that the fractal dimension may be related to the average number of degrees of freedom parameter proposed by Phillips.[13] Many similar problems can be also investigated with this simple technique. We shall address these problems in the forthcoming papers.

## References

1. M. Hamermesh, *Group Theory and Its Application to Physical Problems* (Pergamon, London, 1962).
2. W. Krätschmer, K. Fostiropulos, D. R. Huffmann, *Chem. Phys. Lett.* **170**, 162 (1990).
3. H. Kroto, J. R. Heath, S. C. O'Brien, R. F. Curl, and R. E. Smalley, *Nature* **318**, 162 (1985).
4. T. G. Schmalz, W. A. Seitz, D. Klein, and C. E. Hiley, *J. Am. Chem. Soc.* **110**, 1113 (1991)
5. R. Kerner, K. H. Bennemann, and K. Penson, *Europhysics Lett.*, (1992)
6. R. Kerner, *Computational Materials Science* **2**, 500 (1994)
7. R. Kerner, K. H. Bennemann, K. Penson, *Fullerene Science and Technology* **4**, 1279 (1996).
8. J. L. Morán-López, contribution to this volume, p. 177.
9. R. Kerner, *J. Non-Cryst. Solids* **135**, 155 (1991).
10. D. M. dos Santos-Loff, R. Kerner, Mand . Micoulaut, *Europhys. Lett.* **28**, 573 (1994).
11. R. Kerner, *Physica B* **215**, 267 (1996).
12. R. Kerner, M. Micoulaut, *J. Non-Cryst. Solids* **210**, 298 (1997).
13. J. C. Phillips, *J. Non-Cryst. Solids.* **34**, 153 (1979).

# Ring Statistics in Glass Networks

Matthieu Micoulaut

*Laboratoire GCR-UFR de Physique*
*Paris VI, CNRS-URA 769, Université Pierre et Marie Curie*
*Tour 22, Boîte 142, 4, Place Jussieu, 75005 Paris*
*FRANCE*

## Abstract

We present in this paper a statistical model which gives elements of the intermediate-range order (IRO) structure in some current glass networks. Starting from typical short-range order (SRO) clusters, we construct multiplets of growing size by agglomeration of the SRO clusters and compute a corresponding probability. The assumption that a rapid stabilization of the fraction of atoms trapped inside rings is obtained with size-growing clusters, determines the involved ring-formation energies and the computed ring statistics is then compared to experiment.

## I. Introduction

Full information about the structure of an amorphous solid can never be elucidated, nor it is possible to find with experimental techniques. In contrast with perfect crystals where the structure can be determined on all length scales by repeating periodically a unit cell of a certain number of atoms, amorphous networks have no periodic symmetry so that the structure on large distances can not be inferred. Nevertheless, in a number of papers devoted to the subject, Galeener and others[1] have shown that it can be useful to classify the elements of order in amorphous solids on different scales. On a very small scale [called short-range order (SRO)], the chemical bonding imposes approximate rotational symmetries of nearest neighbors about a given atomic species. In most of the situations, the elements of SRO are the same in amorphous or crystalline solids (*e.g.* the $GeO_{4/2}$ tetrahedra in a-$GeO_2$ and c-$GeO_2$).

When the attempt to determine the structure on its shortest length scales is accomplished, it is possible to discern additional order [intermediate-range order (IRO)] on increasingly larger scales of distance, as it is now suggested in numerous and various glass-forming materials.[2,3] For example, vitreous $B_2O_3$ is supposed to be mainly composed of boroxol rings (composed of three connected SRO $BO_3$ triangles). In tetrahedral based glasses, the rate and the number of edges that a tetrahedron shares with its neighbors can be derived from the chemical shift in nuclear magnetic resonance (NMR) spectroscopy patterns. This has given information about a certain degree of order, which is larger than SRO.[4] All these structural investigations, realized both on theoretical and experimental basis, have led to a quite precise delineation of three scales of distances, termed as *short*, *intermediate*, and *long*. The latter one (LRO) is representative of rotational or translational repetition of an element of IRO over several length scales.

We shall present in this paper the application of a statistical model on amorphous solids which gives a connection between SRO and LRO. The approach which has been first introduced by Kerner, Bennemann and others in order to elucidate the formation of fullerenes,[5] is based on the statistical construction of micro-clusters which have clear and unambiguous experimental evidence (such as carbon pentagons and hexagons in $C_{60}$) and on the statement that the rate of significant substructures will converge very rapidly to a limit value when size-growing clusters are considered. In fullerenes, these limit values are obvious, since $C_{60}$ contains 12 pentagons and 32 hexagons. The model relies also on a few parameters (but with physical meaning) and allows an extensive structural and thermodynamical description of these materials in a simple fashion. Recently, it has been suggested that the model might be adequate for the description of the ring statistics in amorphous silica.[6] In this paper, we shall focus our attention on the structure of three currently studied glasses, which are $B_2O_3$, $B_2S_3$ and $GeO_2$.

## II. Statistical Model

The model starts with the geometrical agglomeration of a star-like entity with a central atom which corresponds to the lowest possible structural level having explicit experimental evidence (a *singlet*) in an amorphous network. In v-$B_2S_3$ and v-$B_2O_3$, such a singlet can be very well represented by a $BX_3$ (X= O, S) triangle, since each central boron atom has three neighboring oxygen or sulphur atoms. The experimental evidence of such structural units is revealed by the diffraction patterns of these materials, which exhibit sharp and characteristic peaks at the corresponding bond lengths.[7] The singlets which have a coordination number of $m$ will form bridges with the next central atoms, in order to give larger clusters with two, three, four central atoms, etc. Although it is obvious that the network can be entirely tiled with such singlets, they may represent a very spurious estimation of IRO. We assume that the structure of the amorphous networks we are investigating, can be tiled with larger clusters (*doublets* with two central atoms, *triplets* with three central atoms,...), each step of agglomeration yielding bigger clusters and a better approximation about the IRO and LRO, because the final structure we shall produce contains more significant information about typical substructures of the network (such as rings) than the initial singlets, and the way in which these rings are connected. During the agglomeration, the energy cost of formation, or the energy stored in a given cluster, can be treated as an additive quantity and evaluated by summation of elementary steps of agglomeration consisting in the creation of a new oxygen or sulphur bond. At each step of agglomeration, we can compute a normalized probability for each particular cluster being produced and the normalizing factors can eventually imitate the statistical sum at the glass transition temperature.[8]

We have to take into account during the first step of this construction both formation of corner-sharing and edge-sharing doublets, constructed out of two singlets, the latter one being also a two-membered ring (dimer). Long or ramified chains, which

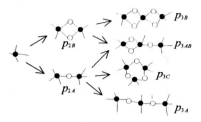

**Figure 1.** From the initial singlet to a set of triplets with the symbols used in the computation of the probabilities.

appear by pure corner joining, will involve an energy cost of $E_1$, edge-sharing units will use an energy cost of $E_2$ and the creation of three and four-membered rings which are produced during the forthcoming steps, will need an energy cost of $E_3$ and $E_4$. The involved probabilities have different multiplicity which corresponds to the number of ways in which a given doublet can be constructed, and can be regarded as the degeneracy of the corresponding stored energy. Indeed, given the coordination number of the central atom $m$ and labeled covalent bridges, a doublet chain has the multiplicity $m^2$ whereas for a dimer we can count $\binom{m}{2} = \frac{1}{2}m(m-1)$ in three dimensions. The simplest way to evaluate the probabilities of these two new structural states with the energies $E_1$ and $E_2$ is the description by means of canonical ensemble summations. The symbols $p_{2A}$ and $p_{2B}$ refer to the ones displayed in Fig. 1,

$$p_{2A} = \frac{2m^2 e_1}{2m^2 e_1 + m(m-1)e_2}, \tag{1}$$

$$p_{2B} = \frac{m(m-1)e_2}{2m^2 e_1 + m(m-1)e_2}, \tag{2}$$

with $e_1 = \exp(-E_1/kT_g)$ and $e_2 = \exp(-E_2/kT_g)$. The next step leads to four different multiplets with three central atoms, among which there is a new type of structure, namely a three-membered ring using the energy $E_3$. The multiplicities and the Boltzmann factors $e_1$, $e_2$, and $e_3$ of each pathway of production are shown in Fig. 1, and the normalized probabilities are given below:

$$p_{3A} = \frac{2m(m-1)e_1 p_{2A}}{2m(m-1)e_1 + m(m-1)^2(m-2)e_2 + m(m-1)^3 e_3}, \tag{3}$$

$$p_{3AB} = \frac{m(m-1)^2(m-2)e_2 p_{2A}}{2m(m-1)e_1 + m(m-1)^2(m-2)e_2 + m(m-1)^3 e_3}$$
$$+ \frac{4m(m-2)e_1 p_{2B}}{4m(m-2)e_1 + m(m-1)(m-2)(m-3)e_2}, \tag{4}$$

$$p_{3B} = \frac{m(m-1)(m-2)(m-3)e_2 p_{2B}}{4m(m-2)e_1 + m(m-1)(m-2)(m-3)e_2}, \tag{5}$$

$$p_{3C} = \frac{m(m-1)^3 e_3 p_{2A}}{2m(m-1)e_1 + m(m-1)^2(m-2)e_2 + m(m-1)^3 e_3}. \tag{6}$$

The next step yields a growing number of multiplets (eleven *quadruplets*) and a new kind of ring appears, namely the four-membered ring, sharing four central atoms,

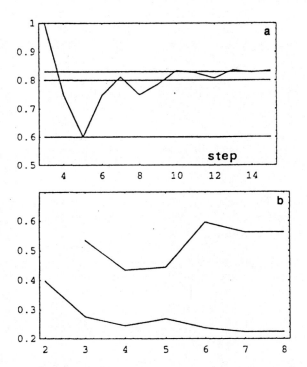

**Figure 2.** (a) The obtained boroxol rate in v-$B_2O_3$. The horizontal lines represent the values established by Jellison et al., Johnson et al., and Hannon et al.[10] (b) The trimer and dimer rate in v-$B_2S_3$.

obtained from the triplet $3A$. The construction has been continued up to the sixth or seventh step leading to numerous nonredundant clusters (about a hundred) with seven or eight central atoms each, some of which containing two-, three- or four-membered rings or several of them.

The size of the clusters seems to be already sufficient in order to characterize a certain IRO, compared to the involved structures described in the some typical glass-formers.[3] Once given the probabilities, it is easy to compute the fraction of atoms trapped inside the rings (we shall respectively denote $B^{(l)}$, $C^{(l)}$ and $D^{(l)}$, the 2-, 3- and 4-ring fraction obtained from structures with $l$ central atoms) by summing up all the corresponding contributions, e.g., $C^{(3)} = p_{3C}$ and $B^{(3)} = \frac{2}{3}p_{3AB} + p_{3B}$.

Now, what we need is a simple principle that would enable us to establish the equation to be solved.[5] We can reasonably assume that the growth process is not only governed by energetical and combinatorial factors (which appear in the probabilities) but also by a certain "self-similarity," when the agglomeration is well established. We mean by this statement that the ring fraction should not evolve any more when a critical cluster size is reached. This can be expressed by:

$$B^{(l)} \simeq B^{(l-1)}, \qquad C^{(l)} \simeq C^{(l-1)}, \qquad D^{(l)} \simeq D^{(l-1)}. \tag{7}$$

Equations (7) are a very strong condition and they should be satisfied only for sufficient large structures where step fluctuations of the ring fraction are vanishing. Previous examples show that condition (7) is fairly satisfied for multiplets sharing at least 15 central atoms [e.g., Fig. 2(a) or Ref. 5]. Nevertheless, for smaller clusters, one can imagine that the process is still governed by approximate self-similarity, identified with a minimization of the step fluctuations, because the number of multiplets sharing rings grows very rapidly, following nearly a geometrical progression law (in the forthcoming application on $B_2O_3$, the number of "ring"-multiplets *versus* the multiplet size is 1-1-2-5-8-20-44-98-...). For each step, we shall minimize with respect to the variables $e_i$ the following expression, inspired from a least-squares minimization

$$\mathcal{F}^{(l)}(e_i) = \left[B^{(l)} - B^{(l-1)}\right]^2 + \left[pC^{(l)} - C^{(l-1)}\right]^2 + \left[D^{(l)} - D^{(l-1)}\right]^2. \tag{8}$$

The solutions $e_i$ obtained below are close to the ones we could derive from other convergence criteria, such as exponential convergence or a geometrical progression law, discussed at this conference by Kerner.[9] Let us now see how the model can be applied to the description of the ring statistics in several glass formers.

## III. Ring Statistics in Particular Glass Networks

### III.1 Boron Oxide

The structure of vitreous boron oxide is a typical example of well-characterized IRO. Indeed, there is a strong experimental evidence[7] for the existence of larger structural groups, namely boroxol rings $B_3O_3$ (BR), made up of three connected SRO $BO_3$ triangles. The network is then constructed of randomly oriented BR connected via oxygen atoms. The interpretation of different spectroscopic data in terms of BR have yielded the approximate value of 80–83% of borons trapped inside BR.[10] An extensive study of this glass with the statistical model has been reported elsewhere.[8]

The possibility of formation of a two-membered ring, for which there is no experimental evidence in x-ray results, can be excluded because it would produce in the diffraction pattern a sharp and characteristic peak which is not observed.[7] The four-membered ring can be also excluded on the basis of angular considerations, therefore $e_2 = e_4 = 0$ and $m = 3$.

The construction has been performed up to the 10th step of agglomeration and has produced 288 different multiplets. The evaluation of the ring fractions has been realized and the convergence criterion applied, in order to determine the Boltzmann factors $e_3/e_1$ satisfying condition Eq. (8). This led to the energy difference $E_1 - E_3 = 0.21$ eV $= 5.1$ kcal/mol with the mean value of $T_g = 500$ K. The value is in very well agreement with previous energetical estimates derived from thermodynamical considerations (5.3 kcal/mol, Ref. 8) and experimental data (5 kcal/mol, Ref. 11).

With the energy difference established and the glass transition temperature fixed, one can compute the rate of BR and the result exhibit a converging behavior to the limit value of 84%, which is consistent with various theoretical and experimental results presented above, especially with the model of Jellison *et al.*[10] who suggested that 83% boron atoms are contained in BR.

### III.2 Germanium Oxide

The second glass studied with the statistical model is the germanium oxide glass $GeO_2$. This system is of great interest because of its unique physical properties, among which the so-called "density anomaly" corresponding to a density maximum around 15–16% added $Na_2O$ in the binary alloy.[12]

Structural studies based on SRO concepts ($GeO_6$ octahedra) have been accomplished,[13] although it is now suggested that the germanate anomaly should either

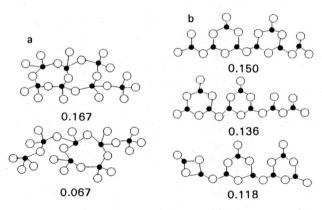

**Figure 3.** The most frequent multiplets with their computed probability (a) v-GeO$_2$ (b) v-B$_2$S$_3$.

result from an alternative structural reorganization. Micro-Raman experiments have been performed and results have been obtained in this sense: The anomaly is supposed to be strongly related to the existence of 4- and 3-membered rings, the growing proportion of the latter ones with increasing Na$_2$O being responsible of the density anomaly.[14] Galeener[15] has suggested that GeO$_2$ may be composed of nearly planar 3- and 4-rings, according to spectroscopic investigation. This proposal has been also made by Barrio and coworkers[1] on the basis of a Bethe lattice model, but a precise and quantitative evaluation of the ring statistics is lacking in the literature.

The agglomeration has been performed up to the 6th step, yielding 46 nonredundant aggregates with seven germanium atoms. 28 structures share at least a three-membered ring and 15 a four-membered ring. Edge-sharing tetrahedra have been excluded for the same reason as mentioned in the previous subsection, because there is no evidence of such rings in the diffraction data,[13] therefore $e_2 = 0$ and $m = 4$. The minimization of $\mathcal{F}^{(l)}$ in terms of the variables $e_3/e_1$ and $e_4/e_1$ for the ultimate step of agglomeration ($l = 7$) leads to the values $E_4 - E_1 = 0.138$ eV and $E_3 - E_1 = 0.205$ eV (with $T_g \simeq 800$ K). The fraction of rings can be computed and shows converging behavior up to 20.6% three-membered rings and 40.3% four-membered rings. The important ring fraction is in agreement with qualitative energy minimization arguments[15] and with Raman spectra analysis of v-GeO$_2$ which assigns the usually observed bands at 425 and 515 cm$^{-1}$ to symmetric stretching of Ge-O-Ge bonds associated with 4- and 3-membered rings.[16] In comparison with the Raman spectra of vitreous silica, the predicted fractions are in a much higher proportion, consistent with.[1,6] The most frequent multiplets are displayed in Fig. 3(a), confirming that v-GeO$_2$ seems to be mainly composed of 4-membered rings.[14,15]

### III.3 Boron Sulphide

The last system we shall investigate is the vitreous boron sulphide, which has received little attention, in contrast with its oxide analogue. Reports of Raman and NMR spectra exist in the literature and the unit cell of the crystalline compound is well characterized (having a rate of 0.75 three-membered rings and 0.25 two-membered rings).[17] Nevertheless, until recently, very few studies have been provided in order to give a basis for understanding the structure and physical properties of the corresponding glass.[18] With the singlet defined (a $BS_3$ triangle, $m = 3$), we have constructed the multiplets up to structures with eight boron atoms (120 different multiplets), among which appear thioboroxyl groups ($B_3S_6$, the analogue of the boroxol group, i.e., a trimer). The multiplets contain also some edge-sharing units (two-membered rings or dimers), following a situation which is rather common in chalcogenide based glasses and which is observed in c-$B_2S_3$. For the same reasons as mentioned above (the S-B-S angle is about 120°), the four-membered rings may not exist in vitreous boron sulphide, therefore $e_4 = 0$. Performing the same minimization of $\mathcal{F}^{(l)}$ for the different steps of agglomeration yields the values $E_2 - E_1 = 0.054$ eV and $E_3 - E_1 = -0.083$ eV (with $T_g = 570$ K). Very close values are obtained with condition (7), so that we can assume that the critical size is fairly reached in this system for small structures. We observe that the energy differences have not the same sign, i.e., the formation of three-membered rings is energetically preferred and easier than the formation of single bonds (leading to chains) and two-membered rings. The fraction of rings *versus* the size of the multiplets is shown in Fig. 2(b) and is consistent with very recent NMR work on v-$B_2S_3$ inferring that the structure should be composed of 75% trimers and 25% dimers.[19] The most frequent structures, typical examples of LRO, are composed of connected trimers and dimers, as displayed in Fig. 3(b).

## References

1. F. L. Galeener, *Diffus. Defect. Media* **54-55**, 305 (1988); R. A. Barrio, F. L. Galeener, E. Martinez, and R. J. Elliott, *Phys. Rev. B* **48**, 15 672 (1993).
2. J. Krogh-Moe, *Phys. Chem. Glasses* **6**, 46 (1966).
3. L. F. Gladden and S. R. Elliott, *Phys. Rev. Lett.* **59**, 908 (1987).
4. R. Dupree, D. Holland, P. W. MacMillan, and R. F. Pettifer, *J. Non-Cryst. Solids* **68**, 399 (1984).
5. R. Kerner et al., *Z. Phys. D* **29**, 231 (1994); R. Kerner, *Comp. Mat. Science* **2**, 39 (1994).
6. R. Kerner, *J. Non-Cryst. Solids* **182**, 9 (1995).
7. R. L. Mozzi and B. E. Warren, *J. Appl. Cristallogr.* **3**, 153 (1970).
8. M. Micoulaut, R. Kerner, and D. M. Dos Santos-Loff, *J. Phys. Condens. Matter* **7**, 8035 (1995).
9. R. Kerner, these proceedings p. 323.
10. G. E. Jellison, L. W. Panek, P. J. Bray, and G. B. Rouse Jr., *J. Chem. Phys.* **66**, 802 (1977); P. A. V. Johnson, A. C. Wright, and R. N. Sinclair, *J. Non-Cryst. Solids* **50**, 281 (1982); A. C. Hannon et al., *J. Non-Cryst. Solids* **177**, 299 (1994).
11. G. E. Walrafen, M. S. Hokmabadi, P. N. Krishnan, and S. Guha, *J. Chem. Phys.* **79**, 3609 (1983).
12. M. K. Murthy and J. Ip, *Nature* **201**, 285 (1964).
13. M. Ueno, M. Misawa, and K. Suzuki, *Physica B* **120**, 347 (1983).
14. G. E. Henderson and M. E. Fleet, *J. Non-Cryst. Solids* **134**, 259 (1991).
15. F. L. Galeener, *The Physics of Disordered Materials*, edited by D. Adler, H. Fritzsche, and S. R. Ovshinsky, (1985).
16. G. S. Henderson, G. M. Bancroft, M. E. Fleet, and D. J. Rogers, *Am. Mineral.* **70**, 946 (1985).
17. B. Krebs, *Angew. Chem. Int. Ed. Engl.* **22**, 113 (1983).
18. S. W. Martin and D. R. Bloyer, *J. Am. Ceram. Soc.* **73**, 3481 (1990).
19. S. W. Martin, in preparation.

# GaAs Layers Grown by the Close-Spaced Vapor Transport Technique Using Two Transport Agents

E. Gómez,[1,*] R. Valencia,[1] R. Silva,[1] and F. Silva-Andrade[2]

[1] *Instituto de Física "Luis Rivera Terrazas"*
*Universidad Autónoma de Puebla*
*Apartado Postal J-48, Puebla, Pue., 72570*
*MEXICO*

[2] *Centro de Investigación en Dispositivos Semiconductores*
*Instituto de Ciencias, Universidad Autónoma de Puebla*
*Apartado Postal 1651, Puebla, Pue., 72000*
*MEXICO*

## Abstract

A parallel study is made on two sets of GaAs homolayers grown with a close-spaced vapor transport system and different transport agents. One set was grown leaving fixed all growth parameters except for the water vapor concentration. The second set was grown using atomic hydrogen as the only reactant, here just the spacer thickness between the source and the substrate was varied. Scanning electron microscopy and energy dispersive spectroscopy were used to characterize the layers. For the first set it was observed that the low water vapor concentrations used during the growth process produced As-rich surfaces, while larger ones produced Ga-rich surfaces. For the second set, grown with atomic hydrogen, it was found a phenomenological relationship among the spacer thickness, the growth rate and the surface morphology, these films came out to be Ga-rich surfaces, probably because of the high temperature effect and the atomic hydrogen interaction with the source and the substrate during the growth process.

## I. Introduction

The close-spaced vapor transport (CSVT) is a modified chemical vapor deposition (CVD) technique since a transport agent is needed to achieve an epitaxial growth.[1] It consists on a sandwich assembly of a solid source and a substrate separated from each other by a spacer typically of about 1 mm of thickness. A temperature gradient is kept between them and the transporting agent reacts with the source producing volatile compounds, which migrate to the substrate where the reverse reaction takes place forming the epitaxial layer.[2-4] Thus the driving force of this technique is a temperature gradient between the source and the substrate and the reaction of the transport agent with the surface of the source and the substrate allows the material transport. In particular, GaAs layers grown by CSVT using water vapor as the transport agent have been extensively investigated[5-7] to elucidate the growth mechanism. Recently atomic hydrogen instead of water vapor has been used to grow GaAs homo-layers by the same CSVT technique, some partial results about this subject have been already reported.[8,9]

The aim of this work is to make a parallel study of two sets of GaAs homolayers grown by the CSVT technique and two different transport agents. In one set the water vapor concentration was varied, while in the second one atomic hydrogen was the only reactant and additionally some growth parameters were varied. The surface morphology and stoichiometry of the "as grown" GaAs layers were analyzed using a JSM-5400LV (JEOL) scanning electron microscope (SEM) in combination with a (Voyager-Noran) energy dispersive x-ray spectrometer (EDS).

## II. Experimental CSVT Setups

In both CSVT systems the purification of hydrogen is carried out by means of a catalytic filter, a water filter, and a palladium cell hydrogen purifier. To carry out the growth process with water vapor, it is necessary first to remove oxygen and water from the reactor, then introduce water vapor in a controlled concentration, here molecular hydrogen is used only as the carrier gas. In this system the thermal gradient is produced by a graphite heater. In the second system two tungsten filaments produce the gradient temperature between the source and the substrate, besides one of these filaments produces the atomic hydrogen by high temperature dissociation. The temperature of both filaments are adjusted by independent variable transformers and estimated by their change in resistance, for more details, see Ref. 8. The source-substrate assembly is very similar in both systems, it consists of a quartz plate with a hole in the center, which holds the GaAs source. This is approximately 6 mm × 350 $\mu$m, their 6 mm long edges are cut to 3.2 degrees bevel angle to provide an inlet-like for volatile compounds to reach the substrate surface, and a quartz O-ring is used as the spacer between the source and the substrate.

## III. Sample Preparation

In this experiment undoped (100) GaAs wafers with $n = 10^{16}$ cm$^{-3}$ were used as sources and Cr-doped semi-insulating GaAs wafers with $(100) + 2°(110)$ orientation, $\rho = 10^8$ $\Omega$ cm and EPD = $5 \times 10^4$–$7 \times 10^4$ cm$^{-2}$ as substrates. Both materials were bought from the MR Semicon Inc. All wafers used were first degreased by the ultrasonic cleaning (2 min. each) in acetone, xylene, acetone, and rinsed in deionized (DI) water. Then, they were etched in 3:1:1 ($H_2SO_4 : H_2O : H_2O_2$) etchant at $T = 80°C$ and finally each sample was dipped into 1:1 ($HCl : H_2O$) solution to remove residual surface oxides,[10] rinsed in DI water and spin-dried just prior introduction into the reaction chamber.

## IV. Results and Discussion

The first set of GaAs homo-layer were grown keeping fixed the parameters values to: source temperature $T_s = 800°C$, spacer thickness $\delta = 1$ mm, thermal gradient between source and substrate $\Delta T = 80°C$, deposition duration $t = 3$ min., and the water vapor concentration was varied to $C_{H_2O} = 150, 800$, and $4000$ ppm. The reaction proposed in the literature[2,5,6] for the transport of GaAs in a CSVT system with water vapor as the transport agent is the following:

$$2\text{GaAs} + \text{H}_2\text{O} \rightleftharpoons \text{Ga}_2\text{O} + \text{H}_2 + \text{As}_2, \tag{1}$$

where $\text{Ga}_2\text{O}$ is assumed to be the transporting species.

One experimental result is that the growth rate (GR) (defined as GR $[\mu m/min] = d\,[\mu m]/t[min]$, where $d$ is the layer thickness[6]) is proportional to the square root of the water vapor pressure. This behavior is similar to the one reported by Gottlieb and Carboy.[5]

During the growth process of the second set of GaAs layers, that is, with atomic hydrogen as the reactant, the deposition parameters were kept fixed to: $f_{H_2} = 425$ cm$^3$/min, $T_s = 650°C$, and deposition duration $t = 30$ min., while the spacer thickness was varied to $\delta = 0.025, 0.6, 1$, and $2$ mm. In our experiment it is found that, for $\delta = 0.6, 1.0$, and $2.0$ mm the parameter decreases with increasing the source to substrate separation in a similar way, as for the GaAs layers grown with water.[6] The growth mechanism for the GaAs layers deposited with atomic hydrogen as the only reactant up to now is not well understood.

### IV.1 SEM and EDS Analysis

Figure 1 shows the surface morphology of the GaAs layers grown with different water vapor concentrations, $C_{H_2O} = 150, 800$, and $400$ ppm. These SEM images show smooth textures, the first two of them are kind of "orange peel" while the third one shows small furrows.

Figure 2 shows the variation on the surface morphology of the GaAs layers grown with atomic hydrogen and different spacer thickness $\delta$, for (a) $\delta = 0.025$, (b) $\delta = 0.6$, (c) $\delta = 1$, and (d) $\delta = 2$ mm.

It has been reported[11] that, thermodynamic considerations and studies of the initial stages of epitaxy have led to the distinction among three different growth modes, which can be adopted by a film growing under (near-) equilibrium conditions. In the Frank- van der Merwe (FM) mode the overlayer grows continuously in a two-dimensional (2D) layer-by-layer way. In the Volmer-Weber (VW) mode nucleation occurs by immediate formation of three-dimensional (3D) islands, and in the Stranski-Krastanov (SK) mode, the initial nucleation occurs in a 2D way but above a certain thickness, $t_{SK}$ island formation occurs. According to the stated above, the first three SEM images of Fig. 2 correspond to different (SK) growth modes: in (a) is shown a kind of stepped surface, in (b) a pyramidal growth, in (c) a wavy texture, and (d) shows a polycrystalline morphology.

EDS analysis were performed on different representative regions of each sample under study. The surface of GaAs films grown with water vapor concentrations of $C_{H_2O} = 150$ and $800$ ppm are As-rich. This result is in accord with Côté et al.,[6] who reported that the GaAs layers with GR in the range of 0.2 to 2.0 $\mu m/min$ always occur in an excess As condition. However the surface of the film grown with water concentration of $4000$ ppm resulted Ga-rich. The GR for our samples are in the

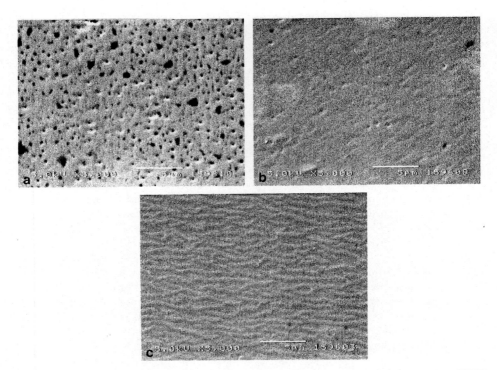

**Figure 1.** SEM images of the surface morphology of the GaAs layers deposited with a CSVT system and different water vapor concentration: (a) $C_{H_2O} = 150$ ppm, (b) $C_{H_2O} = 800$ ppm, (c) $C_{H_2O} = 4000$ ppm.

range of 0.49 to 0.86 $\mu$m/min. The EDS spectra of these samples also show signals corresponding to the residual contaminants C, O, and Si.

The EDS analysis carried out on GaAs films grown with atomic hydrogen show that all surfaces of this set are Ga-rich. The spectra of these sample show only a very weak signal corresponding to the residual contaminant C. This signal is slightly large in layers with polycrystalline surface morphology.

## V. Conclusions

Two sets of GaAs films were grown with a CSVT system and two different transport agents: Water vapor and atomic hydrogen. For GaAs films grown with water vapor, it was found that: The GR is proportional to the square root of the water vapor pressure. The surface morphology of these layers is smooth. Small water vapor concentration during the growth process produced As-rich surface, while the largest concentration used produced Ga-rich surfaces. EDS analysis showed also that these films contain C, O, and Si contaminants.

For GaAs films grown with atomic hydrogen: It was observed a phenomenological relationship among the spacer thickness, the growth rate parameter and the surface morphology of the grown GaAs layers: smaller spacer thickness produced higher growth rates and films with SK growth mode, the largest spacer thickness (2 mm)

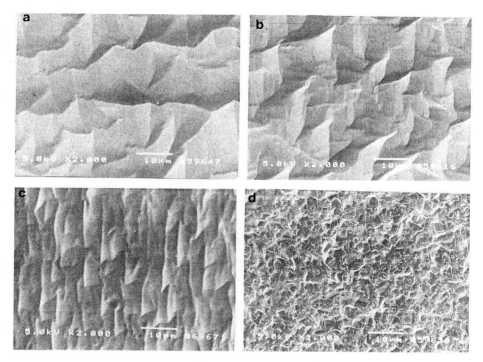

**Figure 2.** Surface morphology (SEM images) of the GaAs layers deposited with a CSVT system and atomic hydrogen, having used spacers with different thickness ($\delta$). (a) $\delta = 0.025$ mm, (b) $\delta = 0.6$ mm, (c) $\delta = 1.0$ mm, and (d) $\delta = 2.0$ mm.

produced the lowest growth rate and a polycrystalline morphology. By EDS analysis we conclude that all GaAs overlayers are Ga-rich surfaces, we infer that the non-stoichiometry of these layers is due to the nature of the CSVT growth process and the atomic hydrogen as the reactant, in which the effects of high temperature and the interaction of the atomic hydrogen with the surface and the substrate take part. Comparing the results of the two sets we observe that the GaAs films grown with atomic hydrogen have less residual contaminants, which is desirable in the manufacture of optoelectronic devices. Both sets present a variety of surface morphologies growth parameters dependent, which requires x-ray diffraction analysis to determine the morphologies with better quality.

# References

* On Postdoctoral position
1. A. Jean, EMIS Datareview, (1990).
2. E. Sirtl, J. Phys. Chem. Solids **24**, 1285 (1963).
3. F. H. Nicoll, J. Electrochem. Soc. **110**, 1165 (1963).
4. P. Robinson, RCA Review **21**, 574 (1963).
5. G. E. Gottlieb and J. F. Corboy, RCA Review, Dec. 585 (1963).
6. D. Côté, J. P. Dodelet, B. A. Lombos, and J. I. Dickson, J. Electrochem. Soc. **133**,

1925 (1986).
7. B. A. Lombos, D. Côté and J. P. Dodelet, M. F. Lawrence, and J. I. Dicson, *J. Cryst. Growth* **79**, 455 (1986).
8. F. Silva-Andrade, F. Chávez, and E. Gómez, *J. Appl. Phys.* **76**, 1946 (1994).
9. E. Gómez, R. Silva, F. Silva-Andrade, *Rev. Mex. Fís.* **43**, 290 (1997) .
10. N. Toyoda, M. Mihara, and T. Hara, *J. Appl. Phys.* **51**, 392 (1980).
11. H. Lüth, *Surface and Interfaces of Solids*, second edition (Springer, Berlin, 1993).

# Index

6-point octahedron, 303
8-point rhombohedron, 303
9-point $bcc$ unit cell, 303
13-point cluster, 303
14-point $fcc$ unit cell, 303
$\eta$ pairing operators, 79

Acetylene group, 148,327
Aerosol electrometer, 152
Agglomeration processes, 323,342
Alkali metal clusters, 125
Alloy analogy approximation, 292
Alternating gradient magnetometer, 214
Aluminum oxide, 144
Amorphous
  carbon, 114
  silica, 340
Anderson
  lattice, 53
  localization, 284
  transition, 295
Andreev reflections, 21
Angular resolved photoemission, 1
Anisotropic magnetoresistance, 228
Atomic sphere approximation, 248,303
Auxiliary bosons, 65
Au/Fe/Cr sandwiches, 213
a-$GeO_2$, 339

Ballistic
  conductance, 228
  limit, 228
Barth-Hedin exchange-correlation potential, 303
$bcc$-$hcp$ allotropic transition, 302
BCS
  superconductors 16

BCS (*cont'd*)
  theory, 2, 79
  wave functions, 84
Bell numbers, 82
Benzene ring, 148, 327
Benzo(a) pyrene, 144
Bethe lattice model, 244,344
Bipartite lattice, 77,81
Bipartite networks, 283
$Bi_2Sr_2CaCu_2O_{6+\delta}$, 1
Block representation, 39
Boltzmann
  equation, 30,228
  factors, 328
Born-Mayer potential, 264,268
Born-Oppenheimer dynamics, 120
Borocarbides, 12
Bragg-Williams approximation, 303
Broyden mixing scheme, 121
$B_2O_3$, 339
$B_2S_3$, 340

C clusters, 144
$CaVO_3$, 96
$CdCl_2$ solutions, 256
CdS films, 255
CdS/CdTe solar cells, 256
$CeFe_4P_{12}$, 54
CeNiSn, 54
Ceperley-Alder exchange-correlations, 121
$Ce_3Bi_4Pt_3$, 54
Charge density waves, 35,195
Charge transfer
  insulator, 96
  transition, 95
Chemical bath deposition, 255
Chemical vapor deposition technique, 237
Chevrel phase, 11

Cluster expansion method, 301
Cluster variation method, 204,219,301
Co
   wires, 189
   work function, 115
Coherent pairing states, 79
Coherent potential
        approximation, 28,62,216,247
Collinear magnetism, 87
Colossal magnetoresistance, 27,291
Conjugate gradient method, 120
Contact spectroscopy, 15
Conventional superconductors, 17
Cooper pairs, 1
Core level photoemission spectra, 96
Correlation functions, 79
Couple channel method, 277
Cr
   clusters, 162,196
   matrix, 166
CrFe alloys clusters, 196
Critical surface interaction, 220
Cr-Cr coupling, 211
Cr-Fe coupling, 211
Cr-Fe interdiffusion, 210
Cubic anharmonicity, 263
Cubo-octahedral clusters, 113
CuO, 96
$CuO_2$ layers, 1
Curie law, 59
CuS, 96
Cu-Au system, 302
Cu-Pd system, 302
Czochralsky method, 18
$C_{2v}$ geometry, 126
$C_{60}$ molecule, 327,340
$C_{780}$, 133
c-$GeO_2$, 339

Daily cigarette exposure
        equivalence, 158
Debye temperature, 302
Defect levels, 256
Dehydrogenation, 148
Dendritic growth, 332
Density matrix renormalization
        group, 45
Diatomic chains, 263,268
Differential mobility analyzer, 151
Diffusive conductivity, 228

Dimerized systems, 35,45
Diophantine fraction, 325
Dipolar response, 135
Dirac matrices, 91
Disordered alloys, 301
Dissipation-fluctuation
        theorem, 120
Dissociative
   adsorption, 275
   scattering, 279
Double exchange mechanism, 28,291
Dynamical antiferromagnetic
        measurements, 6
Dynamical mean field theory, 95
Dyson equation, 62
$D_2$ molecule, 277
$D_{ooh}$ symmetry, 126
$D_{4d}$ symmetry, 122
d-wave gap, 7

Electronic correlations, 47
Eliashberg
   equations, 2
   theory, 2,16
Endohedral fullerenes, 133
Energy dispersive spectroscopy, 347
Ensemble-averaged Green
        function, 293
$ErRh_4B_4$, 14
$f$ electronic systems, 11
$fcc$-like clusters, 197
Fe
   clusters, 113,162,188
   nanoparticles, 188
   wires, 189
   work function, 115
Fermi
   golden rule, 135
   surface, 1
   velocity, 31
Fermion-boson Hilbert space, 65
Ferromagnetic superconductor, 12
$Fe_N$ clusters, 162
Fe/Cr multilayers, 165
Fe/Cr/Fe trilayers, 209,214
Fibonacci chains, 324
First-principles molecular
        dynamics, 119
First Dobinski identity, 82

Fluctuation
   exchange approximation, 2
   functions, 79
Fractals, 323
Frank-Van der Merwe
      nucleation mode, 349
Fridel model, 177
Fullerenes, 133,143,148,323,340
Functional integral formalism, 168

GaAs, 114
GaAs homolayers, 347
Ge clusters, 126
Generalized Hubbard model, 73
GeSe molecule, 119
$GeSe_2$-glass, 332
$GeS_2$-glass, 332
GeTe, 120
GeTe clusters, 119
$GeTe_2$, 126
$Ge_2Te$, 126
$Ge_2Te_2$, 126
$Ge_3$, 126
Giant magnetoresistance, 210,227
Glass
   networks, 339
   transition, 335,340
Golden
   rectangle, 324
   section, 287
Goldstone theorem, 284
Green function, 286
Growth processes, 323
Gutzwiller approximation, 67

H associative desorption, 275
Hall experiments, 32
Harmonic approximation, 307
Harmonic interactions, 263
Hartree-Fock
      approximation, 36,164
Haydock-Heine-Kelly recursion
      method, 241
Heavy Fermion compounds, 11
Heisenberg model, 50,87,167,197
Hessian matrix, 223
High-$T_c$ superconductors, 1,12,79
$HoMo_6S_8$, 11
Honeycomb lattice, 244
$HoNi_2B_2C$, 14

Hubbard
   Hamiltonian, 35,46,79,161,178,185,239
   tight-binding Hamiltonian, 229
Hubbard-Stratonovich
      transformation, 168
Hund rules, 240
$H_2(D_2)/Cu(111)$ system, 275
Hückel limit, 45

Icosahedral clusters, 113,195
Impurity
   band, 67
   model, 28
   scattering relaxation time, 230
Independent-particle susceptibility, 134
Inflation matrix, 325
Infrared adsorption spectrum, 59
Intermediate valence
      Tm compounds, 27
Intermediate-range order, 339
In-plane magnetic anisotropy, 190
Ionization potential measurements, 114
Iron oxide, 144
Ising
   model, 167,228
   square antiferromagnet, 203
Isothermal molecular dynamics
      simulations, 120

Jahn-Teller distortions, 28

KBr systems, 268
Kleinman-Bylander construction, 121
Kohn-Sham
   equations, 120
   Hamiltonian, 135
Kondo
   hole, 53
   hole boundstates, 58
   insulators, 53
   interaction, 18
Korriga-Kohn-Rostoker multiple
      scattering formalism, 247
Korteg de Vries equation, 263
Kotliar-Ruckenstein
   mean field approximation, 53
   operators, 37
Kubo
   criterion, 110
   formula, 30

Kubo-Greenwood formula, 228

$L1_2$ structure, 252
$LaFeO_3$, 96
Lagrangian dynamics, 120
Lanczos
  diagonalizations, 202
  iterative method, 49,170
Landau damping, 137
Langevin molecular dynamics, 119
Latex, 144
$LaTiO_3$, 96
$La_{1-x}M_xMnO_3$, 27
$La_{1-x}Sr_xMnO_3$ crystals, 291
$La_{2-x}Sr_xCuO_4$, 2
LEED spectroscopy, 214
Lifshitz tails, 288
LiI systems, 268
Linear muffin-tin orbital
  method, 251,303
Linear response theory, 134,228
Local density
  approximation, 119,301
  functional approach, 87
Local
  Green function, 30
  spin density theory, 162,187
Lorentzian distribution, 293
Low frequency Raman response, 54
Luminescence spectra, 257
Luminescent devices, 255

Madelung
  correlation, 229
  energy, 249
Magneto-anisotropy, 185
Manganites, 27
Mapping method, 77
Mass spectroscopy, 122,143
Matrix Green function, 56
Mean field approximation, 247
Metal-insulator
  phase diagram, 100
  transition, 20,53,73,95,292
Metal-nonmetal transition, 109
Methane, 145
Micro-Raman experiments, 344
Mn-O sublattice, 28
Mobility edge, 292
Molecular beam experiments, 120

Molecular field approximation, 30
Monoatomic chains, 263
Monte Carlo method, 87,203,309
Mott
  insulator, 77
  transition, 74,95

Na octahedra, 122
NaCl particles, 156
Nagaoka theorem, 171,202,241
NaOH solutions, 256
Naphtalene, 148
  double rings, 327
$Na_{20}@C_{780}$, 133,136
$Na_{25}@C_{780}$, 133
Network glasses, 323
Neutron scattering spectrum, 28
$NH_4NO_3$ solutions, 256
Ni
  clusters, 162
  work function, 115
NiAl alloys, 247
NiCr system, 301
NiO, 96
NiS, 96
$Ni_3Al$, 252
Ni-V system, 302
Ni/Co superlattices, 227
Ni/Cu superlattices, 227
NMR spectroscopy, 340,344
Noncrossing approximation, 99
Nonmagnetic insulator, 65
Noncollinear
  magnetism, 87
  spin states, 196
Nonmagnetic spacers, 227

Off-plane magnetic anisotropy, 190
One band Hubbard model, 96
One-dimensional lattices, 263
One particle Green function, 66
Optical
  conductivity in $LaFeO_3$, 105
  reflectance measurements, 54
Optimally doped systems, 2
Organic compounds, 16
Overdoped systems, 2

Padé approximants, 61
Pair approximation 219

Pairing off-diagonal
  long-range order, 79
Pauli spin matrices, 89
PbNa clusters, 119
PbSe molecule, 119
PbTe molecule, 119
Pd
  aerosol, 143
  clusters, 153,163
Pd-V system, 302
Pd/Ag superlattices, 227
Pd/Ag system, 233
Penrose lattice, 283
Perovskite manganese oxides, 27
Perylene, 143,148
Phase stability, 301
Photoabsorption
  cross section, 135
  spectra, 126
Photoelectric work function, 148
Photoemission studies, 1,143
Photoluminescence spectroscopy, 255
Polyacetylene, 45
Polycyclic aromatic hydrocarbons, 143
Porous silicon, 315
Primary combustion aerosols, 143
Proton- and electron-induced
  Auger-electron spectroscopy, 210
Pseudo fermions operators, 39
Pseudo-potential wave method, 119
Pyrochlore structure, 27
Pyrolitic graphite, 155

Quantum confinement, 316
Quantum
  dots, 316
  well states, 233
  wires, 316
Quartic anharmonic interactions, 263
Quasi-crystals, 283,323

Raman spectra, 344
Random alloys, 250,283
Reaction approximation, 219
Reentrant superconductivity, 11
Relaxation time approximation, 30
Renormalization group theory, 219
Renormalized
  band structure calculations, 96
  perturbation expansion, 293

Respiratory particles, 144
Retarded
  Green function, 236
  one-electron Green function, 135
Rh clusters, 163
RHEED spectroscopy, 214
Ring statistics, 339
RKKY exchange interactions, 18
Rotating wave approximation, 264
Rotational-translational
  energy transfer effect, 275

Saddle-point
  approximation, 168
  slave-boson approximation, 239
Scanning electron microscopy, 256,347
Scanning tunneling
  microscopy, 188,210
Screened CPA, 250
Second Dobinski identity, 82
Second moment approximation, 110
Self-consistent
  Dyson equation, 134
  Hubbard model, 209
  tight-binding approach, 162,186
Self-similarity, 323
SEMPA experiments, 214
Short-range
  correlations, 1
  magnetic order, 167
  order, 301,339
Si clusters, 126
Silica, 144
Sine-Gordon equation, 263
Sinusoidal magnetic phases, 13
SIS-tunneling measurements, 5
Si-Ge system, 302
Slater determinant, 88,196
Slave boson
  approach, 35,104
  mean field approach, 54
Small cell renormalization, 61
$SmB_6$, 53
SmS, 53
Soliton solutions, 263
Sommerfeld constant, 16
Soot
  formation, 143
  particles, 144
Spherical jellium model, 123

Spherically averaged
    pseudopotential, 134
Spin
    coherent states, 80
    density approximation, 162
    density waves, 19,35,195,214
    fluctuations, 1
    Peierls regime, 45
Spin-orbit interactions, 186
Square band approximation, 178
$SrVO_3$, 96
Steering effect, 275
Stern-Gerlach deflection, 162,187
Stirling numbers, 82
STM, 214
Stochastic matrix method, 330
Stoner
    criterion, 55
    theory, 187
Stranski-Krastanov
    nucleation mode, 349
Sulphuric acid aerosols, 144
Superantiferromagnetic order, 204
Supercell model, 316
Superfluid helium, 16
Surface energy, 247

T matrix, 56
Te clusters, 126
Tetragonal Bain distortion, 312
$Te_3$, 126
Thermal evaporation 255
Thermionic emission, 148
Thin film photovoltaic devices, 255
Thioboxil group, 345
Tight-binding
    Hamiltonian, 209,315
    supercell model, 320
Time of flight
    distributions, 276
    mass spectrometry, 145
Time-dependent density
    functional theory, 133
Time-resolved desorption, 143
$Tl_2Mn_2O_7$, 27
TmSe systems, 53, 292
$TmSe_{1-x}La_xS$, 54
Transfer matrix approach, 75

Transition metal
    clusters, 109,161,177,187
    compounds, 95
    wires, 189
Transmission
    electron microscopy, 320
    measurements, 54
Tunneling
    contact spectroscopy, 15
    measurements, 1
Two-band Hubbard Hamiltonian, 95
Two-dimensional Hubbard
    Hamiltonian, 2
$t$-U-V extended Hubbard, 74

Ultracoherent states, 79
Unconventional superconductors, 17
Underdoped cuprates, 2
Unrestricted Hartree-Fock
    approximation, 186,195,229
$URu_2Si_2$, 11,15

V clusters, 163
Van der Waals forces, 150
Van Vleck susceptibility, 54
Van-Hove singularity, 6
Vapor transport system, 347
Virial theorem, 268
Volmer-Weber nucleation mode

Wannier exciton, 74
Wexler formula, 19
Work function, 247

$XY$-model, 288
X-ray
    diffractometer, 256
    spectroscopy, 114

Yang
    eigenfunctions, 80
    method, 80
$YBa_2Cu_3O_{6+\delta}$, 1
YbBiPt, 16

Zeeman splitting, 187
Ziman criterion, 293